Lecture Notes in Bioengineering

Lecture Notes in Bioengineering (LNBE) publishes the latest developments in bioengineering. It covers a wide range of topics, including (but not limited to):

- Bio-inspired Technology & Biomimetics
- Biosensors
- Bionanomaterials
- Biomedical Instrumentation
- Biological Signal Processing
- Medical Robotics and Assistive Technology
- Computational Medicine, Computational Pharmacology and Computational Biology
- Personalized Medicine
- Data Analysis in Bioengineering
- Neuroengineering
- Bioengineering Ethics

Original research reported in proceedings and edited books are at the core of LNBE. Monographs presenting cutting-edge findings, new perspectives on classical fields or reviewing the state-of-the art in a certain subfield of bioengineering may exceptionally be considered for publication. Alternatively, they may be redirected to more specific book series. The series' target audience includes advanced level students, researchers, and industry professionals working at the forefront of their fields.

Indexed by SCOPUS, INSPEC, zbMATH, SCImago.

More information about this series at http://www.springer.com/series/11564

Alejandro Muñoz-López · Redha Taiar ·
Borja Sañudo
Editors

Resistance Training Methods

From Theory to Practice

Springer

Editors
Alejandro Muñoz-López
Departamento de Motricidad Humana y
Rendimiento Deportivo
University of Seville
Calle Pirotecnia, Sevilla, Spain

Redha Taiar
Department of Sport Science
Université de Reims Champagne Ardenne
Reims, France

Borja Sañudo
Department of Physical Education and Sport
University of Seville
Sevilla, Spain

ISSN 2195-271X ISSN 2195-2728 (electronic)
Lecture Notes in Bioengineering
ISBN 978-3-030-81988-0 ISBN 978-3-030-81989-7 (eBook)
https://doi.org/10.1007/978-3-030-81989-7

This Springer imprint is published by the registered company Springer Nature Switzerland AG
The registered company address is: Gewerbestrasse 11, 6330 Cham, Switzerland

Preface

This book provides traditional and modern concepts and guidelines for resistance training practice. The contents span from traditional resistance training methods to the actual modern methods, including several relevant concepts regarding new advances in research, monitoring methods, and programming in resistance training.

Resistance training is one popular method for fitness training and performance preparation. It is well known that force production is the main conditioning factor, even the unique for many authors and professionals. This book covers the use of free weights such as rotary inertial devices in modern resistance training methods, together with modern training paradigms developed on the basis of key physical, physiological, and biomechanical concepts. Moreover, it discusses how technology can help to control and program resistance training. Further, it gives new insights into strength training for special populations (i.e., women, kids, and older adults).

The book is organized in four different sections. Each chapter follows the same structure to guide the reader from foundations to practical guidelines. In the introduction, authors state the main aims and objectives of the chapter. The *from theory* part covers the theoretical foundations that on the *from practice* part are developed from a practical point of view. Next, in the *filling gaps* part, authors explore their point of view on how to reduce the gaps between the theory and the real practice, based on their own experience on the field as experts. Finally, the *take home messages* part summarizes the most important points of the chapter.

The first section *Resistance Training Foundations* introduces the basics for developing robust resistance training programs. It includes four chapters:

- First Chapter "Applied Physics to Understand Resistance Training." Many times, physics become complicated to practitioners. Before, when mostly isoinertial devices were used for strength conditioning, the basics related to Newton's laws were easily applied. However, nowadays, coaches use different types of equipment, including rotary and constant load devices. Besides, the increasing technology to control and monitor resistance training programs provides a huge number of variables such as velocity, power, force, or acceleration. This chapter covers all the applied physics foundations for other chapters.

- Second Chapter "Muscle Strength Determinants and Physiological Adaptations." Any resistance training program can enhance fitness through several body physiological adaptations. This chapter covers the basic physiological adaptations of resistance training but not only focused on general strength development itself but including several aspects about the methods which will be explained on the next block.
- Third Chapter "Kinetic and Kinematic Analysis for Exercise Design: A Practical Approach." Despite force platforms have been extensively used, especially in biomechanics, strength and conditioning coaches can benefit from the understanding of the kinetic and kinematic raw data from the variables explained in Chapter "Applied Physics to Understand Resistance Training," in relation also to the methods which will be covered in the next block. This chapter explains how the raw data analysis via curves (i.e., force-time or force-velocity curves) study can enhance the specific knowledge with different strength training equipments, so practitioners can not only better adapt their daily routines but also provide deeper analyses which will be later explained on Block 3.
- Fourth Chapter "Equipment and Training Devices." The last chapter of Part One offers a preface to Part Two. As the core of the book is on the use of four resistance training paradigms such as free weights, rotary inertial devices, variable resistance training devices, and the use of the own body as a resistance per se, this chapter introduces the equipment and training devices available as solutions to be used during resistance training programs, discussing important scientific and practical knowledge achieved in the last few years.

The second section *Developing and Building Training Paradigms* introduces the four main resistance training paradigms previously explained. It describes the advantages and main applications of each single paradigms, including the latest research findings achieved in sports science. It includes four chapters:

- Fifth Chapter "Resistance Training for the Maximization of the Vertical Force Production: Jumps." In this first chapter, authors describe how to implement isoinertial loads for improving the vertical force production. The use of isoinertial loads is the most traditional approach to enhance strength gains in sport. This chapter describes the main benefits of using weight training and jumps, with practical tips from the experienced authors and how science has told us the most convenient way to integrate them on daily routines, integrated to new concepts and approaches.
- Sixth Chapter "Resistance Training for the Maximization of the Horizontal Force Production." Not only high external loads or complex equipment can be used to enhance strength. The use of the own body weight with the addition of very small external isoinertial loads has been proved to be enough to increase the sprint performance. This chapter describes the main benefits of using isoinertial loads for improving the horizontal force production, with practical tips from the experienced authors and how science has told us the most convenient way to integrate them isolated or with other options, such as those described in this block.

- Seventh Chapter "Resistance Training Using Flywheel Resistance Training Devices." Rotary inertial devices have become increasingly popular in the last ten years. However, many coaches still do not integrate this option in daily routines, despite the higher benefits for some developments which the science has already highlighted, compared to other training options. This chapter will describe the main benefits of using this kind of equipment, with practical tips from the experienced author and how science has told us the most convenient way to integrate them isolated or with other options, such as those described in this block.
- Eighth Chapter "Variable Resistance Training Methods." Variable resistance training mainly covers the use of equipment which vary the load within the range of motion during the concentric and/or eccentric phase of an exercise. It has been shown that, for some specific cases, combining this approach with other traditional ones, such as isoinertial loads, can increase the force development. This chapter will describe the main benefits of using this kind of equipment or how to implement the rotary dynamic correspondence, with practical tips from the experienced authors and how science has told us the most convenient way to integrate them isolated or with other options, such as those described in this block.

The third section *Monitoring Training and Testing* gathers recommendations on how to apply each of the four previously introduced paradigms, showing how to use the available technology for training and assessment. It includes three chapters:

- Ninth Chapter "Velocity-Based Training for Monitoring Training Load and Assessing Training Effects" deals with one of perhaps, the greatest advance in strength training. This approach has radically changed the way on how to develop and set up resistance training programs in practice. This chapter will describe, from a practical point of view, yet considering also the last scientific advances, how to use the velocity measured by using different technologies for controlling and planning resistance training exercises.
- Tenth Chapter "Measuring and Testing with Flywheel Resistance Training Devices." Rotary inertial devices can also be monitored. A main difference with isoinertial equipment is that there is no repetition maximum to program the training load during resistance training exercises. However, scientists and coaches have used several different approaches to monitor these devices in resistance training programs. For example, the use of the so-called eccentric overload has been particularly used in these devices. This chapter covers how to monitor these devices during training and testing procedures, with details achieved from the authors' practice and experience.
- Eleventh Chapter "How to Use Force Sensors for Resistance Training in Daily Practice." Not only velocity or power derived from velocity measurements is used for monitoring or testing in resistance training programs. Other testing equipments such as force platforms, strain gauges, dynamometers or, more

recently, accelerometers have been used. The main usage of these devices is that force can be directly measured, rather than being estimated from velocity, like when using isoinertial devices. So, this chapter covers how to implement those devices in daily practice, especially for testing and assessment.

The fourth section *Program Design and Periodization: Combining Strategies*, continues reporting on more specific approaches that are related to programming and training routine design. This part sets the basics for using the four paradigms combining science with practical experience. It includes four chapters:

- Twelfth Chapter "Basics of Programing and Periodization in Resistance Training." Before introducing some relevant specific strategies on how to set up and program resistance training exercises, a basic theory covering the basics of programming and periodization in resistance training is mandatory. This chapter not only reports on traditional strategies, but it also combines the author own results from practice with the most advanced knowledge on resistance training programs.
- Thirteenth Chapter "Programing and Periodisation for Team Sports." Strength training for team sports such as football is full of myths and recommendations. In-pitch or in training centers can both elicit gains in muscle strength and power. In this chapter, the authors show different strategies to implement resistance training programs in team sports. Authors show what kind of strategies are scientifically proved and how they can be applied in real environments for different kind of team sports options.
- Fourteenth Chapter "Programing and Periodization for Individual Sports." Another important application of resistance training programs is concerned with individual sports. For many years, strength training has been used on individual disciplines such as athletics. However, recent and modern advances have changed the way resistance training programs are nowadays applied in these specific contexts. This chapter gives novel insights into how individual athletes may benefit from resistance training.
- Fifteenth Chapter "The Role of Resistance Training in Strategies to Reduce Injury Risk." One of the main aims when a resistance training program is developed, especially for sports, is to reduce the injury risk. Different human structures such as bones, muscles, tendons, ligaments and other soft tissues can benefit from resistance training. In this chapter, which is focused on sports in general, authors discuss how to implement daily strength and conditioning routines to reduce the injury risk in sports, using the techniques and equipment previously described in the book.

The fifth section *Special Considerations in Resistance Training* describes how to implement resistance training using the previously described paradigms on special populations. In addition, authors discuss the main factors influencing strength and the relevant supplementation for the respective population. This part includes four chapters:

- Sixteenth Chapter "Resistance Training in Older Adults" describes how resistance training programs can be implemented in this population group. In the past, those programs have been mainly based on light loads without any kind of control. Nowadays, it is proved that a systematic and supervised resistance training can enhance the quality of life in the elderly, reduce the risk of falls and, among others, even prevent pathologies such as Alzheimer. This chapter covers the existing knowledge concerning how older adults can benefit from strength training and how to implement specific programs for them.
- Seventeenth Chapter "Resistance Training for Children and Adolescents." During maturity, especially for adolescents, there exist interesting maturity points, known as sensible phases, which can enhance the response to a given stimulus. Hence, it is very important to understand how resistance training can improve the potential genetic adaptations in this population for further enhancements during adult life. This chapter covers the existing knowledge about how children and adolescents can benefit from strength training and how to implement specific programs on them.
- Eighteenth Chapter "Resistance Training in Women." Women, athletes or not, are different compared to men in relation to how hormones can affect their performance. Many of these differences are related to the menstrual cycle. Also, it has been proved that women can benefit from resistance training programs during pregnancy. This chapter reviews recent scientific and practical knowledge relating to resistance training in women, filling an important gap in the current literature.
- Nineteenth Chapter "Supplementation and Ergogenic Aids for Enhancing Muscular Strength Production." The last chapter covers how to combine supplementation strategies and ergogenic aids with resistance training routines. While the existing literature has traditionally be focused on the use of these strategies to enhance muscle mass, authors in this chapter provide a unique and original point of view on how to maximize other potential strength determinants instead of working on the structural factor.

This book offers extensive and timely information on resistance training methods to strength and conditioning coaches, physicians, and doctors. Mainly intended to support them in their daily practice, it will also offer a source of inspiration and an authoritative guide for improving the quality of their interventions. At the same time, it also addresses athletes or other individuals who want to gain insights into on how to improve their performance, prevent and recover from injuries, and treat some pathologies. The book includes many examples that will help trainers to set up training programs and the relevant quantifications for specific sports and populations, and to achieve the optimal development of the different manifestations of strength, specific for the needs of the single sport, athlete, or individuals.

Seville, Spain — Alejandro Muñoz-López
Reims, France — Redha Taiar
Seville, Spain — Borja Sañudo

About This Book

This book provides traditional and modern concepts and guidelines for resistance training practice, combining new training methods and technologies developed in the last years, together with traditional methods. The contents cover from traditional resistance training methods to the actual modern methods, including several relevant concepts regarding new advances in research, monitoring methods, and programming in resistance training.

Resistance training is one of the main elements of fitness and performance preparation. It is well known that force is the main conditioning factor, even the unique for many authors and professionals. From the use to weights to modern resistance training methods such as rotary inertial devices, readers will find modern training paradigms, following the main physical, physiological, and biomechanical concepts, besides with brief information which includes how technology can help to control and program resistance training, including detailed information about special populations.

The core of the book can help strength and conditioning coaches, physicians, doctors, and even athletes who have just started to study how to improve sports performance, injury prevention, injury recovery and some pathologies treatments. On the same manner, it may help all the experienced audience which wants an update on their daily practice to improve the quality of their interventions.

Contents

Editors and Contributors

About the Editors

Alejandro Muñoz-López has a M.Sc. in High Performance in Sports, in soccer performance and injury rehabilitation, and a Ph.D. in Sport Sciences. He is a former strength and conditioning coach of some professional and amateur Spanish soccer teams and of the Latvian Football Federation. His main research line is monitoring the training load of field and strength exercises applying new advances in sports technology. He was the co-founder of Science2improve, a company with the main aim to help coaches and practitioners to apply the new advancements in sports science to their daily routines. He is working as a professor at the University of Seville (Spain) in the sports training subject. Despite being young, the continuous application of the theory to the practice in the strength and conditioning topic over the last ten years realized him about this book idea. He is also an author and co-author of relevant sports science papers. At the moment, he is more focused on the development of a new training methodology to implement the flywheel resistance training devices on resistance training programs. As part of this purpose, he is a sports scientist and consultant of SmartCoach Inc.

Prof. Redha Taiar has Ph.D. in Biomechanics and now is a professor at the University of Reims Champaign, France. He was the head of the European Master of Biomechanics, Ergonomics, and Clinical Research, was the head of the University diploma of Podiatrics, and was the head of the University diploma of Ergonomics. He was also the head of the Laboratory of Biomechanics at the University of Reims and the vice-head of department of Sport Science. He was the head of redha Taiar Biomechanical Engineering (RTBE) Society developed for sport and medical advice for industry and the vice-head of the congress of Physical Medicine and Rehabilitation Society (SOFMER) from October 2013 in Reims city. His researches focus on the industry engineering for medicine and high level sport. He is engineer for different industries like Arena for the high level sport and Sidas, Medicapteur, for the medical development. For industry workers, his last work was for the Notrax society Conception and validation of anti-fatigue mats. For sport, his last works focus on the development of swimsuit for triathlon and swimming for Brazil Olympic Games (2016) and the suit fabrics for skiing Olympic Games at Sotchi in 2014. He is a specialist on Biomechanics of health disease and rehabilitation.

Borja Sañudo received the B.S. and M.S. degrees in physical activity and health from the International University of Andalusia in 2007 and the Ph.D. degree in sport science from University of Seville, Spain, in 2009. From 2010, Borja is full professor at the Department of Physical Education and Sport (University of Seville, Spain). His major research interests are centered on the role of exercise and other lifestyle factors for promoting improvements in physiological function, quality of life and disease-free survival in clinical populations and improvements in athletes' performance. Since 2005, he has been working on a number of studies with clinical populations (mainly fibromyalgia, Type 2 diabetes and obesity) and athletes (whole body vibration and exercise and the use of technologies in sports). He is the author of five books, more than 100 articles in peer-reviewed journals, and more than 80 international presentations.

Contributors

Helvio Affonso APPTO Assessoria Esportiva, APPTO Physiology Laborratory of Excerise, Nutrition and Sports and Tranning, Universidade de Villa Verde, Espirito Santo, Brasil

Luis Suarez Arrones Sport and Informatics Department, University of Pablo de Olavide, Seville, Spain

Beatriz Bachero-Mena University of Seville, Sevilla, Spain

Cristina Casals Faculty of Education Sciences, University of Cádiz, Puerto Real, Cádiz, Spain

Filipe Almeida Viana Conceiçao Center of Research, Education, Innovation and Intervention in Sports (CIFI2D), Faculty of Sports, LABIOMEP, Porto Biomechanics, Universidade do Porto, Porto, Portugal

Víctor Cuadrado Autonomous University of Madrid, Cantoblanco University City, Madrid, Spain

Moisés de Hoyo Lora Physical Education and Sports Department, University of Seville, Seville, Spain

Raúl Domínguez Departamento de Motricidad Humana y Rendimiento Deportivo, Universidad de Sevilla, Sevilla, Spain

Avery D. Faigenbaum Department of Health and Exercise Science, The College of New Jersey, Ewing, USA

Oliver Gonzalo-Skok Sevilla FC, Seville, Spain

Fernando Hernández-Abad Faculty of Health Sciences, Universidad Europea de Canarias, La Orotava, Spain;
FSI, Sport Research Lab, Granada, Spain

Franco Impellizzeri School of Sport, Exercise and Rehabilitation, Faculty of Health, University of Technology Sydney, Ultimo, Australia

Pedro Jiménez-Reyes Centre for Sport Studies, Rey Juan Carlos University, Madrid, Spain

Irineu Loturco NAR—Nucleus of High Performance in Sport, São Paulo, Brazil

John J. McMahon Directorate of Sport, Exercise and Physiotherapy, University of Salford, Salford, UK

Susana Moral-González European University of Madrid, Madrid, Spain

Alejandro Muñoz-López Departamento de Motricidad Humana y Rendimiento Deportivo, Universidad de Sevilla, Seville, Spain

Alireza Naderi Department of Sport Physiology, Islamic Azad University, Tehran, Iran

Fabio Yuzo Nakamura Research Center in Sports Sciences, Health Sciences and Human Development, CIDESD, University Institute of Maia, ISMAI, Maia, Portugal

Francisco Javier Nuñez Universidad Pablo de Olavide, Physical Performance & Sports Research, Seville, Spain

Fernando Pareja-Blanco Physical Performance and Sports Research Center, Pablo de Olavide University, Seville, Spain

Jesús G. Ponce-González Faculty of Education Sciences, University of Cádiz, Puerto Real, Cádiz, Spain

Marco Pozzo SmartCoach Technologies Inc, Pleasanton, CA, USA

Rodrigo Ramírez-Campillo Department of Physical Activity Sciences, Universidad de Los Lagos, Santiago, Chile

Alejandro Rodríguez-Fernández Faculty of Sport Sciences, University of Leon, León, Spain

Michael E. Rogers 106G Heskett Center, Wichita State University, Wichita, KS, USA

Eduardo Sáez de Villarreal Universidad Pablo de Olavide, Physical Performance and Sports Research Center, Seville, Spain

Antonio Jesús Sánchez-Oliver Departamento de Motricidad Humana y Rendimiento Deportivo, Universidad de Sevilla, Sevilla, Spain

Javier Sanchez-Sanchez Pontifical University of Salamanca, Salamanca, Spain

Alberto Sánchez-Sixto Departamento de Deporte., Centro de Estudios Universitarios Cardenal Spinola CEU, Bormujos, Sevilla, España

Borja Sañudo Departamento de Educación Física y Deporte. Universidad de Sevilla, Sevilla, Spain

Pierre-Jean Vazel Athlétisme Metz Métropole, Metz, France

Emilio Villa-González Department of Physical and Sports Education, University of Granada, Granada, Spain

Resistance Training Foundations

Applied Physics to Understand Resistance Training

Marco Pozzo and Franco Impellizzeri

Abstract Professionals in strength training deal on a daily basis with physical entities such as force, speed, power and work, and use them to assess performance of their athletes and tune their training routine. These variables are related to each other through fundamental laws and principles of Physics. Knowledge of their meaning and how they correlate to each other is key for their use in resistance training. The purpose of this chapter is to provide the reader with a basic knowledge of the fundamental concepts of Mathematics, Physics and Information Theory applied to Sport Physiology, as well as a description of the working principles of the technologies used in resistance training devices and their practical implications. This knowledge is necessary for the full comprehension of practically any concept in strength training, as well as for the understanding of all the subsequent chapters of this book.

Keywords Physics · Force · Speed · Power · Training technology · Data acquisition · Filtering · Eccentric overload

1 Introduction

Learning the rudiments of Physics and information technology is a necessary effort in understanding the science behind strength training, especially when selecting exercises, choosing equipment and evaluating the strength of the claims behind new tools or devices.

The purpose of this chapter is to provide such basic knowledge to understand the physics, advantage and limitations of strength training equipment, and the measures

M. Pozzo (✉)
SmartCoach Technologies Inc, Pleasanton, CA, USA
e-mail: marco.pozzo@smartcoach.tech

F. Impellizzeri
School of Sport, Exercise and Rehabilitation, Faculty of Health, University of Technology Sydney, Ultimo, Australia

A. Muñoz-López et al. (eds.), *Resistance Training Methods*, Lecture Notes in Bioengineering, https://doi.org/10.1007/978-3-030-81989-7_1

that are used to monitor this kind of training. They rely on a variety of technologies to provide resistance, but can be classified into the following macro categories:

- *Gravity-Dependent Equipment*: load is provided mainly by the gravitational force acting upon a mass (the athlete's body, an external weight or a combination of the two). This category includes weight stack machines and free weights, but also equipment where the athlete body slides onto inclined rails
- *Flywheel Resistance Training Devices (FRTDs)*: load is provided mainly by the inertial force of a rotating mass (Berg and Tesch 1994)
- *Elastic Resistance Equipment*: machines in which resistance is provided by an elastic component, such as one or more springs or rubber bands
- *Fluidodynamic Machines*: they use the pressure from a liquid (hydrodynamic) or gas (pneumatic machines), such as Keiser® air resistance equipment
- *Robotic Machines*: they use an electrical motor controlled by a computerized control unit to provide resistance.

The physics of these devices and their practical implications will be discussed in the next section.

2 From Theory

This section covers some generic principles of Physics, Mechanics and Information Theory and their applications to strength training technology.

In Physics and here, any physical entity is assumed to be expressed in *SI* (*International System*) standard units: mass in kg, force in Newtons (N), displacement in m, speed in m/s, angular velocity in rad/s, power in Watts (W), energy (work) in Joules (J), and so on. If a variable is not expressed in SI units, it must be converted first; for example, a mass expressed in lbs must first be transformed in kg (1lbs ≅ 0.454 kg).

For conciseness, mathematical symbols are used in the following: "$\approx$" means "approximately equal to"; "$\equiv$" means "coincides with", and "$\ll$" means "much less than". The dot "$\cdot$" represents multiplication. Variants of the same formula have the same number followed by one or more apostrophes (e.g. (6), (6′), (6″)).

2.1 Fundaments of Mechanics

Training physics is based on just two independent physical entities: force *(F)* and velocity *(v)*. Any other training variable is obtained by calculation from them (DiLisi 2019):

- *Power* is the product of force and velocity:

$$P = F \cdot v \tag{1}$$

- *Work* (energy), is the integral of power:

$$E = \int_t P \cdot dt \tag{2}$$

- *Acceleration* is the derivative of speed:

$$a = \frac{dv}{dt} \tag{3}$$

- *Displacement* is the integral of speed:

$$x = \int_t v \cdot dt \tag{4}$$

The last two operations are used extensively in this chapter, and it's important to understand their physical meaning: the *derivative* of a variable is the rate of variation (change) over time; the *integral* is the area beneath its graph versus time. Integral and derivative are each other's inverse operations, e.g. the integral of speed (the area below its graph) is displacement. And the derivative of displacement is speed. You can find online several easy calculus short courses and videos explaining derivatives and integrals if you want to know more.

Any performance index is derived from one or more of the main variables: average concentric power is the average of power P over the entire concentric phase; peak power is its maximum instantaneous value; eccentric overload is the ratio of the two variables, and so on.

An often misunderstood concept is the difference between force and mass, which are two completely different variables. Mass is a physical property, i.e., the quantity of matter, measured in kg. Force is an interaction between objects, and is measured in N. Weight is just one specific force—the vertical attraction exerted by the planet on any object; astronauts have the same mass everywhere, but different weight on Earth or Moon, and no weight at all in space. It is as common as so painfully wrong to use expressions such as "*a force of x kg*" or "*a weight of x kg*". While tolerated at the grocery store, they should be banned in sport science.

An entity derived from mass is the *momentum* (p) of an object, which is the product of mass times velocity and represents its quantity of motion:

$$p = M \cdot v \tag{5}$$

Momentum and force are correlated by the fundamental *II Newton's Law* (for linear motion):

$$F = \frac{dp}{dt} \tag{6}$$

which states that force applied to an object changes its momentum. Mass could change in time (like in rocket or dragster car as they burn fuel) but, in sport applications, it is usually constant. So, the (6) can be rewritten by carrying mass out of the derivative:

$$F = \frac{dp}{dt} = \frac{d}{dt}(M \cdot v) = M \cdot \frac{dv}{dt} = M \cdot a \tag{6'}$$

which is the classical, renowned expression where a_{tot} is the total body acceleration:

$$F = M \cdot a_{tot} \tag{6''}$$

Another variable used in sport science is the *impulse* (J), which measures the total amount of exchanged force (for example, an athlete landing after a jump):

$$J = \int_t F \cdot dt \tag{7}$$

By replacing F with its expression from the II Newton's Law (6), and remembering that the integral is the inverse operation of derivative, so they cancel each other out, it's easy to observe that the impulse is the total change of momentum:

$$J = \int_t F \cdot dt = \int_t \frac{dp}{dt} \cdot dt = p\Big|_{t_{initial}}^{t_{final}} = p_{final} - p_{initial}$$

which is mass times the change in speed:

$$J = p_{final} - p_{initial} = M \cdot (v_{final} - v_{initial}) \tag{7'}$$

In the example of the athlete landing, one could compute the landing speed from the jump height, and compute the impulse without the need for a force plate.

2.2 Fundaments of Energy Physics

Another fundamental concept that has heavy implications in sport science is the *I Law of Thermodynamics*, which states that the change of energy (ΔU) in a system is the difference between heat received (Q) and work done (E) (Fig. 1A):

$$\Delta U = Q - E \tag{8}$$

In practical terms, it simply means that energy cannot be created or destroyed, just transformed among different forms (*Principle of Conservation of Energy*), such as:

- *Gravitational potential* (E_g): due to the height of a mass relative to a zero position
- *Kinetic*: energy due to movement of a moving mass (E_k) or spinning flywheel (E_{FW})
- *Elastic* (E_e): energy accumulated in a spring or rubber band under tension
- *Heat* (Q): energy due to friction
- *Work* (E_w): energy exchanged with the athlete

The athlete-machine conjoint is (usually) a *closed system*, i.e. there is no energy exchange with the outside world ($\Delta U = 0$), hence all the energy is converted into heat ($E = Q$). This is the case of aerobic training devices (Fig. 1B), where the concentric work is purposely dissipated in friction. In strength training devices, friction is usually absent or negligible, so the net energy balance is zero: practically all the concentric work done by the athlete is instead accumulated in the machine in one or more of the above forms (kinetic, elastic, etc.):

$$E_w \approx E_k + E_e + E_g \tag{9}$$

and then returned completely in the eccentric phase (Fig. 1C):

$$E_{CON} \approx E_{ECC} \tag{10}$$

The best-known example is in FRTDs, where all the concentric work is transformed in flywheel kinetic energy ($E_{FW} = E_{CON}$), and returned to the athlete in the eccentric phase ($E_{ECC} = E_{FW}$). *Active* training devices are not closed systems, as they have an external source of energy (Fig. 1D), whose implications of are discussed further on.

2.3 *Physical Working Principle of Training Devices*

The most common type of strength training equipment are *weight machines*, where resistance is provided by a mass displaced in a linear motion (vertically or inclined).

Their principle of operation is summarized by the II Newton's Law for linear motion (6″), where M is load and a_{tot} is the total acceleration applied to it. If the movement is vertical, the total acceleration is the sum of acceleration a (derivative of speed) and gravity acceleration $g = 9.81\ \text{m/s}^2$ (Fig. 2A, B).

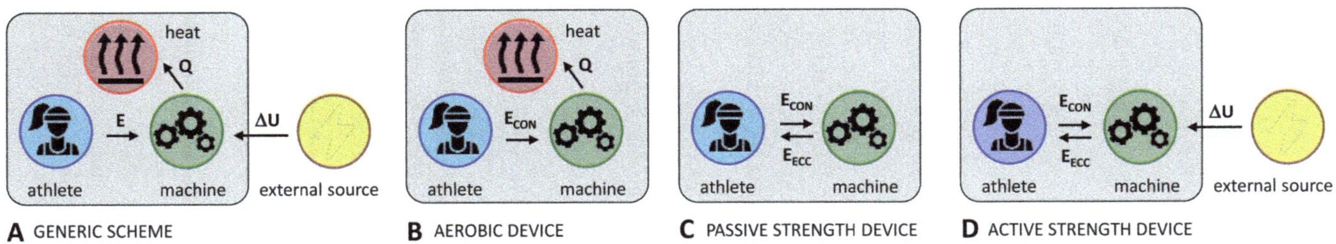

Fig. 1 General representation of the I Law of Thermodynamics in the generic case (**A**) and specific applications: aerobic device, where all the concentric work is dissipated (**B**); mechanical strength device, where all the concentric work is accumulated in the machine and returned in the eccentric phase (**C**); active strength device with an external energy source (**D**)

$$F = M \cdot a_{tot} = M \cdot (g + a) = M \cdot g + M \cdot a \tag{11}$$

If, in the general case, the path is not vertical but inclined by an angle θ, only the component of g parallel to the motion intervenes in the equation:

$$F = M \cdot g \cdot \sin(\theta) + M \cdot a \tag{11'}$$

The total force is the sum of two components: a constant *gravitational* force F_g, and an *inertial* force F_a, due to the inertia of the load (its mass M):

$$F = F_g + F_a \text{ where } F_g = M \cdot g \cdot \sin(\theta) \text{ and } F_a = M \cdot a \tag{11''}$$

This simple formula shows clearly why weight machines should not be called *isotonic* (which means "constant force"): force is approximately constant ($F \approx F_g$) only if the inertial force F_a is negligible ($F_a \ll F_g$); that is, when acceleration is low and movement is slow. If acceleration is high, the inertial force component can add substantially to the total resisting force (Fig. 2A, B).

If the movement is horizontal (θ = 0°), the gravitational component F_g becomes null and, as paradoxical as it might sound, weight resistance become purely inertial ($F \equiv F_a$). In other words, resistance in weight machines is never fully isotonic, but a combination of isotonic and inertial force; which one prevails, depends on the machine geometry (angle of motion θ) and execution (speed and acceleration). Physics prevails on semantics.

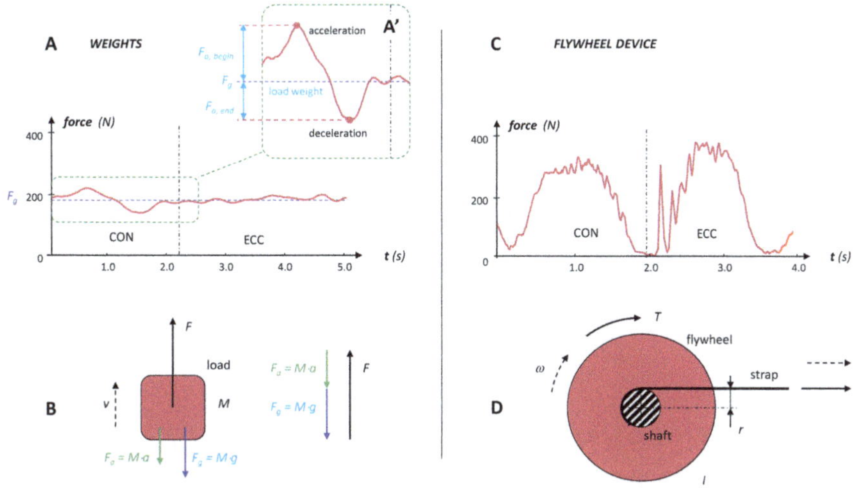

Fig. 2 Force time course of one bout of unilateral knee extension on a weight stack (**A**) versus flywheel device (**C**), and corresponding force diagrams (**B** and **D**). In zoom **A'**, note the inertial force surge $F_{a,\ begin}$ at the beginning of the CON force, due to the initial upwards acceleration of the weight, and the final dip ($F_{a,\ end}$) due to deceleration at the end of the ROM, that sum to the gravitational force F_g

In *flywheel training devices*, resistance is provided by a mass (e.g. a set of discs or radial masses) put into rotation. Their behaviour is described by the II Newton's Law for angular motion:

$$T = I \cdot \alpha \tag{12}$$

This is formally identical to the (6″) for linear motion, where their angular counterparts are used: T (torque, Nm) instead of F, I (*angular moment of inertia*, kg·m^2) instead of M, and α (angular acceleration, rad/s^2) instead of a. Since the weight of the flywheel is borne by its shaft and ball bearing, there is no gravitational force component.

Torque T is the angular expression of a force F applied at a distance r (*lever arm*) from the centre of rotation (Fig. 2C, D):

$$T = F \cdot r \tag{13}$$

In these machines, the force F is applied to a rope or strap, rolled onto the flywheel shaft at a distance r (the shaft radius) from the axis of rotation. Linear (v) and angular speed (ω, rad/s) and accelerations (a, α) are also correlated with each other by the radius of rotation r:

$$v = \omega \cdot r \text{ and } a = \alpha \cdot r \tag{14}$$

Substituting these expressions into (12) yields to its alternate expression in linear terms:

$$F = \left(\frac{I}{r^2}\right) \cdot a \tag{12′}$$

We can now define a new variable M_{eq}, the *flywheel equivalent mass (FEM or FWM):*

$$M_{eq} = \left(\frac{I}{r^2}\right) \tag{15}$$

and rewrite the physical law of flywheel machines as:

$$F = M_{eq} \cdot a \tag{12″}$$

which is identical to the II Newton's Law for linear motion (6″). The implications of this mathematical fun are now evident: from a physical point of view a flywheel device is, no more and no less, equivalent to a weight machine with load M_{eq} displaced horizontally (so to elide the gravitational force component), and the concept of flywheel equivalent mass allows correlating the physics of weights and flywheel devices.

In spring and rubber band devices (*elastic resistance equipment*) resistance is given by the Hooke's Law:

$$F = S \cdot \Delta x \tag{16}$$

where Δx is the displacement (elongation) and S is the *stiffness* measured in N/m. In these devices, force increases with displacement, i.e. is higher at the end than at the beginning of the range of motion (*ROM*). In these machines, the mass of moving parts, and their consequent inertial forces, are negligible compared to resistance provided by the elastic component.

The last category is that of *pneumatic machines* (Keiser®) which use a pressurized air cylinder to provide resistance. Since the only moving parts are a small amount of compressed air and a piston, of negligible mass altogether, the inertial force component is practically absent. The combined effect of the pressurized cylinder and kinematic linkages provides a substantially constant force (F_{air}) throughout the entire ROM:

$$F = F_{air} \tag{17}$$

The load can be set by increasing or decreasing the air pressure by acting on two buttons, and is shown on a display in *equivalent mass M* where the force equals $F = M \cdot g$. The load is expressed in kg in European machines and in lbs for the US market.

2.4 Fundaments of Data Acquisition

In modern sport science, training variables are measured and used for training monitoring, requiring a basic understanding on how such data is acquired and processed; this is the purpose of the next few paragraphs.

In the previous formulas, variables have assumed to be *time continuous*, such as analogue signals recorded on a magnetic tape, but this is generally not the case. Instead, physical variables (e.g. force measured with a transducer) are acquired and converted into *time discrete* data, i.e. a series of values stored on computer software for processing and analysis. This process is called *digitization* and consists of two well separate processes often confused with each other:

- *Sampling,* the process of taking consecutive measurements at equally spaced time intervals (*discretization in time*). Each measurement is called a *sample* and the time difference Δt between samples is called *sampling time*. Its reciprocal:

$$f_{samp} = \frac{1}{\Delta t} \tag{18}$$

is called *sampling frequency* (Hz) and represents the data rate per second
- *Quantization,* the conversion of samples into digital data on a finite number of bits (*discretization in values*)

In real life, any physical variable has a certain maximum rate of variation in time. The faster a signal changes in time, the larger is its *bandwidth B*, i.e. its occupation in the so-called *spectrum* or *frequency domain*. Incidentally, this is exactly what happens in a high-speed internet connection: digital data is sent at a high transmission rate, meaning more bit transitions and a fast-changing signal, therefore a large bandwidth; hence its name.

The fundamental theorem of Nyquist-Shannon ensures that *sampling is a loss-less process*, meaning that it does not introduce any error (Cunningham 2003), if the sampling frequency is at least twice the signal bandwidth (Fig. 3):

$$f_{samp} > 2 \cdot B \tag{19}$$

The optimal sampling rate ($f_{samp^*} = 2B$) is called the *Nyquist frequency*. Signals varying faster in time have larger bandwidth and require higher sampling rate. If this condition is not respected, a distortion called *aliasing* occurs. A relatable example is in movies: the movement of an object, e.g. a spinning wheel, is rendered correctly until its speed is relatively low. As it speeds up, the perception is that the wheel is slowing down or even turning in the opposite direction. In this case, a high-speed camera (a higher sampling rate) would be required.

Samples are then converted (*analogue-to-digital conversion*, A/D) (Kehtarnavaz 2008), into their digital representation on a finite number of bits; the number of bits, *N*, is called *resolution*. Digital data can only have so many different values, from zero to D_{max} (Fig. 4):

$$D_{max} = 2^N \tag{20}$$

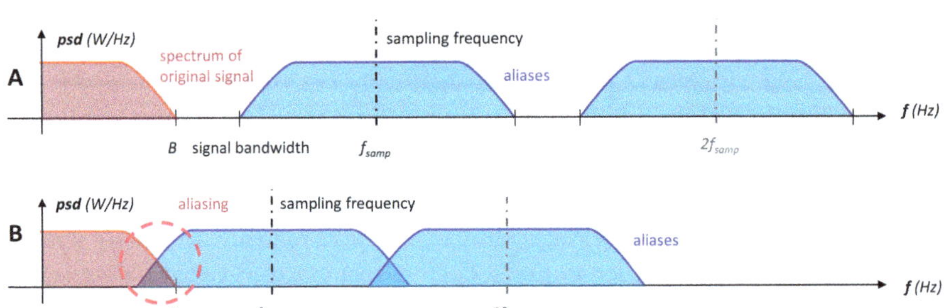

Fig. 3 Spectrum (representation in frequency domain) of a signal sampled with sufficient (**A**) and insufficient sampling rate (**B**). Sampling has the effect of creating *aliases* (replicas) of the original spectrum at multiples of the sampling frequency f_{samp}. If the Nyquist condition is not respected, aliases overlap the original spectrum, resulting in *aliasing* distortion

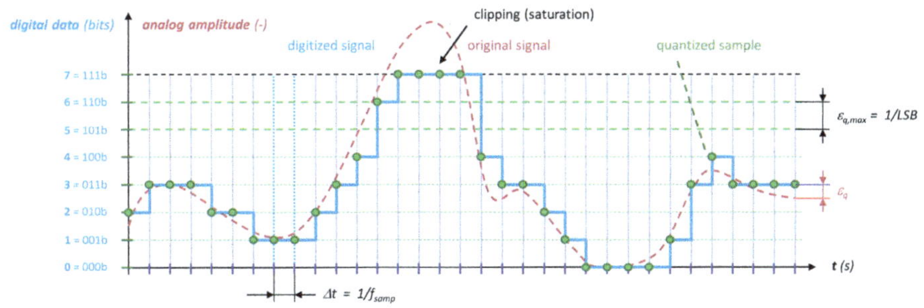

Fig. 4 Time-continuous, analogue signal (red) and its digitized version (blue) on 3-bit resolution ($D_{\max} = 7$), to exaggerate the effect of quantization error (ε_q). To fully exploit the available resolution, the input signal must be *conditioned* (attenuated or amplified) to cover the entire vertical scale (*ADC input range*), but not exceeding it to avoid *saturation* (clipping of signal peaks)

The effect of quantization is an error, due to the approximation of the exact sample value with its closest digital number, which is computed as:

$$\varepsilon_q = \frac{1}{2^N} \tag{21}$$

So, it is quantization (the resolution) and not sampling (the sampling rate), that can introduce inaccuracy in measured training data. The next section will explain how to deal with these limitations.

2.5 Data Filtering

Filtering is often applied to training data for a number of purposes, including the reduction of noise. The topic is vast and complex and deserves a book on its own. The purpose of this section is to briefly introduce the topic and warn on their use without a basic knowledge, which can lead to disastrous effects and fallacious data interpretation.

Filters can be of the following types:

- *Low-pass filters*: they cancel the frequencies above a given *cut-off frequency* (f_c). They are mainly used to reduce and smoothen out the background noise and are, by far, the most used in sport science
- *High-pass filters*: they cancel the frequencies below their cut-off frequency
- *Notch filters*: they cancel a narrow range of frequencies around a central value (f_c) and are often used to filter out interference from power lines (50 Hz in Europe, 60 Hz in US)

Combining low- and high-pass filters it's possible to build *band-pass* and *band-reject* filters, which selectively allow or reject frequencies in a given band.

In actual filters, the further from the cut-off frequency, the higher is the attenuation expressed by the filter *profile*, i.e. its characteristic in the frequency domain (Butterworth, Chebychev, Bessel, etc.) and *order*, i.e. a number that indicates how steep the transition is from pass to rejected frequencies. Digital filters have two different implementations, so-called *FIR* (*Finite-*) or IIR (*Infinite-Impulse Response*).

The above variety alone should already suggest caution in the unaware application of filtering. In addition to that, one important implication is that filters introduce a delay, which is greater with higher filter orders. One consequence is shown in Fig. 5 where speed (v) is acquired, while acceleration (a) is computed by derivation, and then filtered (a_{filt}) to reduce noise. The consequence is the introduction of a lag that alters the temporal relationship with speed, leading to erroneous interpretation while analyzing both on the same time scale. The correct way would be to also filter speed with an identical filter, which will introduce an equal lag and therefore maintain the temporal relationship. This simple example should hopefully already discourage from the random use of filtering.

3 From Practice

This section will discuss the practical implications of the notions explained previously, starting from how to acquire training variables used in training monitoring.

3.1 *Measuring Training Variables*

Force and velocity are the two main independent variables from which all the others can be derived (1–4). Those formulas are irrespective of machine type, technology

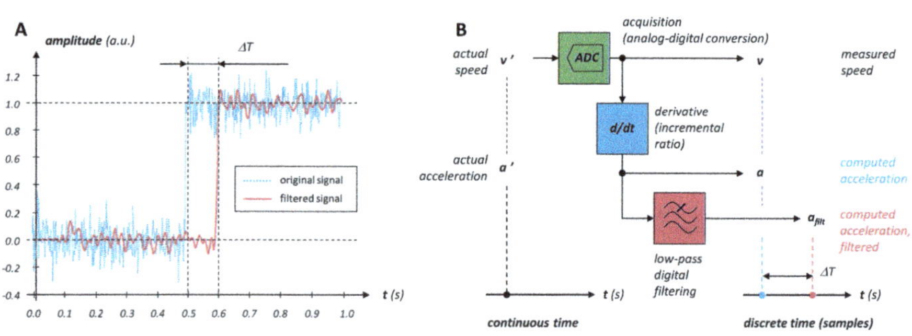

Fig. 5 **A** Effect of filtering on a sample noisy signal sampled at 500 samples/s and filtered with order n = 100 low-pass FIR filter. The introduced lag ΔT corresponds to 50 samples, i.e. half the filter order. **B** Example of incorrect data handling procedure, where acceleration is computed from acquired speed and then filtered, which alters the temporal relationship between the two variables

or even load; so, if both force and speed can be measured, this is all it's needed to compute everything else: displacement, power, work.

However, force transducers (load cells) are, usually less practical and more expensive than speed sensors. An alternative, common used technique is that of only *measuring* speed directly, and *computing* force instead from the machine's physics using formulas (11–17). This approach is as good as the closer the device is to its ideal behaviour.

The best option for speed is to measure it directly. Among speed sensors there are analogue tacho generators, which require periodical calibration. Since this process is time consuming, and limits accuracy to how precisely calibration is done, it is preferable to use digital, calibration-free incremental encoders instead. Alternatively, speed can be computed by derivation from a displacement sensor. In the time-discrete domain (sampled data), derivation and integrals are calculated respectively as incremental ratios and integrals as cumulative sums, e.g.:

- Acceleration: $a = \frac{dv}{dt}$ becomes:

$$a_n = \frac{\Delta v}{\Delta t} = \frac{v_n - v_{n-1}}{\Delta t} = \frac{v_2 - v_1}{\Delta t} \tag{3'}$$

- Displacement: $x = \int_t v \cdot dt$ becomes:

$$x_n = \sum\nolimits_n (v_n \cdot \Delta t) = \Delta t \cdot \sum\nolimits_n v_n = \Delta t \cdot (v_1 + v_2 + \ldots) \tag{4'}$$

where v_n, is the generic nth value and v_{n-1} is the previous one.

A third option for speed is to compute it by integration from acceleration. While in theory it works perfectly, integral calculations are risky business in practice: since the integral is a cumulative sum, it also accumulates background noise which results in a growing drift in time, which is why they work best on a short period and where noise is relatively lower (i.e. higher acceleration).

3.2 Acquiring Training Variables

One of the first aspects when acquiring training data is that of its sampling rate. The Nyquist-Shannon theorem states that there is an optimal sampling frequency, above which the sampling process is error-free. Such frequency varies for each physical phenomenon, depending on the nature of the measured signal (its rate of change in time, i.e. its bandwidth *B*): around 100 Hz for speed in weight exercises (Bardella et al. 2017); 1–2 kHz for raw surface electromyography (*sEMG*) signals (Ives and Wigglesworth 2003; Sadhukhan et al. 1994) and so on.

The theory only tells us about the minimum frequency, not the maximum. A common, often abused misbelief, is that the higher the sampling frequency, the better; but no: a sampling rate much higher than its optimal value is not only meaningless, but also counterproductive, as it will degrade the signal quality (Fig. 6). This is because beyond the Nyquist frequency, the entire signal spectrum is already fully captured, and further augmenting f_{samp} only has the effect of picking up more background noise (and also, unnecessarily increasing the data size). Noise has a constant *spectral density* N_0 (measured in *W/Hz*) at any frequency (*white noise*), and the total noise power is given by:

$$P_{noise} = N_0 \cdot f_{samp} \tag{22}$$

The unnecessarily higher is f_{samp}, the higher is the noise power, and the worse is the signal quality. This is measured by the *signal-to-noise ratio* (*SNR*), i.e. the quotient between signal and noise power:

$$SNR = \frac{P_{signal}}{P_{noise}} = \frac{P_{signal}}{N_0 \cdot f_{samp}} \tag{23}$$

One might think of reducing noise by filtering, such as averaging consecutive samples. But this is actually equivalent to reducing the sampling rate in the first place; and two wrongs do not make a right. Also, since averaging or filtering smoothens the signal spikes, it might alter its peak value (but not it's average), which is also the reason why mean values (e.g. mean concentric power) are generally more stable indexes than peak ones. And one more reason to avoid the use of filtering without a basic knowledge on their theory.

The second practical factor in data acquisition is resolution. But this must be put in context too: accuracy only makes sense relative to the fluctuation of the variable being measured. With a proper *conditioning* (amplification) of the signal, even a resolution as low as 10-bit ($D_{\max}$ = 1024 steps) results in a quantization error below 0.1% which is, by far, below the intrinsic variability of any human-related training variable.

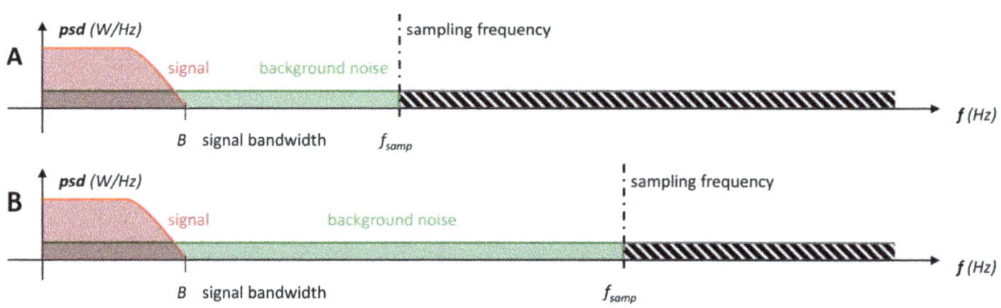

Fig. 6 Effect of optimal (**A**) and excessive (**B**) sampling frequency. In both cases, the signal is sampled losslessly but in B, a larger portion of background noise is captured, resulting in a higher noise power (green area) for the same signal power (red area) and a worse signal/noise ratio

In fact, higher resolutions (12–14 bits) should, and are, actually used to guarantee sufficient precision even when the signal amplitude is low, as this result in less *effective* resolution and loss on precision. Without entering too much into detail, this happens when using an instrument with a measurement range much larger than needed (e.g. a 2000 N load cell to measure forces of 100–200 N), which is never a good idea.

3.3 Practical Considerations on Machine Technologies

Formulas (11–17) represent ideal physical models of different training equipment. How closely the actual machine is to its model, depends on several factors. In essence, all or most of the energy conversion must occur between the athlete's work and the main source of resistance (weight, flywheel, spring or rubber band, pneumatic cylinder, etc.).

With reference to (9), in a weight machine for example it is expected that energy conversion occurs mainly between the athlete's work and gravitational energy of the load ($E_w \approx E_g$); which is usually the case. Instead, elastic and pneumatic devices should have no preponderant moving masses so that inertial or gravitational forces, if any, are negligible; and normally, this is also a fairly good approximation.

Likewise, it should happen in flywheel equipment. Closeness to ideal behaviour here is even more paramount, since not only the amount of resistance, but the entire working principle of the machine—and its capability of generating overload—is based on this assumption. That is, all of the athlete's concentric work is ideally transformed in flywheel kinetic energy, and integrally returned in the eccentric phase (10). Said in formulas:

$$Ew = E_{CON} \approx E_{FW} \approx Ecc \tag{24}$$

Friction is the most obvious cause of nonideality: the rope or strap going through several pulleys, and the use of low-quality ball bearings are a recipe for energy dissipation. But even in its absence, machine structure also plays a role. In devices like in Fig. 7A, the only moving part in addition to the flywheel is the small harness pulley; hence, the majority of the work goes into flywheel kinetic energy, and only a negligible amount is wasted in friction and in moving the pulley (Fig. 7B).

However, there are machines (Fig. 7C) which have a considerable number of bulky moving parts. There, only some of the concentric work is converted into flywheel kinetic energy and returned in the eccentric phase, whereas a non-negligible fraction it wasted in setting these parts in motion, which therefore constitute a so-called *parasitic inertia*. With reference to the energy flow in Fig. 7D, out of the total athlete's concentric work (E_{CON}), only a fraction is converted into kinetic energy stored in the flywheel (E_{FW}) and returned in the eccentric phase. Part of it is used to lift and rotate the leverage (leg flexion), which acts both as a gravitational load instead of inertial, and as a parasitic rotating inertia. Far from

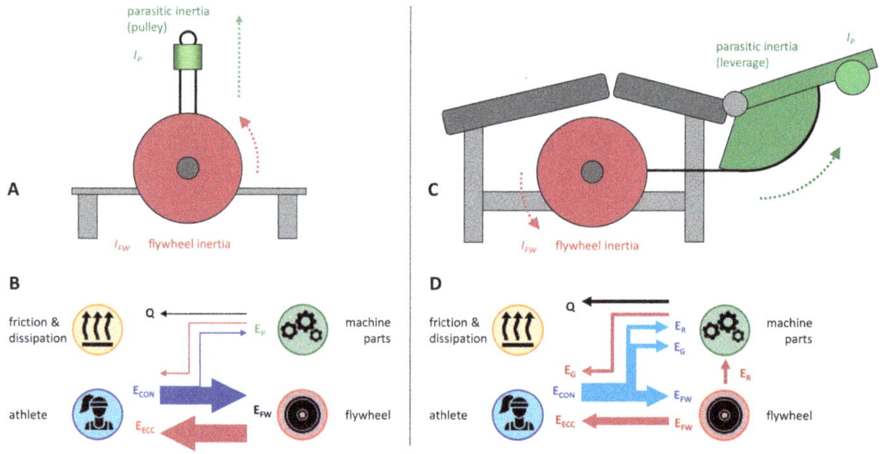

Fig. 7 Two sample flywheel devices with negligible (squat, **A** and **B**) and not negligible parasitic inertia (leg curl, **C** and **D**) and related energy and energy flow

aiming at a detailed analysis of these parasitic effects, the overall effect is that, for the same concentric energy, only a fraction is returned and made available to produce overload. Parasitic inertias are more preponderant when smaller, lower inertia flywheels are used, such as during exercises with injured players, which is when eccentric overload is known to be less subjectively perceived.

3.4 *Physics of Eccentric Overload*

The expression "eccentric overload" is (too) often associated with FRTDs training, to the extent of using the expression "eccentric training" interchangeably, and erroneously, with "flywheel training". Per se, overload means no more and no less than a higher amount of a given variable with respect to another. One classical definition of eccentric overload is the use of loads above 100% of 1RM on the eccentric phase (Aagaard 2010; Hoppeler 2016). It appeared associated for the first time with flywheel devices, and computed as the ratio of eccentric/concentric peak force, in (Berg and Tesch 1998), while other authors have used the ratio of average (Sabido et al. 2018; Illera-Domínguez et al. 2018) and peak power (Norrbrand et al. 2008; Fernandez-Gonzalo et al. 2014) instead.

Therefore, in principle, eccentric overload can be computed as the ratio of eccentric/concentric of any variable: force, speed or power, peak or average:

$$\lambda_{Fpeak} = \frac{F_{ECC,peak}}{F_{CON,peak}} \quad \lambda_{vpeak} = \frac{v_{ECC,peak}}{v_{CON,peak}} \quad \lambda_{Ppeak} = \frac{P_{ECC,peak}}{P_{CON,peak}} \tag{25}$$

$$\lambda_{Favg} = \frac{F_{ECC,avg}}{F_{CON,avg}} \quad \lambda_{vavg} = \frac{v_{ECC,avg}}{v_{CON,avg}} \quad \lambda_{Pavg} = \frac{P_{ECC,avg}}{P_{CON,avg}} \tag{26}$$

In lack of a standardization of the definition, it is more formally correct to specify on which variable eccentric overload is computed onto.

The principle of overload is, actually, the same for all three variables. This is the moment where all the previous theory falls into place at once:

- Energy (*E*), which is the integral of power *P*, is the same for concentric and eccentric since, in theory, no dissipation occurs (10)
- Displacement (ΔL), which is the integral of speed *v*, is the same for concentric and eccentric, since the initial and final positions after one repetition are identical
- Impulse (*J*), which is the integral of force *F*, is the same for concentric and eccentric since it represents the net variation of momentum (7′) which is zero, because initial and final speed is zero, hence momentum (5) is zero.

Incidentally, it is meaningless to compute eccentric overload on the above three variables, exactly because they are identical for concentric and eccentric, and their ratio is unitary.

As known, the integral of a variable is the area beneath the curve, i.e. the variable multiplied by time. Said empirically, is the total "quantity" of the variable over time. To produce eccentric overload on a given variable, it is sufficient to express it for a shorter time in the eccentric than in the concentric phase; in fact:

- expressing the same energy *E* in less time *T*, results in higher power: $P = E/T$
- travelling the same distance ΔL in less time *T*, results in higher speed: $v = \Delta L/T$
- exploiting the same impulse *J* in less time *T*, results in higher force: $F = J/T$.

Note that the above are generic formulas and, so far, no assumption has been made on a specific device. So, as far as no energy dissipation (e.g., friction) occurs, this concept applies to any technology, machine and technique (Fig. 8A), not just flywheel.

As a matter of fact, a plyometric barbell jump squat from a box is a textbook eccentric exercise: during the entire downwards flight time, no force is produced because the athlete is free falling at the same speed and acceleration (*g*) of the barbell. All of the force is generated on the short landing time *T*, where the entire athlete's momentum is dissipated. The shorter the time, the higher is the force (which is the derivative, i.e. the change of momentum in time (6)):

$$F = \frac{dp}{dt} \approx \frac{\Delta p}{\Delta t} = \frac{M \cdot v_c}{T} \tag{27}$$

where v_c is the speed at the time of impact, computed from jump height *h*, using the law of free-falling body under constant acceleration *g*:

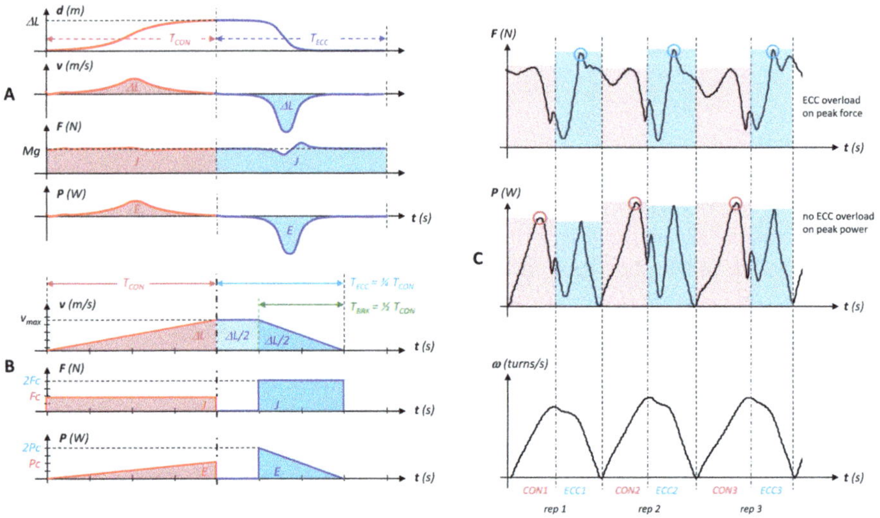

Fig. 8 Concept of eccentric overload. **A** displacement, speed, force and power in a weight exercise showing the integral (i.e. area beneath each curve) being equal to displacement (ΔL), impulse (J) and energy (E) and identical for concentric (red) and eccentric (blue). **B** Model of eccentric overload generation on force and power on a flywheel device. **C** Force, power and flywheel speed in actual flywheel exercise, showing eccentric overload on peak force but not peak power

$$v_c = \sqrt{2 \cdot g \cdot h} \tag{28}$$

Simply, on some machines or with certain techniques, some types of overload are easier to attain or control than others; or just not possible at all. For example:

- In flywheel devices, the flywheel keeps accelerating throughout the entire concentric phase, then decelerating on the eccentric one, So the peak speed is the same for both phases and no overload can be attained on peak speed ($\lambda_{V,peak} = 1$)
- In elastic and pneumatic machines, there is theoretically no moving mass, hence no momentum can be built in concentric and consumed in eccentric. Consequently, the principle of overload on force cannot apply, and no overload can be generated on force.

To explain how to attain overload on a flywheel device, a model is represented in Fig. 8B where, for simplicity, force is assumed constant and applied throughout the entire concentric phase. The result is a constant acceleration, with speed increasing linearly until v_{max} at the end of the ROM, where the full strap length ΔL has been unwound. At that point, the flywheel keeps spinning by virtue of its inertia and starts rewinding the strap, initiating the eccentric phase. The flywheel is let go for a portion of the ROM (half in the example: $\Delta L/2$), then a constant force is applied, decreasing speed linearly for the remaining $\Delta L/2$.

Remembering that the area beneath speed (its integral) is displacement, which is equal for both phases, it is easy to demonstrate using areas of triangles and rectangles, that $T_{ECC} = \frac{3}{4}T_{CON}$.

In general, the later the braking force is applied on a flywheel, the shorter the relative duration of the eccentric phase:

$$T_{ECC} \leq T_{CON} \tag{29}$$

and proportionally even shorter the duration of the braking action over which force and power are generated. This behaviour is peculiar of flywheel machines since they are the only where initial eccentric speed is maximal, and higher speed means same displacement (ΔL) in less time. Also, since energy E is supposedly the same in both phases, and average power is the ratio of energy and time:

$$P_{CON,avg} = \frac{E}{T_{CON}},\ P_{ECC,avg} = \frac{E}{T_{ECC}} \tag{30}$$

the consequence is that average eccentric power can be higher than concentric:

$$P_{ECC,avg} \geq P_{CON,avg} \tag{31}$$

In practical cases, however, the difference is minimal. And since energy is dissipated in the braking phase (T), which is a fraction of the eccentric phase (T_{ECC}), peak values are more pronounced than average ones which are computed over the entire eccentric duration. This is evident in Fig. 8B where it is easy to compute areas and eccentric overload indexes:

$$\begin{array}{lll} \lambda_{Favg} = 4/3 = 125\% & \lambda_{vavg} = 4/3 = 125\% & \lambda_{Pavg} = 4/3 = 125\% \\ \lambda_{Fpeak} = 2Fc/Fc = 200\% & \lambda_{vpeak} = v_{\max}/v_{\max} = 100\% & \lambda_{Ppeak} = 2Pc/Pc = 200\% \end{array}$$

Whether it is better to compute eccentric overload on force or speed does not have a firm answer; however, since power is the product of force times speed, eccentric overload might occur in force but not power (Fig. 8C), because force production capability is lower at higher speed, which suggests power being a more conservative approach. It is also the most practically feasible if a force transducer is not available.

No matter on which variable it is attained, since the quantity of eccentric energy and momentum are constrained to be, at most, equal to concentric one, eccentric overload is always a trade-off between quantity and time: the more overload, the shorter its duration (Fig. 9A–C).

The only way to obtain eccentric overload without sacrificing duration is to have more eccentric than concentric energy, which is possible with robotic, active training devices (e.g. SmartCoach Exentrix) where the excess of eccentric energy is

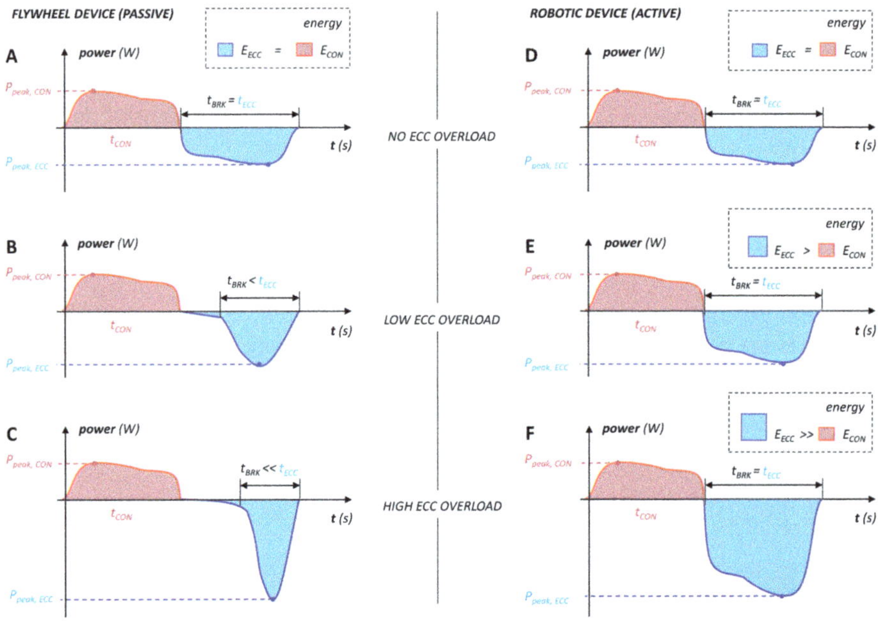

Fig. 9 ECC overload in passive (left) versus active training devices, at increasing levels of ECC overload. In **A**, **B**, **C** the red area (integral of CON power = CON energy) is the same as the blue one (ECC energy), just spread differently over time: depending on execution, the shorter the braking phase (t_{BRK}), the higher peak ECC power and overload. On the right, active device with increasing levels of ECC energy for same CON work, which allows attaining the desired ECC overload over the entire ROM, without sacrificing duration

adjustable and taken from an external source: the electrical power supply (Fig. 9D–F). This also allows eccentric overload on energy and momentum, physically impossible on any purely mechanical device.

3.5 *Physics of Pulleys*

Last but not least, pulleys deserve a mention because of their extensive use and because, despite their simplicity, their working principle is a great example of practical application of the main Laws of Physics. A basic pulley is shown in Fig. 10A; one end of the rope is tied to an anchor point ⓿, while forces are applied to ends ❶ and ❷. Since ropes transfer forces by changing their direction but not magnitude, the same force F_1 is present at both ends ⓿and ❶. And since the I Newton's Law states that "*the sum of all the forces acting on a body, including inertial ones, is zero*", that means that the force at end ❷ must equal the total of forces on the opposite sides (two identical forces F_1):

$$F_2 = 2F_1 \tag{32}$$

What about speed? The I Law of Thermodynamics states that input and output energy must be equal, and so must be their amount per time unit, i.e. power (2). In point ⓿there is no power since speed is zero, so power (which is force times speed (1) is transferred only between points ❶ and ❷:

$$P_1 = P_2 \Rightarrow F_1 \cdot v_1 = F_2 \cdot v_2 \tag{33}$$

which yields the relationship between speeds:

$$v_2 = \frac{1}{2} F_1 \tag{34}$$

So, a simple pulley multiplies force by two, but divides speed by two, keeping power (their product) unchanged. Since displacement is the integral of speed (4), this is also divided by two. Since pulleys are reversible, they can be used to double the resistance (Fig. 10B) at half speed and displacement, or vice versa (Fig. 10C), depending on which side athlete and load are connected.

In general, the more the pulleys the higher the transmission ratio K (for example, $K = 4$ in Fig. 10D) but the concept is the same: given ❶ and ❷ the (interchangeable) points of application, force and speed on one side are respectively K and $1/K$ times those on the other side:

$$\begin{aligned} F_2 &= K \cdot F_1 \\ v_2 &= \frac{1}{k} \cdot v_1 \end{aligned} \tag{35}$$

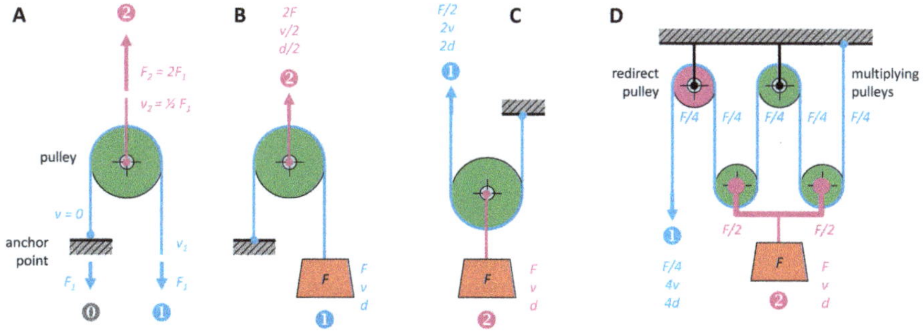

Fig. 10 Working principle of a pulley (**A**) used to multiply resistance (**B**) or speed (**C**), and general principle of a multiple pulley (block and tackle, **D**)

4 Filling Gaps

Now that physics of training technologies and their implications have been analyzed, some practical conclusions can be drawn.

To begin with, no machine has purely isotonic, inertial, or elastic behaviour; it also depends on speed, execution, and the actual machine structure. One example is shown in Fig. 11, where weight and pneumatic machines are compared at different speeds: at low velocity (Fig. 11A), both exhibit a fairly constant resistance; the variability versus angle is purposely due to the physical construction, and it is meant to counteract the *sticking point*.

At higher speed (Fig. 11B), the inertial component F_a prevails in weights, opposing to speed build-up. This is precisely the purpose of pneumatic resistance device: allowing high speed training with constant, controlled load. Also, elastic device has the same behaviour, but their resistance also increases with ROM.

The above is also the reason of existence of conical shaft flywheel devices (e.g. VersaPulley™): at the beginning of CON, the rope is wound around a higher radius, so the force required to initiate the movement is lower than in plain shaft machines, and more speed can be built in the flywheel. Also, conical flywheel devices exhibit longer CON-ECC coupling time, because the path of the rope to change side of the shaft at the end of ROM is longer. Whether a smoother or a more explosive coupling time is desired, depends very much on the application.

In air machines at high speed (Fig.11B), one can also observe a peculiar behaviour where the path done by force versus angle is lower in ECC than in CON (*hysteresis*). This is due to fluid-dynamic friction in the pneumatic cylinder, which increases with speed, and is the reason why pneumatic machines are not well suited to produce eccentric overload at high speed. It can be demonstrated that the area enclosed in the force versus angle path represents the energy loss.

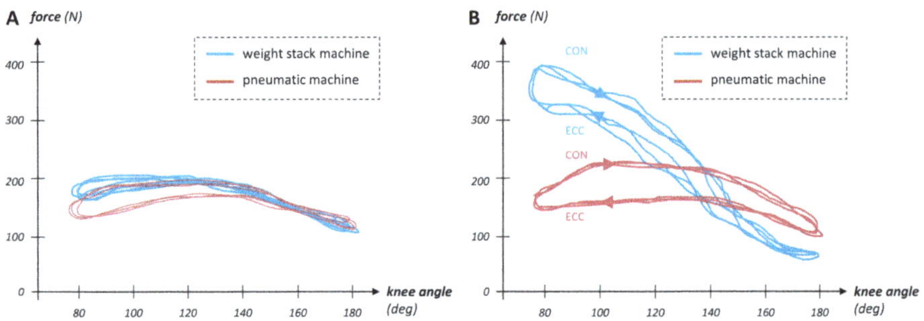

Fig. 11 measured machine force versus knee angle (180° = full extension) in one-legged exercise in weight stack machine (blue) compared with pneumatic machine (red), at different execution speeds: **A**, slow (≈45 deg/s) and **B**, fast (≈ 90 deg/s). In weight, at the beginning of ROM, the inertial force component ($F_a = M{\cdot}a$) required to accelerate the weight upwards, adds to the weight load ($F_g = M{\cdot}g$) and opposes to increase in speed. This does not happen in pneumatic machines, because of the negligible moving masses. Reproduced from Keiser®, with permission

Table 1 Summary of the main characteristics of different technologies

Parameter	Technology			
	Weights	Flywheel	Elastic	Pneumatic
Resistance	Roughly constant (at low speed only), proportional to load	Highly variable, proportional to acceleration and inertia	Proportional to displacement and stiffness	Fairly constant, proportional to pressure
Load variability	Low	High	Medium	Low
Dynamic load	Yes	Yes	No	No
Speed range	Low to medium	Medium to high	Low to medium	Medium to high
Ecc overload	No	Yes	No	No
Sticking point	Yes	No	Yes	Yes
ROM	Free	Max ROM fixed	Free	Free

Anyway, resistance variability per se is not a disadvantage; flywheels' variable resistance is one of their key features believed to elicit higher degree of motor unit recruitment/de-recruitment at neural level (Pozzo et al. 2006). On the other hand, that calls for the need for a feedback system to control it and tune the desired amount of overload.

In fact, flywheel devices are the only machines with no minimum resistance to overcome to initiate the movement. This is why there is no flywheel equivalent of a 1RM concept: no matter how huge the inertia is, even the smallest force can initiate the movement, just slower (7″). This is one of flywheel's greatest values and worst defect, since it makes normalization difficult.

A summary of the main characteristics of different technologies is shown in Table 1.

5 Take Home Messages

Far from being exhaustive, this chapter provided an insight into physics of training technology and performance monitoring. The main conclusion is that, in general, no training device or any data acquisition equipment has pure, ideal behaviour.

Some among the less obvious conclusions that can be drawn are:

1. Weight exercises are not purely isotonic: wherever there is a mass, there is always an inertial force component; it just negligible or not depending on specific circumstances.
2. When compared at same static, gravitational load than weights ($F_g = M \cdot g$) pneumatic and elastic training devices are better suited for explosive training as they lack the added inertial component that opposes to speed build-up.
3. At high speed, pneumatic devices exhibit hysteresis in their force–displacement curve which makes them poorly suited to produce eccentric overload.

4. Inertia is a physical property which opposes to whatever movement: in linear movement, it is expressed by mass; in rotational motion it is the body's *moment of inertia* (or rotational inertia).
5. From a semantic standpoint, *isoinertial* literally means "constant inertia". Although used to define flywheel devices, this is a misnomer since also weights have a constant inertia: their mass.
6. Parasitic inertias in flywheel devices deviate their behaviour from the ideal physical model, and reduce the capability of producing ECC overload.
7. Conical shaft flywheel devices are designed to mitigate the inertia at the beginning of the concentric phase, to help build momentum; their larger effective radius also results in a longer coupling time.
8. From a physical point of view, eccentric overload can be accomplished with any device capable of storing and returning energy; it's just easier to attain in flywheel devices, although it calls for use of some feedback to control it.
9. Because of the absence of a constant force component (minimum force threshold to overcome to initiate a movement), there cannot be a flywheel equivalent of 1RM.
10. Comparing flywheel devices based purely on their inertia is meaningless, as the shaft diameter plays a major role in the resistance. The concept of flywheel equivalent mass (FWM) allows a more formally correct comparison. Also, since it has the same physical dimensions of 1RM, it is a potentially candidate index to compare weights and flywheels.
11. Higher sampling frequencies do not improve measurement accuracy: a sampling rate much above the optimal value does not improve the measurement precision and instead results in additional noise.
12. Filtering introduces delays in data that can alter the temporal relationship between variables. Caution is advised against the random use of filtering unbeknownst of their theory.
13. To minimize resolution errors, the measurement device range shouldn't be too high compared to the order of magnitude of the measured variable.

To summarize all of the above in one sentence: nothing is universally true nor ideal. And more is not necessarily better: everything shall be put into context and relativized to the actual application.

References

Aagaard P (2010) The use of eccentric strength training to enhance maximal muscle strength, explosive force (RDF) and muscular power—consequences for athletic performance. Open Sports Sci J 3(1):52–55

Bardella P, Carrasquilla García I, Pozzo M, Tous-Fajardo J, Saez de Villareal E, Suarez-Arrones L (2017) Optimal sampling frequency in recording of resistance training exercises. Sports Biomech 16(1):102–114

Berg HE, Tesch A (1994) A gravity-independent ergometer to be used for resistance training in space. Aviat Space Environ Med 65(8):752–756

Berg HE, Tesch PA (1998) Force and power characteristics of a resistive exercise device for use in space. Acta Astronaut 42(1–8):219–230

Cunningham MJ (2003) Electrical Measurement. In: Laughton MA, Say MG (eds) Electrical engineer's reference book, 16th edn. Elsevier

DiLisi GA (2019) Tools classical mechanics, vol 1: Tools and vectors. Morgan & Claypool Publishers

Fernandez-Gonzalo R, Lundberg TR, Alvarez-Alvarez L, de Paz JA (2014) Muscle damage responses and adaptations to eccentric-overload resistance exercise in men and women. Eur J Appl Physiol 114(5):1075–1084

Hoppeler H (2016) Moderate load eccentric exercise; a distinct novel training modality. Front Physiol 16(7):483

Illera-Domínguez V, Nuell S, Carmona G, Padullés JM, Padullés X, Lloret M, Cadefau JA (2018) Early functional and morphological muscle adaptations during short-term inertial-squat training. Front Physiol 9:1265

Ives JC, Wigglesworth JK (2003) Sampling rate effects on surface EMG timing and amplitude measures. Clin Biomech 18(6):543–552

Kehtarnavaz N (2008) Analog-to-digital signal conversion. In: Digital signal processing system design, 2nd edn

Norrbrand L, Fluckey JD, Pozzo M, Tesch PA (2008) Resistance training using eccentric overload induces early adaptations in skeletal muscle size. Eur J Appl Physiol 102(3):271–281

Pozzo M, Alkner B, Norrbrand L, Farina D, Tesch PA (2006) Muscle-fiber conduction velocity during concentric and eccentric actions on a flywheel exercise device. Muscle Nerve 34 (2):169–177

Sabido R, Hernández-Davó JL, Pereyra-Gerber GT (2018) Influence of different inertial loads on basic training variables during the flywheel squat exercise. Int J Sports Physiol Perform 13 (4):482–489

Sadhukhan AK, Goswami A, Kumar A, Gupta S (1994) Effect of sampling frequency on EMG power spectral characteristics. Electromyogr Clin Neurophysiol 34(3):159–163

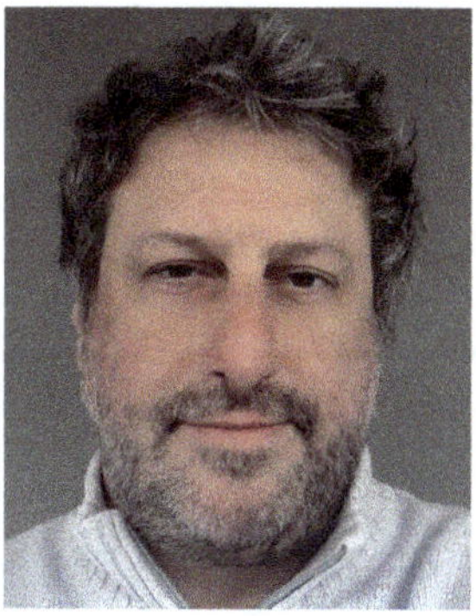

Marco Pozzo is an engineer with expertise in multichannel surface EMG and technical aspects of sports physiology. As a fellow of the European Space Agency (ESA), he worked at Karolinska Institute in Stockholm with Prof. Per Tesch on the application of flywheel technology in space. There he founded the company SmartCoach Europe, expanded to SmartCoach Technologies (CA, USA) in 2018, to develop tools for performance monitoring in resistance training. He has been collaborating with NASA, several top sports club, elite athletes, and Olympic champions. He is constantly engaged in research and is author of several scientific papers and book chapters.

Franco Impellizzeri is a professor in Sport and Exercise Science and Medicine at the University of Technology Sydney (UTS, Australia). He started his career at the MAPEI Sport Research Center (Italy), where he was taking care of the training and testing for elite and top-professional level athletes. He has also worked ten years in clinical setting as at the Schulthess Clinic (Zurich, Switzerland) focusing on clinical outcomes and clinimetrics. He has more than 170 publications and is a member of the boards of various scientific journals. He has also trained elite athletes participating at the Olympic Games in various disciplines.

Muscle Strength Determinants and Physiological Adaptations

Jesús G. Ponce-González and Cristina Casals

Abstract Strength training, also known as weight or resistance training (RT), has become one of the most popular forms of exercise, not only for sport performance but also for improving health-related fitness. A wide variety of physiological adaptations achieved through RT have been documented in the short, medium, and long term. These improvements include changes in body composition, muscle hypertrophy, strength, power and motor performance; as well as other health benefits such as changes in blood pressure, insulin sensitivity, lipid profile, endocrine system, and better performance in daily life activities, among others. This chapter will cover the basic physiological adaptations of RT discussing neurological, musculoskeletal, cardiorespiratory, and endocrine responses and adaptations according to current scientific literature. These physiological concepts will be applied in following chapters in which specific methods and technologies for RT are presented.

Keywords Strength · Skeletal muscle · Hypertrophy · Motor units recruitment · Body composition · Anabolic signaling pathways · Growth factors · Exercise training

1 Introduction

Resistance training (RT) leads to the disruption of homeostasis (the body's dynamic state of equilibrium), posing a stress to which the body responds and adapts, which sets the basis of exercise and sport physiology that are crucial for understanding the principles of training.

The RT-induced responses are the acute effects of this type of exercise that, usually, return to regular values (homeostasis) in a few minutes, hours or 1–2 days

J. G. Ponce-González (✉) · C. Casals
Faculty of Education Sciences, University of Cádiz, C. Republica Saharaui, 12, 11519 Puerto Real, Cádiz, Spain
e-mail: jesusgustavo.ponce@gm.uca.es

A. Muñoz-López et al. (eds.), *Resistance Training Methods*, Lecture Notes in Bioengineering, https://doi.org/10.1007/978-3-030-81989-7_2

after physical exercise; while adaptations are chronic effects, leading to gradual but long term changes, such as a gain in body muscle mass. Therefore, physiological responses are those changes after an acute exercise and physiological adaptations are those achieved by regular physical training.

In this chapter, a critical review of the current scientific literature in exercise physiology and RT is presented, in the hope that the reader can increase the knowledge of the underlying causes of RT-related improvements in performance, strength, hypertrophy, etc. Therefore, this chapter details neurological, musculoskeletal, cardiorespiratory and endocrine responses and adaptations to RT.

2 From Theory

2.1 Neurological Responses and Adaptations

Changes in the nervous system contribute to the development of muscle strength after a period of RT (Sale 1988). Indeed, physical strength can be improved in the absence of hypertrophy (Enoka 1997), by enhancing the brain's ability to recruit muscles to contract and produce a desired movement. Therefore, an interesting question in the RT is related to the physiological mechanisms capable of explaining the disparity between muscle mass and strength gains.

In this line, it has been reported that most of the strength gain during the initial weeks of RT are due to adaptations in facilitatory and inhibitory neurological pathways acting at different levels in the nervous system (Moritani and deVries 1979). A recent meta-analysis (Siddique et al. 2020), conducted with randomized controlled trials that focused on neural adaptations to resistance-training, concluded that the training-related increase in muscle strength is due an increment of motoneuron activation as a consequence of both cortical and subcortical adaptations. Furthermore, RT lead to changes in synaptic efficacy within the motoneuron pool that increase motor unit short term synchrony (Carroll et al. 2001), which play a fundamental role in the execution of voluntary movement and strength gains.

These neurological adaptations to RT are difficult to reveal and there are multiple neurological elements that contribute to both the quantity and quality of muscle fibre activation. Some key points related to the neural adaptations to RT are the following statements: *(i)* increased activation of agonist muscles, *(ii)* reduction in the co-activation of antagonists and *(iii)* better co-activation of synergists (Sale 1988).

Agonist muscle activity results in a movement in the desired direction, while antagonist activity opposes that motion. Thus, it is not only important a higher activation of the agonist muscle, moreover, the reduction of the co-activation of antagonist muscles allows higher strength levels of agonist muscles (Gabriel et al. 2006). Co-activation is a mechanism adopted to provide higher stabilization of joints; co-activation is higher, the greater uncertainty of the action to be taken (Enoka 1997, 1988). It is up-regulated by the central nervous system and in reflex responses at the level of the spinal cord, but also the activation of Renshaw cells excites the interneurons lb of the Golgi organs (Enoka 1997; Gabriel et al. 2006).

In summary, neural adaptations that lead to greater strength are partly due to the disinhibition of inhibitory mechanisms such as: *(i)* Golgi tendon organs are sensory receptors, which, with excessive tension, elicit a reflex inhibition of the muscle contraction; *(ii)* Renshaw cells are inhibitory connecting neurons located in the spinal cord, this interneurons temp the rate of discharge of alpha motor neurons to avoid tetanic contraction and its related damage; and *(iii)* Supraspinal inhibitory signals that consist in conscious or unconscious inhibitory signals released from the brain (Bompa and Buzzichelli 2015). Co-activation is too high in beginners but can be attenuated with training and movement familiarity (Solomonow et al. 1988).

The RT may enhance performance by reducing the extent of cortical activation and therefore the activation of neural elements that interfere with the optimal execution of movement (Carroll et al. 2001). This highlights that, in early phases of training, the improvements in inter-muscular coordination can be more determinant of strength gains than the proper force-generating capacity of the muscle (Tallent et al. 2021); where main factors are: *(i)* Synchronization, the capacity to contract motor units simultaneously or with a minimum latency (<5 ms); *(ii)* Recruitment, the capacity to recruit motor units simultaneously; and *(iii)* Rate coding, the capacity to increase firing rate (motor unit discharge rate) in order to express more strength (Bompa and Buzzichelli 2015).

Neural adaptations can explain why despite similar growth in muscle mass, those who lift heavier weights enjoy greater strength gains than low-load lifters (Jenkins et al. 2017). Subtle changes within the nervous system lead to increases in muscular strength, since the mobilization of the motor unit is a determining factor of strength and would explain most of the strength gains achieved in the absence of hypertrophy (Enoka 1997). Thus, the rapid gain in muscular strength during the early phases of a RT program is mainly due to the increased ability to activate muscle fibres. However, although neural factors appear to play a greater role in initial strength increases, hypertrophic processes also commence at the onset of training and contribute in long-term adaptations (Folland and Williams 2007) (Fig. 1).

Fig. 1 Neural and muscular adaptations to strength training over time

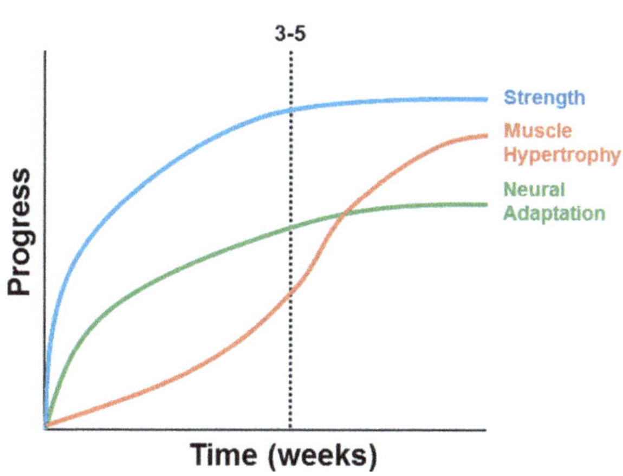

2.2 *Musculoskeletal System Responses and Adaptations*

The RT produces several benefits related to specific musculoskeletal tissues in response to specific mechanical loading. The musculoskeletal system, which forms a framework for the body, is subdivided into two broad systems: (i) skeletal system which incorporates bones, cartilage, and ligaments; and (ii) muscular system which is formed by skeletal muscle and tendons. Although this chapter will focus on skeletal muscle modifications, the rest of the adaptations of musculoskeletal system will be briefly detailed below.

2.2.1 Skeletal System (Bone, Cartilage, and Ligament)

Bone Mass

The mechanical loading, carried out with RT especially with impact, induces osteocyte mechano-transduction and, hence, bone remodeling process. This occurs in part to increase dynamic flow of the pericellular intersticial fluid in the lacunar-canalicular system (Maestroni et al. 2020; Weinbaum et al. 2011). Zhao et al. (2015) performed a meta-analysis including 1769 postmenopausal women, who are the most likely people to develop osteopenia and osteoporosis losing up to 5% of bone mass annually. This meta-analysis concluded that a clinically significant gains in spine (almost 2.4%) and hip (almost 1.8%) bone mineral density (BMD) could be achieved by combining RT with impact exercise as jumping, skipping, hoping, etc. Similar results were reached about benefits on bone mass in youth (childhood/adolescent) (Beck et al. 2017); and it appears to translate across the life span, reducing future risk of fracture and osteoporosis (Warden et al. 2014). Moreover, the level of muscle strength has positively been associated with BMD, but also with bone mineral content and bone area (Clark et al. 2011).

Cartilage and Ligaments

Fibers (elastic and collagenous fibers), ground substance, and cells are the main components for all connective tissue immersed in the body water. Structures such as tendons and ligaments are characterized by collagen fibers arranged in an orderly parallel fashion, giving it tensile strength in one direction; being type I collagen fibers its main matrix element (Amiel et al. 1984). However, the matrix of cartilage is made up of glycosaminoglycans, proteoglycans, elastin and collagen fibers, mostly type II collagen fibers (90–95%) (Sophia Fox et al. 2009). Cartilage is line of connective tissue that works as padding between the bones, whereas a ligament is a band of tissue that connects bones to each other, and ensures the joint is stable.

Cartilage seems to be reduced with immobilization in a range of 5–7% of thickness (Vanwanseele et al. 2003). Due to the avascularity of cartilage, the ability to repair it is limited (Bohndorf 1999). However, it has been found a positive association between muscle cross-sectional area (CSA) and cartilage thickness (Hudelmaier et al. 2003); hence muscle hypertrophy obtained through RT could favor cartilage morphology. Consistent with this, knee extensor muscle weakness is a risk factor for knee osteoarthritis (OA) (Oiestad et al. 2015; Hunter et al. 2020), being RT one of the key element included in clinical guidelines of OA management (Maestroni et al. 2020; Hunter et al. 2020; Lin et al. 2020). In concordance, it has been shown that RT could benefit the stiffening of pericellular and inter-territorial matrix, cartilage volume and glycosaminoglycan, and the protective response of muscle strength against cartilage loss (Maestroni et al. 2020; Bricca et al. 2017, 2019; Amin et al. 2009).

In turn, experiments in rat models showed that training animals had heavier ligaments, higher junction strength and strength/body weight ratios compared to non-trained animals (Tipton et al. 1974, 1975). In humans, deep knee bends with weights improved the stability of the ligaments of the knee (Tipton et al. 1975); later confirmed with magnetic resonance imaging (MRI) in professional weightlifters who started about the age of puberty, showing higher hypertrophy in patellar and cruciate ligaments than controls (Grzelak et al. 2012a, b). Hence, RT could increase blood flow and the availability of anabolic hormones, as testosterone, to induce strength of ligaments.

2.2.2 Muscular System (Tendons and Skeletal Muscle)

Tendons

As explained above, a tendon is a connective tissue, predominated by type I and III collagen fibers, which transmits the tensile loading exerted by the muscle to the skeleton. Collagen synthesis in tendon seems to be upregulated by the release of growth factors (i.e., IGF-1, IL-6) after RT (Heinemeier and Kjaer 2011).

It has been shown that mechanical loading may exert alterations of the tendon stiffness by *(i)* changes in tendon material (i.e., increase of Young's modulus about the elasticity in tension); and *(ii)* changes in morphological properties (i.e., increase in CSA) (Maestroni et al. 2020; Bohm et al. 2015). Kraemer et al. pointed out that there have been great strides in the examination of tendon changes to RT during the last two decades (Kraemer et al. 2017). The RT over 12 weeks induced increases in muscle strength with a concomitant alteration in Young's modulus, which implies a change in the composition of tendon structure. However, tendon CSA remained unchanged (Reeves et al. 2003; Kubo et al. 2006). In contrast, Kongsgaard et al. (2007) showed an increase in a region-specific of patellar tendon CSA, without changes in modulus after 12 weeks of RT training. Higher patellar tendon CSA was also observed when were compared the lower extremities of athletes with more loading in one leg than the contralateral side by the sport characteristics (i.e., fencing or badminton) (Couppe et al. 2008).

Skeletal Muscle

There are several acute and chronic responses into the muscle cells when a person performs RT. The resistance exercise-induced stimuli lead to alterations in gene expression trough different intramuscular signaling pathways that produce the translation of transcripts (mRNA) into proteins by ribosomes (Fig. 2). In addition to these increases in protein synthesis rates and translational machinery, there is also a robust post-transcriptional regulation of protein abundance after exercise training that increase protein translation capacity. However, it has been shown that ubiquitin proteasome and autophagy lysosome pathways are upregulated after resistance exercise increasing also protein degradation (Fry et al. 2013), that could promote the dissociation between proteome and transcriptome. Hence, exercise-induced mRNA does not necessarily translate into proteomic changes (Miller and Konopka 2016). Moreover, Mitchel et al. (2014) showed that acute responses produced on protein synthesis after one bout of first RT were not associated with muscle hypertrophy obtained after 16 weeks of training.

The muscle growth will be determined when a positive balance between synthesis and breakdown of muscle protein occur. This is the main biological phenomenon that induces an increase in muscle CSA after repeated bouts of RT increasing myofribillar volume of individual myofibers, also known as muscle hypertrophy, which produces an increase of muscle strength and power (Goodman et al. 2011). The maximal force generated by muscle is directly proportional to its CSA (Hornsby et al. 2018; Taber et al. 2019) and by the type of muscle fibers, generating greater power per unit of CSA the type II fibers (Hughes et al. 2018; Cormie et al. 2011; Haff and Stone 2015; Suchomel et al. 2018).

Hypertrophic response to RT has traditionally been considered greater in fast-twitch fibers (Type II) compared to slow-twitch fibers (Type I) (Fry 2004). In concordance, type I fibers have relatively small CSA compared to type II fibers (all isoforms) (Bekedam et al. 2003; Gregory et al. 2001; Wessel et al. 2010). However, an abundance of type 1 myosin heavy chain (MHC) mRNA has been associated with higher rate of amino acid uptake, protein synthesis rates and total RNA content, while an abundance of MHC-IIA mRNA has been associated with lower protein synthesis rates and total RNA content (Wessel et al. 2010; Toth and Tchernof 2006; Hood and Terjung 1987). In addition, oxidative fibers show also higher ribosomal protein content, myonuclei per volume cytoplasm and percentage of myonuclei that belong to satellite cells compared to glycolytic fibers (Wessel et al. 2010; Tseng et al. 1994; Grgic et al. 2018). Based on the previously explained, it seems that type I fibers have higher hypertrophic potential but CSA is lower compared to type II fibers, showing an evident muscle fiber-type size paradox (Wessel et al. 2010). It has been hypothesized that it may be due to the increase of protein degradation pathways in oxidative compared to glycolytic fibers, as autophagy process, cathepsin levels (lysosomal proteolysis factor), ubiquitin proteasome system, Calcium-dependent proteolysis, reactive oxygen species

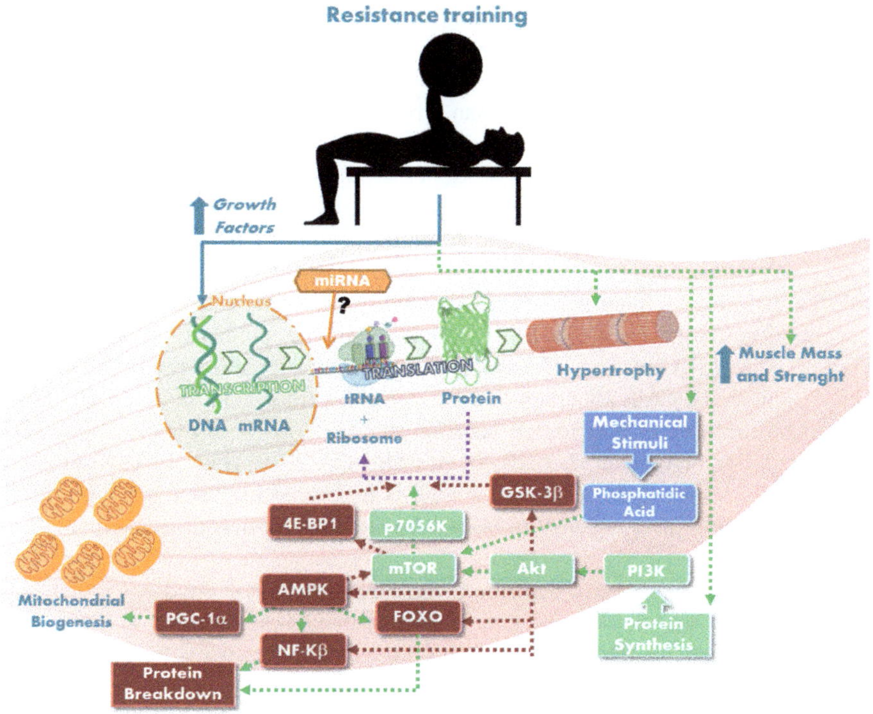

Fig. 2 Physiological responses to resistance training (RT) that increase muscle mass and strength. Green boxes represent anabolic signaling pathways activated by RT; blue boxes represent mechanical stimuli activated by RT, whereas red boxes represent catabolic signaling pathways which are downregulated by RT. Arrows show the direction of the relationships among proteins. Green arrows indicate activating downstream events, whereas red arrows indicate inhibitory downstream events. PI3K, phosphoinositide 3-kinase; AKt, protein kinase B; GSK-3β, glycogen synthase kinase 3-Beta; mTOR, mammalian target of rapamycin; p70S6K, p70S6 kinase; 4EBP1, eIF4Ebinding protein 1; AMPK, 5′adenosine monophosphate-activated protein kinase; FOXO, forkhead box O; NF-κβ, nuclear factor kappa-light-chain-enhancer of activated Beta cells; PGC-1α, peroxisome proliferator-activated receptor gamma coactivator 1-alpha. Part of the figure has been designed by macrovector, brgfx/Freepik and with BioRender.com

(ROS) and its metabolites (Wessel et al. 2010; Parreno et al. 2001; Soori et al. 2016; Steinbacher and Eckl 2015; Powers and Kavazis 2007). Moreover, it has recently discovered that microRNA (miRNA) can also be differentially regulated among the type of fibers; which can influence protein synthesis by affecting mRNA degradation (transcription), or by blocking mRNA binding sites to inhibit translation (Wessel et al. 2010) (Fig. 2). Grgic et al. (2018) suggested that future research should focus in the application of new interventions that can upregulate the protein synthesis without largely increasing protein degradation in type I muscle fibers.

Molecular Response to Resistance Training for Skeletal Muscle Hypertrophy: mTOR Signaling Pathway

A protein kinase called the mammalian target of rapamycin (mTOR) is a central regulator of muscle hypertrophy increasing ribosome biogenesis, protein synthesis, cell growth, and neurite plasticity (Wessel et al. 2010; Ogasawara et al. 2019). The PI3K-Akt-mTOR signaling pathway is activated by growth factors as insulin or IGF-1 (insulin-like growth factor-1) that are released during and after RT (Wessel et al. 2010; Ogasawara et al. 2019). Moreover, the mechanical loading that occurs with RT can also regulate per se mTOR activity by increase in phosphatidic acid concentration (Bigard 2019). This anabolic pathway produces muscle protein synthesis due to activation of several downstream signaling molecules as the GSK-3β induced by Akt phosphorylation; or 70-kDa ribosomal S6 kinase (p70S6k) and eukaryotic initiation factor 4E-binding protein 1 (4EBP1) induced by mTOR phosphorylation (Wessel et al. 2010; Ogasawara et al. 2019) (Fig. 2). This signaling cascade increases the rate of protein synthesis by promoting the translation of the mRNAs encoding the ribosomal proteins and the transcription of ribosomal RNAs (Bigard 2019).

Several reports have been described that when the protein AMPK (5′adenosine monophosphate-activated protein kinase) is activated, it can inhibit mTOR activity. The main functions of AMPK are to turn on the catabolic process to obtain energy supply as fat oxidation or authophagy, and to shut down the anabolic process as protein synthesis by inhibiting the mTOR protein. In fact, AMPK has been linked to protein degradation by the activation of NF-κβ, and MuRF. Moreover, AMPK can stimulate mitochondrial biogenesis by peroxisome proliferator-activated receptor gamma coactivator 1-alpha (PGC-1α) (Wessel et al. 2010). However, high-load RT seems to downregulate AMPK pathway leading to a dilution of the mitochondrial content relative to myofibrillar volume utilizing electron microscopy in young men (Parry et al. 2020). In agreement, citrate synthase activity (CS), as marker of mitochondrial content, is reduced from 3 weeks to 6 months of RT (Parry et al. 2020). Despite that, some authors reported divergent results, showing that RT does not decrease and may even increase VO_2max. Indeed, mitochondrial respiratory function is increased after 12 weeks of high-load RT in young untrained subjects (Porter et al. 2015). It is unlikely that mitochondrial dilution and increased aerobic capacity occur at the same time. One possible explanation is that mitochondrial content measurements are usually expressed normalized by the total amount of protein. As there is greater hypertrophy after RT (greater total amount of protein), it could be masking the real result on mitochondrial content. Furthermore, the most studies have used high-load RT, remaining unknown the effect of low-load RT with high-volume on mitochondrial adaptations (Parry et al. 2020). The fatiguing low-load RT entails higher metabolic stress that may accentuate mitochondrial adaptations and metabolic enzymes changes. The previous training status is also an important factor.

2.2.3 Cardiorespiratory Responses and Adaptations

Although the cardiorespiratory fitness is mainly improved by aerobic training, the RT can also lead to acute and chronic cardiovascular changes with a positive impact on health. In this line, a meta-analysis of randomized, controlled trials (Cornelissen et al. 2011), reported that dynamic resistance exercise training leads to significantly decrease blood pressure, increase peak VO_2 (oxygen uptake), and decrease body fat and plasma triglycerides.

As an acute response to resistance exercises, there is a higher pressor response which is attenuated in trained athletes compared with beginners (Fleck 1988), suggesting that strength training produces a decrease in peripheral resistance. This phenomenon can be explained by an increase in angiogenesis (the process through which new blood vessels—or capillaries—form from pre-existing vessels), greater capillary density and a diameter of the arteries and a central inhibition of sympathetic efferent pathways (Braith and Stewart 2006). Moreover, in the heart, the main adaptations are in the left ventricle with increased absolute left ventricular wall thickness and left ventricular mass; although these adaptations are not significant when expressed relative to body surface area (Fleck 1988).

2.2.4 Endocrine Responses and Adaptations

The endocrine system plays an important role in molecular signalling and many physiological processes. Several hormones can alter the balance between anabolic and catabolic stimuli in muscle in response to RT; highlighting testosterone, growth hormone (GH) and insulin-like growth factor-1 (IGF-1) as anabolic hormones, and cortisol as a catabolic hormone.

- Testosterone is a steroid hormone made from cholesterol, it is mainly synthesized in the Leydig cells of the testes, although small quantities of testosterone can be also produced in ovaries and adrenal glands. Therefore, there is a significant sexual dimorphism of testosterone, not only at rest, but also in response to RT (Kraemer et al. 1991).
- GH stimulates growth, cell reproduction, and cell regeneration, via GH receptor dimerization by also by stimulating the IGF-1 production. GH also regulates metabolism through the stimulation of lipid mobilization and oxidation (Manini et al. 2012).
- IGF-1 is a hormone similar in molecular structure to insulin which has anabolic effects in adults; its signalling pathways include PI3K-Akt-mTOR, MAPK extracellular signal–regulated kinases (ERK), and possibly Ca^{2+}-dependent calcineurin (Schoenfeld 2013).

The dynamic nature of hormones makes difficult to stablish its relation with strength and hypertrophy, especially because the effects of hormones depend entirely on receptors and their interactions with nuclear DNA mechanisms of target cells (Kraemer et al. 2017). Although it has been shown that hypertrophy can occur in the relative absence of anabolic hormones elevations in response to RT (Wilkinson et al. 2006), these acute increments may potentiate hypertrophy leading to the hormone hypothesis (Schoenfeld 2013; Hansen et al. 2001).

It has been shown that the hormonal responses to RT depend on the intensity, volume, load, and rest periods (Kraemer et al. 1990, 1993). Moreover, the secretion of anabolic hormones usually varies throughout the day and is affected, among others, by variations in sleep, diet, taking alcohol or drugs, stress, and fatigue.

Regarding catabolic hormones, cortisol is a steroid hormone, or glucocorticoid, produced by the adrenal gland, which is released in response to stressful situations. Therefore, the testosterone/cortisol ratio has been proposed as a useful outcome that reflects the anabolic/catabolic balance making possible an estimation of the athlete's physiological strain (Urhausen et al. 1995), that will be fully detailed in the following practice section of this chapter.

3 From Practice

3.1 Neurological Responses and Adaptations

Resistance training induces not only increases in the activation of agonist muscles, but also a reduction in the co-activation of antagonist muscles, which together with the optimal activation of the synergist muscles, are the main factors related to the strength gains in the early phases of training by leading a higher force production of the agonist muscles even in the absence of hypertrophy.

This statement highlights the importance of the principle of training specificity in strength gains, since it is the way to improve the neurological adaptations. Thus, it has been shown that the type of muscle action used in strength training and in evaluation could be interrelated where dynamic RT is more effective in increasing dynamic than isometric strength, while isometric RT produces higher increments in isometric strength but not so much in dynamic strength.

In practice, the muscle activation can be assessed by electromyography (EMG), where surface EMG is a useful tool in monitoring the athlete since it is not invasive (Gabriel et al. 2006). This tool allows to some interesting research about neural responses and adaptations to RT than can be applied to the exercise prescription. It has been shown that high-load RT results in greater neural adaptations than low-load RT, leading to rapid increases in muscle strength despite similar gains of hypertrophy between both RT programs (Jenkins et al. 2017).

3.2 Musculoskeletal Responses and Adaptations

Some studies have found muscle strength gains after only 2–4 weeks of RT, although there are significantly increased after 8–12 weeks (Hughes et al. 2018). This early increase in strength is likely caused by neuromuscular and connective tissue adaptations, whereas the early increases in CSA seems to be by œdema (fluid retention) (Maestroni et al. 2020). Lighter loads could be used until failure to maximize motor unit recruitment, and increase muscle size and strength. Loads higher than 70% of 1RM seem to optimize strength gains; however, recent findings have shown that changes in muscle hypertrophy were similar between high-load and low-load RT, when the last was performed until failure.

There are several factors related to training design that could improve or attenuate the degree of hypertrophy, as the type, volume, intensity, and recovery of RT. For example, it has been suggested that to optimize muscle growth and strength gains, the loads of RT should be higher than 70% of 1RM (repetition maximum) (American College of Sports Medicine 2009). In agreement, Schoenfeld et al. (2017) showed in a recent meta-analysis in 2017 that maximal strength benefits are obtained from the use of heavy loads. However, muscle growth could be also achieved across a broad range of medium to low loading zones, but performed to momentary muscular failure, which requires higher exercise volume (work) and time compared to high-load training. Moreover, the type of RT (low vs high loads) could affect the specificity of hypertrophy across the muscle fibers.

Recent findings suggest that preferential hypertrophy for a specific fiber depends also on range of loading zones. High-load RT (>60% of 1RM) primarily increases cross-sectional area of type II muscle fibers, while type I fibers are increased with low-load RT (<60% of 1RM), but momentary muscular failure is needed (Vinogradova et al. 2013; Netreba et al. 2013).

The load and duration of exercise training seem more relevant for tendon adaptation than the type of muscle contraction (Bohm et al. 2015). High loading intensities (85–90% of maximal voluntary isometric contraction [MVIC]) applied in five sets of four repetitions for 3 s each and 2-min of rest between sets showed greater adaptations in maximal strength, tendon stiffness, Young's modulus, and tendon CSA compared to low intensities (Bohm et al. 2015; Reeves et al. 2003; Mersmann et al. 2017; Geremia et al. 2018).

Watson et al. (2019) showed greater benefits in BMD with high-intensity resistance and impact training (5 sets of 5 rep at >80% of 1RM with jumps and drops) compared to a low-intensity exercise training (10–15 rep at <60% of 1RM). Moreover, the mechanical stress produced by RT could improve the strength of the cartilages and ligaments and its junction with bone, becoming stronger and heavier ligaments.

3.3 *Cardiorespiratory Responses and Adaptations*

From a practical point of view, cardiorespiratory benefits are greatly achieved by adding aerobic training to the RT. Notwithstanding, exclusively attending to RT types, dynamic RT, has shown higher cardiorespiratory adaptations than static or isometric RT (Cornelissen et al. 2011). Therefore, it would be favorable to use weight machines or dynabands in the RT that can be consider as adjuvant therapy for the prevention and treatment of hypertension and high blood pressure, where blood pressure improvements can be achieved over a broad range of exercise intensities, from 30 to 100% of 1RM (Cornelissen et al. 2011). Notwithstanding, emerging evidence highlights fewer increments in (intra-)arterial blood pressure and cardiac output of high-intensity dynamic RT programs ($\geq$70% of 1RM), but with significant increments in strength, compared to low-intensity RT (Hansen et al. 2019). Finally, high-intensity circuit-based RT has been proposed as a better protocol to increase cardiorespiratory and metabolic responses than traditional RT (Marin-Pagan et al. 2020), although these processes could interfere with hypertrophy mechanisms.

3.4 Endocrine Responses and Adaptations

Overtraining is a common problem in RT and it leads to a loss of strength and performance, increased perceived efforts, excessive fatigue, and other symptoms such as irritability, insomnia, loss of concentration, loss of appetite, etc. From a physiological point of view, this syndrome can be early detected through the estimation of the strain training by measuring testosterone and cortisol calculating an anabolic/catabolic balance (Urhausen et al. 1995).

The testosterone/cortisol ratio significantly decreases when there is a very high demand and the training is not being well assimilated, although it should be note that small variations are common in response to resistance exercise. Several factors (e.g. workout design, nutrition, genetics, training status and type) can acutely modify both hormones, therefore, it can impact on performance and the RT adaptations (Urhausen et al. 1995). Thus, individual hormonal data may be used to better prescribe RT and to assess the athlete's strain.

4 Filling Gaps

Physiological responses and adaptations to RT can set the basis of strength and conditioning programs; however, the measurement of some outcomes requires complex instruments and analyses, hindering the knowledge of underlying mechanisms and physiological processes. In addition, studies analyse different RT methodologies, adding greater difficulty to apply theoretical conclusions to practice due to the high variety of muscle stimuli.

According to the scientific literature, during the first weeks of training neural factors appear to play a greater role than hypertrophic processes; explaining higher strength levels achieved with high-load training compared to low-load training programs despite similar muscle mass gains. However, muscle hypertrophy is similar between high-load and low-load RT, when the last was performed until failure.

Interestingly, nerve adaptations always accompany strength gains that result from RT, but hypertrophy may or may not be present. Intermuscular coordination implies an increased activation of agonist muscles, a reduction in the co-activation of antagonists and a better co-activation of synergists in response to RT. These neural adaptations highlight the importance of the principle of specificity in sport and RT.

Regarding hypertrophy, current literature has reported that high-load RT (>60% of 1RM) primarily increases cross-sectional area of type II muscle fibers. Nevertheless, if a low-load RT (<60% of 1R) is performed with momentary muscular failure, hypertrophy of type I fibers occurs. The molecular pathways underlying the hypertrophic adaptation in type I fibers remain poor studied to date. Metabolic stress produced by exercise could play an important role in increasing the size of type 1 fibers, more than type and duration of the training. Moreover, some studies have shown that there are some people who have a low muscle hypertrophic response to RT compared to others with similar characteristics. In this sense, some aspects to consider such as genetics and nutrition could have a potential effect on possible adaptations to RT.

Similarly, despite the crucial role of the endocrine system on physiological mechanism, there is little evidence about the importance of the hormone hypothesis in hypertrophy and muscle strength. The dynamic nature of hormones, and the fact that their effects depend entirely on receptors and their interactions with nuclear DNA mechanisms of target cells, dampen their understanding and their possible application to training design.

5 Take Home Messages and Practical Resources

1. During the first weeks of training neural factors appear to play a greater role in strength gains than hypertrophic processes.
2. Intermuscular coordination implies an increased activation of agonist muscles, a reduction in the co-activation of antagonists, and a better co-activation of synergists in response to RT.
3. Neural adaptations highlight the importance of the principle of specificity.
4. Surface EMG is used to measure muscle fibre activation in the practice and is a non-invasive quantitative assessment.
5. High-load training results in greater neural adaptations than low-load training programs.
6. Muscle hypertrophy is similar between high-load and low-load RT, when the last is performed until failure.
7. High-load RT (>60% of 1RM) primarily increases cross-sectional area of type II muscle fibers.

8. Low-load RT (<60% of 1R) increases type I fibers hypertrophy, although momentary muscular failure is needed.
9. High-intensity resistance and impact training improve bone mass density.
10. Although RT has little impact on cardiorespiratory adaptations compared with aerobic training, dynamic RT decreases blood pressure, increases peak VO_2, and decreases body fat and plasma triglycerides.
11. The testosterone/cortisol ratio significantly decreases when there is a high athlete's strain indicating that the training is not being well assimilated and may lead to overtraining.

References

American College of Sports Medicine (2009) American college of sports medicine position stand. Progression models in resistance training for healthy adults. Med Sci Sports Exerc 41(3): 687–708

Amiel D, Frank C, Harwood F, Fronek J, Akeson W (1984) Tendons and ligaments: a morphological and biochemical comparison. J Orthop Res 1(3):257–265

Amin S, Baker K, Niu J, Clancy M, Goggins J, Guermazi A et al (2009) Quadriceps strength and the risk of cartilage loss and symptom progression in knee osteoarthritis. Arthritis Rheum 60 (1):189–198

Beck BR, Daly RM, Singh MA, Taaffe DR (2017) Exercise and Sports Science Australia (ESSA) position statement on exercise prescription for the prevention and management of osteoporosis. J Sci Med Sport 20(5):438–445

Bekedam MA, van Beek-Harmsen BJ, Boonstra A, van Mechelen W, Visser FC, van der Laarse WJ (2003) Maximum rate of oxygen consumption related to succinate dehydrogenase activity in skeletal muscle fibres of chronic heart failure patients and controls. Clin Physiol Funct Imaging 23(6):337–343

Bigard X (2019) Adaptation of skeletal muscle mass and metabolism to physical exercise. In: Nutrition and skeletal muscle. Elsevier, pp 47–61

Bohm S, Mersmann F, Arampatzis A (2015) Human tendon adaptation in response to mechanical loading: a systematic review and meta-analysis of exercise intervention studies on healthy adults. Sports Med Open 1(1):7

Bohndorf K (1999) Imaging of acute injuries of the articular surfaces (chondral, osteochondral and subchondral fractures). Skeletal Radiol 28(10):545–560

Bompa T, Buzzichelli C (2015) Periodization training for sports, 3e. Human kinetics

Braith RW, Stewart KJ (2006) Resistance exercise training: its role in the prevention of cardiovascular disease. Circulation 113(22):2642–2650

Bricca A, Juhl CB, Grodzinsky AJ, Roos EM (2017) Impact of a daily exercise dose on knee joint cartilage—a systematic review and meta-analysis of randomized controlled trials in healthy animals. Osteoarthritis Cartilage 25(8):1223–1237

Bricca A, Juhl CB, Steultjens M, Wirth W, Roos EM (2019) Impact of exercise on articular cartilage in people at risk of, or with established, knee osteoarthritis: a systematic review of randomised controlled trials. Br J Sports Med 53(15):940–947

Carroll TJ, Riek S, Carson RG (2001) Neural adaptations to resistance training: implications for movement control. Sports Med 31(12):829–840

Clark EM, Tobias JH, Murray L, Boreham C (2011) Children with low muscle strength are at an increased risk of fracture with exposure to exercise. J Musculoskelet Neuronal Interact 11 (2):196–202

Cormie P, McGuigan MR, Newton RU (2011) Developing maximal neuromuscular power: part 1–biological basis of maximal power production. Sports Med 41(1):17–38

Cornelissen VA, Fagard RH, Coeckelberghs E, Vanhees L (2011) Impact of resistance training on blood pressure and other cardiovascular risk factors: a meta-analysis of randomized, controlled trials. Hypertension 58(5):950–958

Couppe C, Kongsgaard M, Aagaard P, Hansen P, Bojsen-Moller J, Kjaer M et al (2008) Habitual loading results in tendon hypertrophy and increased stiffness of the human patellar tendon. J Appl Physiol (1985) 105(3):805–810

Enoka RM (1988) Neuromechanical basis of kinesiology. ERIC

Enoka RM (1997) Neural adaptations with chronic physical activity. J Biomech 30(5):447–455

Fleck SJ (1988) Cardiovascular adaptations to resistance training. Med Sci Sports Exerc 20(5 Suppl):S146–S151

Folland JP, Williams AG (2007) The adaptations to strength training: morphological and neurological contributions to increased strength. Sports Med 37(2):145–168

Fry AC (2004) The role of resistance exercise intensity on muscle fibre adaptations. Sports Med 34 (10):663–679

Fry CS, Drummond MJ, Glynn EL, Dickinson JM, Gundermann DM, Timmerman KL et al (2013) Skeletal muscle autophagy and protein breakdown following resistance exercise are similar in younger and older adults. J Gerontol A Biol Sci Med Sci 68(5):599–607

Gabriel DA, Kamen G, Frost G (2006) Neural adaptations to resistive exercise: mechanisms and recommendations for training practices. Sports Med 36(2):133–149

Geremia JM, Baroni BM, Bobbert MF, Bini RR, Lanferdini FJ, Vaz MA (2018) Effects of high loading by eccentric triceps surae training on Achilles tendon properties in humans. Eur J Appl Physiol 118(8):1725–1736

Goodman CA, Mayhew DL, Hornberger TA (2011) Recent progress toward understanding the molecular mechanisms that regulate skeletal muscle mass. Cell Signal 23(12):1896–1906

Gregory CM, Vandenborne K, Dudley GA (2001) Metabolic enzymes and phenotypic expression among human locomotor muscles. Muscle Nerve 24(3):387–393

Grgic J, Homolak J, Mikulic P, Botella J, Schoenfeld BJ (2018) Inducing hypertrophic effects of type I skeletal muscle fibers: a hypothetical role of time under load in resistance training aimed at muscular hypertrophy. Med Hypotheses 112:40–42

Grzelak P, Polguj M, Podgorski M, Majos A, Krochmalski M, Domzalski M (2012a) Patellar ligament hypertrophy evaluated by magnetic resonance imaging in a group of professional weightlifters. Folia Morphol 71(4):240–244

Grzelak P, Podgorski M, Stefanczyk L, Krochmalski M, Domzalski M (2012b) Hypertrophied cruciate ligament in high performance weightlifters observed in magnetic resonance imaging. Int Orthop 36(8):1715–1719

Haff GG, Stone MH (2015) Methods of developing power with special reference to football players. Strength Cond J 37(6):2–16

Hansen S, Kvorning T, Kjaer M, Sjogaard G (2001) The effect of short-term strength training on human skeletal muscle: the importance of physiologically elevated hormone levels. Scand J Med Sci Sports 11(6):347–354

Hansen D, Abreu A, Doherty P, Voller H (2019) Dynamic strength training intensity in cardiovascular rehabilitation: is it time to reconsider clinical practice? a systematic review. Eur J Prev Cardiol 26(14):1483–1492

Heinemeier KM, Kjaer M (2011) In vivo investigation of tendon responses to mechanical loading. J Musculoskelet Neuronal Interact 11(2):115–123

Hood DA, Terjung RL (1987) Leucine metabolism in perfused rat skeletal muscle during contractions. Am J Physiol 253(6 Pt 1):E636–E647

Hornsby WG, Gentles JA, Haff GG, Stone MH, Buckner SL, Dankel SJ et al (2018) What is the impact of muscle hypertrophy on strength and sport performance? Strength Cond J 40(6):99–111
Hudelmaier M, Glaser C, Englmeier KH, Reiser M, Putz R, Eckstein F (2003) Correlation of knee-joint cartilage morphology with muscle cross-sectional areas versus anthropometric variables. Anat Rec A Discov Mol Cell Evol Biol 270(2):175–184
Hughes DC, Ellefsen S, Baar K (2018) Adaptations to endurance and strength training. Cold Spring Harb Persp Med 8(6):a029769
Hunter DJ, March L, Chew M (2020) Osteoarthritis in 2020 and beyond: a Lancet Commission. Lancet 396(10264):1711–1712
Jenkins NDM, Miramonti AA, Hill EC, Smith CM, Cochrane-Snyman KC, Housh TJ et al (2017) Greater neural adaptations following high- versus low-load resistance training. Front Physiol 8:331
Kongsgaard M, Reitelseder S, Pedersen TG, Holm L, Aagaard P, Kjaer M et al (2007) Region specific patellar tendon hypertrophy in humans following resistance training. Acta Physiol (oxf) 191(2):111–121
Kraemer WJ, Marchitelli L, Gordon SE, Harman E, Dziados JE, Mello R et al (1990) Hormonal and growth factor responses to heavy resistance exercise protocols. J Appl Physiol (1985) 69 (4):1442–1450
Kraemer WJ, Gordon SE, Fleck SJ, Marchitelli LJ, Mello R, Dziados JE et al (1991) Endogenous anabolic hormonal and growth factor responses to heavy resistance exercise in males and females. Int J Sports Med 12(2):228–235
Kraemer WJ, Fleck SJ, Dziados JE, Harman EA, Marchitelli LJ, Gordon SE et al (1993) Changes in hormonal concentrations after different heavy-resistance exercise protocols in women. J Appl Physiol (1985) 75(2):594–604
Kraemer WJ, Ratamess NA, Flanagan SD, Shurley JP, Todd JS, Todd TC (2017) Understanding the science of resistance training: an evolutionary perspective. Sports Med 47(12):2415–2435
Kubo K, Yata H, Kanehisa H, Fukunaga T (2006) Effects of isometric squat training on the tendon stiffness and jump performance. Eur J Appl Physiol 96(3):305–314
Lin I, Wiles L, Waller R, Goucke R, Nagree Y, Gibberd M et al (2020) What does best practice care for musculoskeletal pain look like? Eleven consistent recommendations from high-quality clinical practice guidelines: systematic review. Br J Sports Med 54(2):79–86
Maestroni L, Read P, Bishop C, Papadopoulos K, Suchomel TJ, Comfort P et al (2020) The benefits of strength training on musculoskeletal system health: practical applications for interdisciplinary care. Sports Med 50(8):1431–1450
Manini TM, Yarrow JF, Buford TW, Clark BC, Conover CF, Borst SE (2012) Growth hormone responses to acute resistance exercise with vascular restriction in young and old men. Growth Horm IGF Res 22(5):167–172
Marin-Pagan C, Blazevich AJ, Chung LH, Romero-Arenas S, Freitas TT, Alcaraz PE (2020) Acute physiological responses to high-intensity resistance circuit training versus traditional strength training in soccer players. Biology 9(11):383
Mersmann F, Bohm S, Arampatzis A (2017) Imbalances in the development of muscle and tendon as risk factor for tendinopathies in youth athletes: a review of current evidence and concepts of prevention. Front Physiol 8:987
Miller BF, Konopka AR, Hamilton KL (2016) The rigorous study of exercise adaptations: why mRNA might not be enough. J Appl Physiol (1985) 121(2):594–596
Mitchell CJ, Churchward-Venne TA, Parise G, Bellamy L, Baker SK, Smith K et al (2014) Acute post-exercise myofibrillar protein synthesis is not correlated with resistance training-induced muscle hypertrophy in young men. PLoS One 9(2):e89431
Moritani T, deVries HA (1979) Neural factors versus hypertrophy in the time course of muscle strength gain. Am J Phys Med 58(3):115–130
Netreba A, Popov D, Bravyy Y, Lyubaeva E, Terada M, Ohira T et al (2013) Responses of knee extensor muscles to leg press training of various types in human. Ross Fiziol Zh Im I M Sechenova 99(3):406–416

Ogasawara R, Jensen TE, Goodman CA, Hornberger TA (2019) Resistance exercise-induced hypertrophy: a potential role for rapamycin-insensitive mTOR. Exerc Sport Sci Rev 47 (3):188–194

Oiestad BE, Juhl CB, Eitzen I, Thorlund JB (2015) Knee extensor muscle weakness is a risk factor for development of knee osteoarthritis: a systematic review and meta-analysis. Osteoarthritis Cartilage 23(2):171–177

Parreno M, Pol A, Cadefau J, Parra J, Alvarez L, Membrilla E et al (2001) Changes of skeletal muscle proteases activities during a chronic low-frequency stimulation period. Pflugers Arch 442(5):745–751

Parry HA, Roberts MD, Kavazis AN (2020) Human skeletal muscle mitochondrial adaptations following resistance exercise training. Int J Sports Med 41(6):349–359

Porter C, Reidy PT, Bhattarai N, Sidossis LS, Rasmussen BB (2015) Resistance exercise training alters mitochondrial function in human skeletal muscle. Med Sci Sports Exerc 47(9):1922–1931

Powers SK, Kavazis AN, McClung JM (2007) Oxidative stress and disuse muscle atrophy. J Appl Physiol (1985) 102(6):2389–2397

Reeves ND, Maganaris CN, Narici MV (2003) Effect of strength training on human patella tendon mechanical properties of older individuals. J Physiol 548(Pt 3):971–981

Sale DG (1988) Neural adaptation to resistance training. Med Sci Sports Exerc 20(5 Suppl):S135–S145

Schoenfeld BJ (2013) Postexercise hypertrophic adaptations: a reexamination of the hormone hypothesis and its applicability to resistance training program design. J Strength Cond Res 27 (6):1720–1730

Schoenfeld BJ, Grgic J, Ogborn D, Krieger JW (2017) Strength and hypertrophy adaptations between low- versus high-load resistance training: a systematic review and meta-analysis. J Strength Cond Res 31(12):3508–3523

Siddique U, Rahman S, Frazer AK, Pearce AJ, Howatson G, Kidgell DJ (2020) Determining the sites of neural adaptations to resistance training: a systematic review and meta-analysis. Sports Med 50(6):1107–1128

Solomonow M, Baratta R, Zhou BH, D'Ambrosia R (1988) Electromyogram coactivation patterns of the elbow antagonist muscles during slow isokinetic movement. Exp Neurol 100(3): 470–477

Soori M, Lu G, Mason RW (2016) Cathepsin inhibition prevents autophagic protein turnover and downregulates insulin growth factor-1 receptor-mediated signaling in neuroblastoma. J Pharmacol Exp Ther 356(2):375–386

Sophia Fox AJ, Bedi A, Rodeo SA (2009) The basic science of articular cartilage: structure, composition, and function. Sports Health. 1(6):461–468

Steinbacher P, Eckl P (2015) Impact of oxidative stress on exercising skeletal muscle. Biomolecules 5(2):356–377

Suchomel TJ, Nimphius S, Bellon CR, Stone MH (2018) The importance of muscular strength: training considerations. Sports Med 48(4):765–785

Taber CB, Vigotsky A, Nuckols G, Haun CT (2019) Exercise-induced myofibrillar hypertrophy is a contributory cause of gains in muscle strength. Sports Med 49(7):993–997

Tallent J, Woodhead A, Frazer AK, Hill J, Kidgell DJ, Howatson G (2021) Corticospinal and spinal adaptations to motor skill and resistance training: potential mechanisms and implications for motor rehabilitation and athletic development. Eur J Appl Physiol 121(3):707–719

Tipton CM, Matthes RD, Sandage DS (1974) In situ measurement of junction strength and ligament elongation in rats. J Appl Physiol 37(5):758–761

Tipton CM, Matthes RD, Maynard JA, Carey RA (1975) The influence of physical activity on ligaments and tendons. Med Sci Sports 7(3):165–175

Toth MJ, Tchernof A (2006) Effect of age on skeletal muscle myofibrillar mRNA abundance: relationship to myosin heavy chain protein synthesis rate. Exp Gerontol 41(11):1195–1200

Tseng BS, Kasper CE, Edgerton VR (1994) Cytoplasm-to-myonucleus ratios and succinate dehydrogenase activities in adult rat slow and fast muscle fibers. Cell Tissue Res 275(1):39–49
Urhausen A, Gabriel H, Kindermann W (1995) Blood hormones as markers of training stress and overtraining. Sports Med 20(4):251–276
van Wessel T, de Haan A, van der Laarse WJ, Jaspers RT (2010) The muscle fiber type-fiber size paradox: hypertrophy or oxidative metabolism? Eur J Appl Physiol 110(4):665–694
Vanwanseele B, Eckstein F, Knecht H, Spaepen A, Stussi E (2003) Longitudinal analysis of cartilage atrophy in the knees of patients with spinal cord injury. Arthritis Rheum 48(12):3377–3381
Vinogradova OL, Popov DV, Netreba AI, Tsvirkun DV, Kurochkina NS, Bachinin AV et al (2013) Optimization of training: development of a new partial load mode of strength training. Fiziol Cheloveka 39(5):71–85
Warden SJ, Mantila Roosa SM, Kersh ME, Hurd AL, Fleisig GS, Pandy MG et al (2014) Physical activity when young provides lifelong benefits to cortical bone size and strength in men. Proc Natl Acad Sci U S A 111(14):5337–5342
Watson S, Weeks B, Weis L, Harding A, Horan S, Beck B (2019) High-Intensity resistance and impact training improves bone mineral density and physical function in postmenopausal women with osteopenia and osteoporosis: the LIFTMOR randomized controlled trial. J Bone Miner Res 34(3):572
Weinbaum S, Duan Y, Thi MM, You L (2011) An integrative review of mechanotransduction in endothelial, epithelial (renal) and dendritic cells (osteocytes). Cell Mol Bioeng 4(4):510–537
Wilkinson SB, Tarnopolsky MA, Grant EJ, Correia CE, Phillips SM (2006) Hypertrophy with unilateral resistance exercise occurs without increases in endogenous anabolic hormone concentration. Eur J Appl Physiol 98(6):546–555
Zhao R, Zhao M, Xu Z (2015) The effects of differing resistance training modes on the preservation of bone mineral density in postmenopausal women: a meta-analysis. Osteoporos Int 26(5):1605–1618

Jesús Ponce-González is full a professor at the University of Cádiz (Spain) and aresearcher in the MOVE-IT Research group and the Biomedical Research and Innovation Institute of Cádiz (INiBICA) Research Unit. He is an expert in Exercise Physiology and Metabolism, who has a multidisciplinary background focused in molecular biology and training methods. He has shown his international projection with more than 50 scientific articles including, among other variables, measurements at the muscular level with biopsies that allow to know the effect of physical exercise from a molecular point of view.

Cristina Casals is an assistant professor at the University of Cádiz (Spain) and a researcher in the MOVE-IT Research group and the Biomedical Research and Innovation Institute of Cádiz (INiBICA) Research Unit. She has developed her research on Exercise Physiology and Nutrition; she has published more than 40 articles in sport sciences, determining different aspects of exercise performance, such us the analysis of exercise-induced muscle damage in response to different types of training protocols.

Kinetic and Kinematic Analysis for Exercise Design: A Practical Approach

Víctor Cuadrado

Abstract This chapter introduces the importance to study movement in the sports context, where human performance focuses on the continuous optimization of the physical condition of athletes in specific situations, which often require to be performed at high intensities. To optimize these actions, it is necessary to prioritize strength training, focused on improving useful strength, understood as the application of strength under specific time and velocity conditions per training to competitive exercise (issues reflected in force–velocity and force–time curves). To carry out from practice, it is necessary to monitor, quantify, adapt and prescribe strength training to understand the existing relationship between the external load proposed for the subject and its organic consequences to achieve the adaptations sought and thus optimize performance. To achieve this, it is important to measure and control movement from a mechanical perspective. In this sense, in this chapter, an initial analysis of different methods of strength training through kinetics and kinematics will be proposed.

Keywords Kinetics · Kinematics · Force–time curve · Force–velocity curve · Flywheels · Force–velocity profile · Vertical jump · Variable resistance

1 Introduction

Human movement has been a constant subject matter of research throughout history. There are endless studies dating back to Classical Greece (carried out by Aristotle; 384–322 BC) (Komi 1994), up to the present time. The need to understand what its characteristics are and how it originates, how it evolves as organic maturity occurs, why it is altered under certain circumstances, and how it can be optimized for improved performance in various situations, has led to significant development and application of scientific knowledge to attempt to answer these questions

V. Cuadrado (✉)
Autonomous University of Madrid, Cantoblanco University City, 28049 Madrid, Spain
e-mail: victor.cuadrado@uam.es

A. Muñoz-López et al. (eds.), *Resistance Training Methods*, Lecture Notes in Bioengineering, https://doi.org/10.1007/978-3-030-81989-7_3

(Zatsiorsky and Kraemer 2006). To this end, the technology advances that have taken place until now help us know more about issues related to the process of nervous activation which originate movement and the mechanical consequences ruled by the laws of physics.

In the area of physical-sports training, the above, together with the principles which must guide performance optimization, require the use of technology to understand, from a mechanical perspective, which variables describe movement and which explain how it happens (through kinetics and kinematics), primarily in strength training. Therefore, monitoring various strength training methods, based on certain mechanical variables, is a useful tool, necessary to personalize and optimize physical performance.

This chapter provides the theoretical and practical basis for the importance of monitoring strength training by means of kinetic and kinematic analysis of the various methodologies described in future chapters.

2 From Theory

To simplify the difficult task of trainers, coaches, and other actors in charge of prescribing strength training, it is necessary to address the essential rationale of said task.

One of the main objectives of strength training should be to increase the subject's ability in practical terms, and thus improve their physical-sports performance. In this sense, strength training should focus on improving useful strength, understood as the application of strength under specific time and velocity conditions pertaining to competitive exercise, (González-Badillo and Serna 2002; González-Badillo 2000). Hence, it is necessary to study the strength applied by the subject. Citing González-Badillo (González-Badillo and Serna 2002; González-Badillo 2000) applied strength is the external expression of internal tension generated in a muscle or group of muscles in a given time or at a given velocity. Thus, the importance of understanding and studying the existing relationships between the different mechanical variables that characterize the ability to apply force and which are represented by the curves force–time (C f–t) and force–velocity–power (C f–v–p).

2.1 Force–time Curve (C f–t)

When discussing applied force, we must understand that this occurs over time, thus giving way to the concept of application of force in a time unit (RFD), which is derived from the force- or torque time curves recorded during explosive voluntary contractions (Maffiuletti et al. 2016). Therefore, in sports training, RFD is considered explosive strength.

The study of RFD helps understand in detail whether any adaptations have taken place and if so, of what kind, after conducting certain strength training, regardless of the method used. Given that in sports, very often, performance is decided by improved velocity at which certain specific actions are carried out (such as running or jumping) (Tillin and Folland 2014), it may be considered that as sports performance is improved, conditions worsen and subjects will have less time to apply strength, as the same action must be carried out at a greater speed (González-Badillo and Gorostiaga 1995). Therefore, to improve sports performance it is necessary to improve the force–time ratio, that is, to produce the same strength in less time or achieve greater strength in the same amount of time. This enhancement is seen in C f–t as it moves upwards and to the left (see Fig. 1). Depending on the characteristics of the action conducted or, in other words, depending on the RFD characterizing the sports gesture carried out by the subject, there may be endless possibilities for the force–time ratio (see Fig. 2) that will depend on the time available to apply strength in a given context and the strength peak achieved, which is closely related to the resistance to be overcome (González-Badillo 2000).

Therefore, based on C f–t, essential issues can be plotted on a graph such as the maximum peak of strength achieved and the time used for this at various moments of the training process, thus assessing the consequences of a given strength training.

2.2 *Force–Velocity (C f–V) and Force–Velocity–Power Curve (C f–V–P)*

Changes occurring in the applied force and shown in C f–t, as stated above, are also shown in another curve that should be studied and known, C f–v–p. Producing the same force in less time is the same as displacing the same resistance at a greater

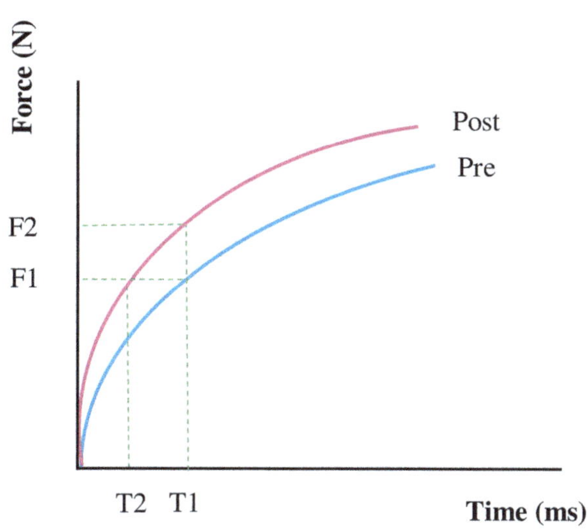

Fig. 1 Improvement in RFD after training

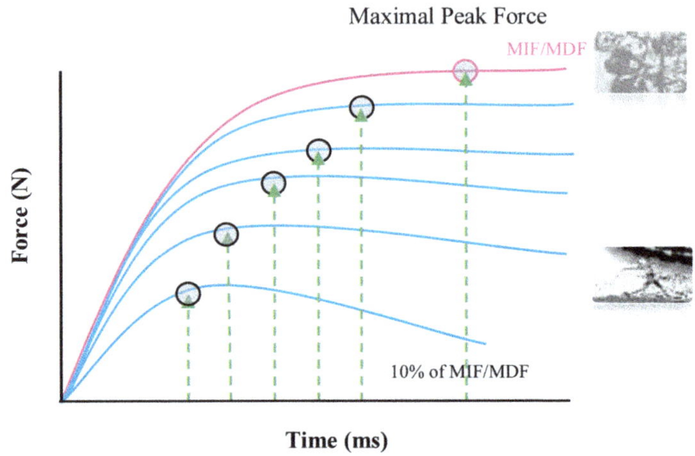

Fig. 2 Different explosive force values based on the external load to overcome. *MIF* Maximal isometric force. *MDF* Maximal dynamic force

velocity and thus achieving greater strength in the same amount of time is the same as displacing greater resistance at the same velocity (González-Badillo and Serna 2002, González-Badillo 2000; González-Badillo and Gorostiaga 1995).

The essence of this curve lies in the graphical representation of the mechanical consequences derived from the neurophysiological "reality" of the skeletal muscle. At the muscular level, the relationship between these two variables is reversed, as muscle shortening speed increases there are fewer cross-bridges available and therefore less force may be exerted (Rassier et al. 1999; Lieber and Ward 2011). In the case of a drop in muscle shortening velocity, there is an increased possibility of forming a higher number of cross-bridges and therefore there is a proportional increase in the force exerted (see Fig. 3) (González-Badillo and Serna 2002; Rassier et al. 1999; Lieber and Ward 2011; Edman 1992). In line with this focus on muscular architecture, it is worth noting that the force–velocity ratio is also conditioned by the percentage of fast or slow fibres, which has a significant impact on C f–v–p. (see Fig. 3).

Based on these core force–velocity ratio considerations, two essential matters must be understood. On the one hand, the possible existing relationship between the two variables depend on the load used by the subject and therefore it is necessary to know which is the right load to use. And on the other, to take into account the development desired graphically, to reflect adequate adaptations following a training period, which imply displacement of the curve towards the right and upwards (see Fig. 4). This displacement need not occur proportionately throughout the curve, it may be more accentuated in certain areas than in others and this will depend on the type of stimulus proposed and the area to be modified based on the subject's needs.

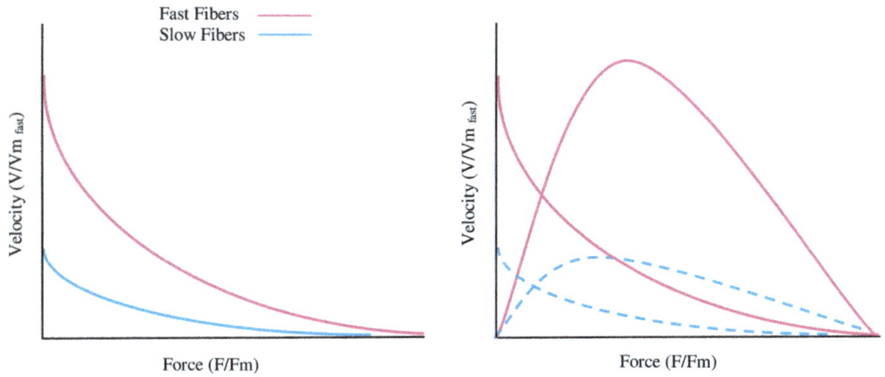

Fig. 3 Influence of muscle fibre type on C f–v–p

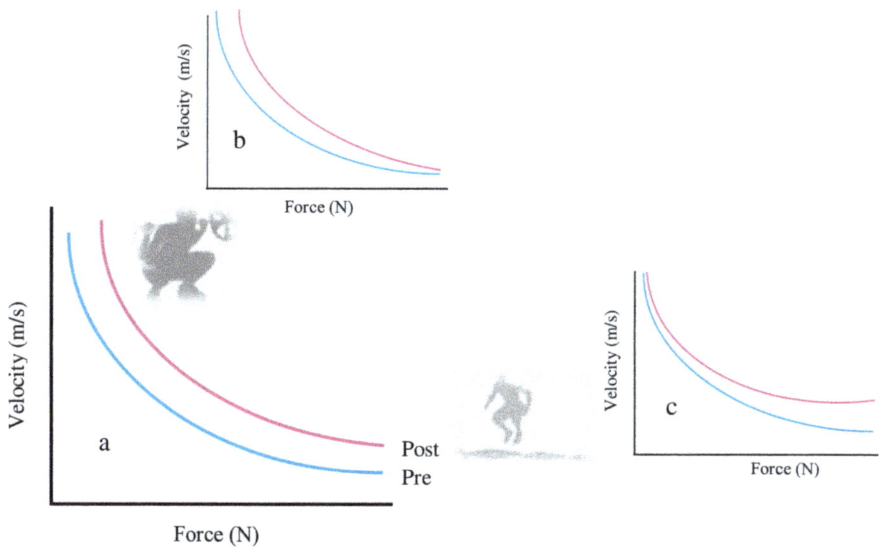

Fig. 4 Changes in C f–v after different training orientations. **a** Improvements across the force–velocity spectrum. **b** Improvements reflected on high force at low speed zone. **c** Improvements reflected on low velocity at high-speed zone

Moreover, from a mechanical perspective, force and velocity determine the power generated, as is graphically shown in the area under the curve (see Fig. 5). Therefore, changes towards the right and above C f–v, as indicated in the above paragraphs, result in an increase in the power generated, and there are thus various possibilities of increasing it. This has triggered controversy in the scientific community primarily in regard to its relationship with sports performance and in determining the loading with which to achieve maximum power in certain exercises

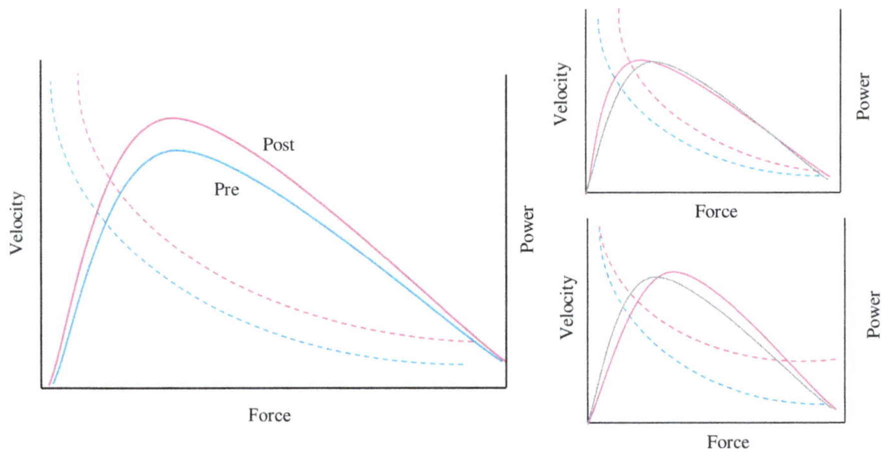

Fig. 5 Graphical representation of C f–v–p and possible changes in power (after training) based on different modification on C f–v

(Kawamori 2004; Soriano et al. 2015; Foster 2016). In this sense, once the necessary power to conduct an exercise or series of exercises is known, the force–velocity curve will provide a very close index of the optimal conditions to achieve maximum sports performance (González-Badillo et al. 2005).

It is thus necessary to control and monitor strength training regarding force–time and force–velocity-power ratios in order to understand which areas should be modified at any given time and based on this, plan the necessary stimuli using the most appropriate training methods.

The next section addresses the various strength training methods (which are amply dealt with in subsequent chapters), based on the mechanical perspective mentioned in the above paragraphs.

3 From Practice

In practical terms, strength training must include actions such as: monitor, quantify, prescribe, adapt, etc. And therefore, it is necessary to understand the existing relationship between the external load proposed for the subject and its organic consequences to achieve the adaptations sought and thus optimize performance. Therefore, an in-depth mechanical analysis of the main strength training methods is essential. This entails characterizing the main training methods covered in this book and specifically addressed in the next chapters, where training strategies for each one of them are provided.

3.1 Flywheel Devices (FW)

Strength training carried out with these devices basically focuses on the use of the momentum of force generated on a flywheel device, with the intention of providing the maximum linear resistance possible through a rope rolled around a cylinder (cylindrical pulley) (see Fig. 6a), or a cone-shaped vertical axis (cone pulley) (Norrbrand et al. 2008) during the various stages of movement (Núñez et al. 2020). To this end, the subject wears a harness that is attached to a rope (in the case of the cylindrical pulley) or pulls on it directly (in the case of the cone pulley). The movement begins until the rope is completely unrolled but the device continues turning due to inertia, making the rope recoil. Efficiency in the use of this training method is based on the subject's intention to apply force at the maximum velocity possible during an acceleration phase or a concentric phase in order to then try to stop the rotating movement during the deceleration or eccentric phase. (Nuñez Sanchez et al. 2017; Maroto-Izquierdo et al. 2017). Therefore, the starting point requires characterizing (based on mechanics) movements conducted with a FW. Below is a graphic representation of the evolution of force and velocity with increased loading (understood as inertial load), using the two devices mentioned above (cone pulley and cylindrical pulley) (unpublished data from our laboratory). As shown in Fig. 6a), the peak velocity achieved both during the concentric and the eccentric phases of the exercise, is much higher when the cone pulley is used. We can also see how the peak velocity during the concentric phase with CP is considerably reduced with the 0.33 kg·m^2 load (this being the largest load proposed for measurement). However, this trend is not so clear during the eccentric phase. On the other hand, we found that the peak velocity is higher in the concentric phase with CP for all loads, whereas when CS is used, the trend is the opposite, especially as the load increases. In relation to the force values (Fig. 8b), we found that with the CP device there was a slight (not very significant) increase in the concentric phase as loading increased, while during the eccentric phase the values were practically the same despite the increased resistance provided by this device. Furthermore, we found that the difference in the peak force in both phases with CP is considerably greater than with CS. On the other hand, when exercise is conducted with CS, the peak force achieved during the concentric phase is greater for all loads and there is a clear trend as the loading provided increases. In relation to the peak force during the eccentric phase with this device, it is greater for each load compared to CP, but it is maintained quite well despite its increase. In practical terms and based on the above, we should consider which device to be used based on the training goals (taking into account the force–velocity ratio). If the goal is to generate greater eccentric overload, as well as the device to be used, we need to consider whether it is necessary or not to increase overload.

Several studies have shown that overloading, with FW, is effective to develop muscle hypertrophy, maximum force, power and even to improve various functional skills, measured both vertically and horizontally (Petré et al. 2018). This fact has recently attracted the interest of the scientific community in monitoring strength

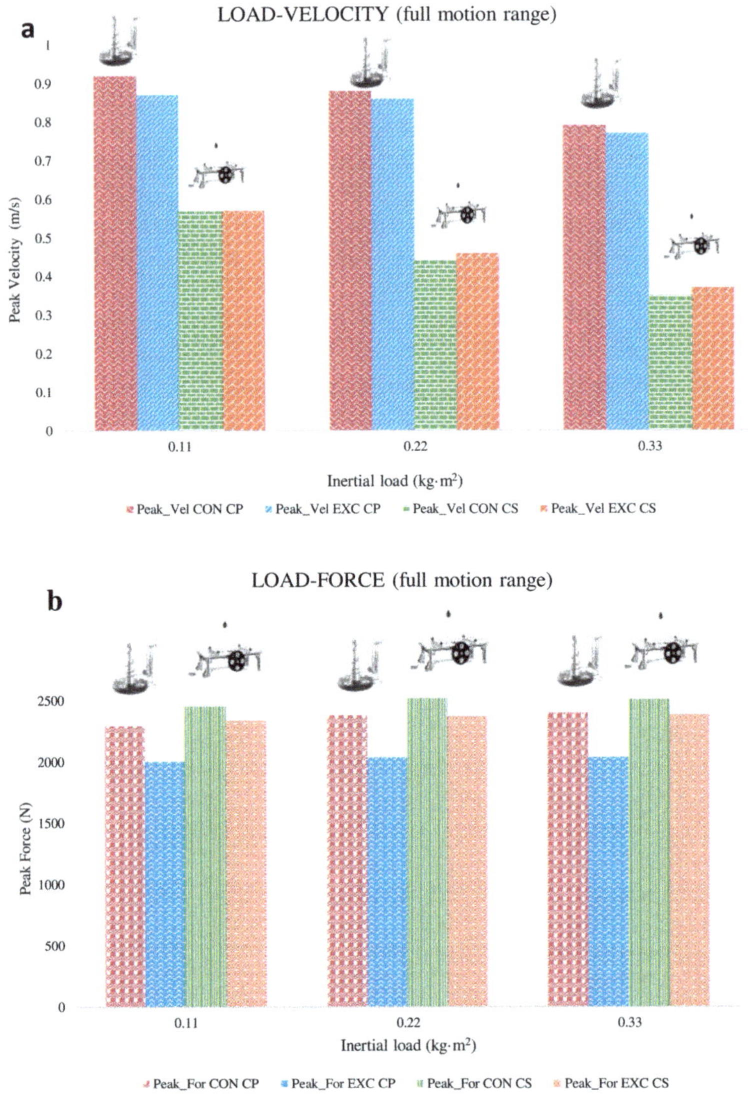

Fig. 6 Half squat **a** velocity and **b** force values achieved in the concentric and eccentric phase under different inertial loads with both FW devices (unpublished private data)

training by evaluating the F–V ratio of an individual under FW conditions and thus obtain data on the mechanical limitations imposed by the muscular capabilities through the FW load spectrum, similar to the work by (Jiménez-Reyes et al. 2017; Samozino et al. 2008) conducted with free weights. Spudić et al. (2020), proved that using four progressive loads (with an average of six repetitions with each one), helps achieve the F–V profile with FW. This allows personalizing FW training to

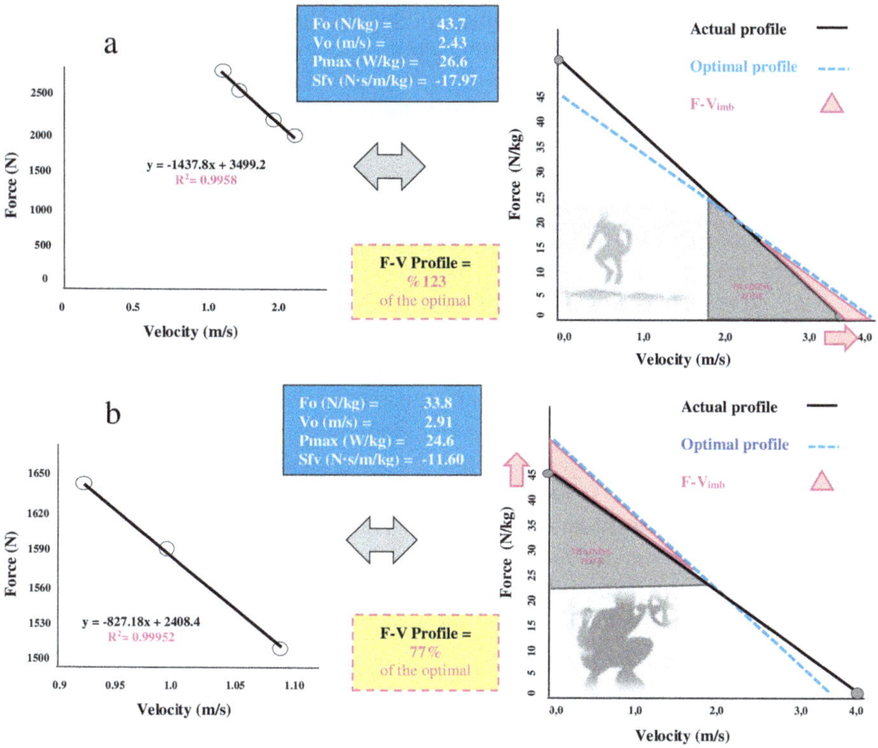

Fig. 7 Force–velocity profile in jump (CMJ) **a** FV_{imb} oriented to velocity and **b** FV_{imb} oriented to force

improve athletic performance aimed at the force, velocity or power skills, taking into account previous considerations based on the use of different devices with varying inertial loads.

There are not many studies on the kinematic analysis of movements carried out with FW devices. Worthy of note is the work by Worcester et al. (2020) which studied the kinematics of doing squats in the sagittal plane with free weights vs an inertial device. On the one hand, they studied the impact of increasing the inertial load when using FW, on ankle, knee and trunk-hip joints (both in position and in the angular velocity achieved). An on the other hand, they compared the position acquired by each joint throughout the squat done with FW vs free weight with a load close to 10% of average power. The results showed that increasing inertial loads does not produce significant changes in the kinematics of movement in the sagittal plane, however, a significant drop was observed in the angular velocity as the inertial load was increased, on all joints except the trunk. When the angulations achieved in these two conditions were compared (standardized based on the percentage of movement), they found that the knee achieved a higher angulation of about 10° with FW (reaching approximately 100°), during practically all of the first part of the path of movement, becoming completely equal in the second part. In the ankle, when 10% of the path was completed, angulation was gradually increased with FW up to half of the path, where there was a difference of 20º, at which point it reached an angulation close to 120º, while during the second part of the path, the difference decreased gradually until angulation was equal when 75% of the path was reached. As for the trunk, the differences are more marked in favour of free weight, both when studied separately or together with the hip). Separately, when it reached 25% of the path, there was a very sharp increase until reaching 50% of the path, where there was a difference of about 20º, finally reaching approximately 60º (when the exercise was conducted with free weight). During the second part of the path, the dynamics were similar but opposite, so that when the path approached 75–80%, angulation was equal. When the study focused on the trunk-hip ratio, the dynamics were similar except for angulation achieved which reached approximately 120º with free weight.

As for angular velocity, it was similar in both conditions during the eccentric phase of the movement, while during the concentric phase, with free weight, angular velocity was considerably higher, possibly because the flywheel requires maximum effort throughout the range of movement, while with free weight, the greatest effort occurs at the beginning until the point of mechanical advantage is reached, after which the force applied increases.

3.2 *Ballistic Push-Off Loaded Actions (BPLA)*

In summary, and in practical terms, this section addresses the vertical jump, which can be considered the main BPLA used in strength training aimed at enhancing sports performance. Specifically, the mechanical considerations related to the force–velocity ratio which have led to conceiving a new paradigm for monitoring and strength training, such as the F–V–P profile for jumps.

Vertical jumps are a common method used by trainers to assess the muscular power of the ability for vertical impulsion. For a long time now, jump performance has become an important part of fitness tests in sports and certain medical areas. Throughout history, various authors have proven that the height of several types of vertical jumps can help assess muscular force and power (Vandewalle et al. 1987) or even the composition of muscle fibres (Bosco et al. 1983). Therefore, for decades, the focus has been more objectively and scientifically on understanding the characteristics of the vertical jump and its relationship with performance (Aragón-Vargas and Melissa 1997). In particular, the driving force of the lower body during a vertical jump is regularly used as a method to assess the explosive characteristics not only of sedentary subjects but also of elite athletes (Hubley and Wells 1983; Cardoso Marques and González-Badillo 2006; Marques et al. 2007). One of the main issues to be resolved therefore by trainers as well as researchers when studying the determining factors for performance during explosive actions, is which mechanical capacity of the neuromuscular system is more important: force or velocity (Samozino et al. 2008, 2012). In other words, is it better to be "strong" or "fast" to achieve maximum performance in ballistic movements? (Samozino et al. 2012). Such an analysis could provide a better understanding of the relationship between the mechanical properties of the neuromuscular system and functional performance, whether to further research animal motor behaviour (James et al. 2007; Jaric and Markovic 2009) or to plan athletic training for humans (Cormie et al. 2010; Cronin and Sleivert 2005; Frost et al. 2010). Therefore, to optimize these actions, studying power is essential. Although power should be understood from a new perspective in which it is not only the result of velocity through force, but rather the interaction of the aspects of force and velocity, highlighting the more important and novel relationship with the muscular characteristics to generate high power (Samozino et al. 2008, 2012). Based on this physiological evidence, in recent years a force–velocity profile has been validated (Samozino et al. 2012; Jiménez-Reyes et al. 2014), which integrates muscle characteristics and is subject-customized to enhance the interaction between force and velocity as factors in the neuromuscular system responsible for optimizing performance during explosive actions. This F–V profile represents a balance between maximum force capacity and maximum velocity potential for the subject. This profile is closely related to performance and most importantly, there is an optimal F–V profile which, firstly, is what allows interaction of the subject's neuromuscular system to optimize power generation according to their individual characteristics and fitness level, and secondly, can be modified with adequate training, thus helping improve any deficiencies detected in any subject, bringing their actual profile closer to the optimal one. Therefore, the mechanical variables achieved after conducting the F–V profile are the following (Morin and Samozino 2016):

- F0 (N/kg): Maximal output of concentric force (per unit of body mass) produced by lower limbs during ballistic push-off actions. This is determined by the full spectrum of the force–velocity ratio, providing more comprehensive information on the force capabilities, than obtaining the load with which RM is achieved. It corresponds to the intersection with the *y* axis of the linear ratio (F–V).
- V0 (m/s): Maximal theoretical velocity of extension of lower limbs during ballistic push-off actions. It corresponds to the intersection with the *x* axis of the linear ratio (F–V). It also represents the capacity to generate force at a very high extension velocity.
- Pmax (W/kg): Maximal capacity of the neuromuscular system to produce power (per unit of body mass) with lower limbs, during the concentric phase of ballistic push-off actions.
- Sfv: Individual balance index of force and velocity capabilities. The more negative the value the greater the profile leans towards force and vice versa.
- Sfv_{opt}: Optimal profile representing the optimal balance between force and velocity capabilities. For a given Pmax, the profile will be related (all else remaining constant) to the highest performance possible for an individual in ballistic push-off actions.

- FV_{imb}: Relative difference between actual and optimal F–V profile. When the value is 100% it means the profile is optimized. If the value is greater than 100%, then a velocity deficit occurs (understood as the application of force to very high speeds). In contrast, if the value is lower than 100%, the opposite deficit occurs.

Figure 7 represents two complete profiles showing all the variables explained in the above paragraph. Figure 7a corresponds to a profile with an imbalance of 23% of application of force at high velocity, making it necessary therefore to work on the low area of the force–velocity ratio (shaded in grey) with the main objective (at that point) of displacing the current profile to the optimal one by going through the area highlighted in red (chart on the right). In contrast, Fig. 7b represents the profile of a subject with an imbalance of 23% in the capacity to generate high levels of force, making it necessary therefore to focus training on the high area of the force–velocity ratio (shaded in grey) with the main objective at that point of displacing the current profile to the optimal one by going through the area highlighted in red (chart on the right). On the left of both figures is a regression analysis of each test (comprised of four and three jumps, respectively), confirming the results given its proximity to 1.

For further reading on individual training needs to optimize the profile, and on the time necessary for this, we recommend the following papers (Jiménez-Reyes et al. 2017, 2020).

3.3 *Variable Resistance Training Systems (VRTS)*

This section aims to show the mechanical effects of using variable resistance training systems, mainly elastic bands and chains. Firstly, it is necessary to understand how these two systems work in order to see how they can be applied to frequent strength training exercises.

In the case of elastic bands, where the anchors are placed is decisive for the effects to be produced during the exercise. If the elastic band is anchored below the bar or element to be displaced (upon which force is applied), an overload will be generated during the concentric phase of the movement, making it a resisted medium acting against gravity. As the upward path advances, the band tautens, generating more resistance (which increases in a curvilinear trend due to the stress–strain ratio), and a gradual decrease of velocity until movement stops (either due to reaching the end of the joint range or the elasticity limit of the band). Once this point is reached, the band will release the elastic energy stored in the opposite direction (in favour of gravity), causing an increase in the acceleration of movement during the descending or eccentric phase. (McMaster et al. 2009). If, on the other hand, the band is anchored above the bar, it will tauten in the descending or eccentric phase storing the corresponding elastic energy, which is then released in the ascending or concentric phase, increasing acceleration, becoming thus a facilitating element to lift the resistance to be overcome. When using an elastic band, bear in mind that given its viscoelastic properties, the stress–strain ratio may vary (Page and Ellenbecker 2005), and thereby the effect on the anchor point, which may consequently impact resistance in the exercise. Therefore, the type of elastic band (tubing, bungee or band), must be taken into account depending on the desired effect.

As for chains, they are considered a resistance medium characterized by applying linear resistance as path progresses due to rising of further chain links, leading to increased force in the concentric phase, which is maximal at the joint end of the range. Next, during the descending or eccentric phase, halting resistance is gradually reduced as more links rest on the ground (McMaster et al. 2009). When using chains, it is necessary to know the structure, density, length and diameter, since these will determine their weight and therefore the intensity of the load. In this regard, some papers propose reference values in order to obtain load details, based on knowing the value represented by each link of the chain lifting from or resting on the ground (Berning et al. 2004).

Using VRTS has generated a lot of interest and controversy in the scientific community. On the one hand, there are papers in which training with VRTS vs traditional training implies enhanced force (increased RM), power, velocity, when loads are greater than 75% of RM combined with the resistance provided by elastic bands (approximately 20% RM), in upper body exercises (Rivière et al. 2017; Baker and Newton 2009), and/or the lower body trying to balance the load (with stress caused by bands) when training was blended (Anderson et al. 2008). Other studies indicate that adding an additional load with elastic bands (of approximately 30% of RM) during the weekly training plan in team sports may enhance force and power of the upper and lower body (Joy et al. 2016). Furthermore, there are papers that show a similar enhancement of force and power (traditional training vs elastic bands, assuming 20–35% of total load prescribed) in the upper and lower body (both with squats and isokinetic assessment at various angular velocities) (Shoepe et al. 2011). In other cases, the use of chains as additional loading until reaching equal total load in both types of training (traditional vs traditional + chains), meant an increase in peak and mean velocity in the concentric phase of 10%, explained in part by the increase in peak velocity at the end of the eccentric phase, assuming a faster stretching-shortening cycle in the next concentric phase (Baker and Newton 2009).

Yet, on the other hand, there is scientific evidence that training with VRTS does not significantly improve force or power performance. There are papers comparing trainings (traditional vs chains as additional loading representing 5 and 10% of the prescribed load in Olympic movements) in which there were no improvements in the peak velocity and the RFD even worsened (Berning et al. 2004). In other studies, when comparing traditional training with elastic bands and chains (assuming in both cases 10% of total prescribed load) in back squats, no significant differences were found either in muscle electrical activity or force (peak or mean), either in the eccentric or the concentric phase (Ebben and Jensen 2002). Thus, and due to the limitations of many studies (level of training and subject anthropometric characteristics, correct quantification of load assumed by VRTS, alternating with other types of training, etc.), has meant that for many authors, the use of these systems has more a rehabilitating purpose rather than to optimize specific performance. Moreover, based on the core theme of this chapter (force applied/force–time ratio/force–velocity–power ratio), the practical use of these training systems would be as stimuli (in addition to traditional ones) that somehow help achieve the desired adaptations, shown in the F–V curves and profile, and if this essential premise is met, then they can be used according to scientific literature, however also being cautious of the limitations reported in order to optimize performance at all times.

4 Filling Gaps

In the area of sports training, it is common to distance oneself from its essence, which is no other than to optimize physical performance with the right stimuli for the subject/body in question, in order to make the necessary adaptations for a given situation. Therefore, it is essential to know the subject's characteristics, as well as those of the stimulus proposed with various exercises and/or training tasks. Going back to that discussed at the beginning of the chapter, optimizing physical and sports performance is based to a great extent, on enhancing the capacity to apply force, as reflected by the appropriate modifications in the force–time and force–velocity–time ratios.

Therefore, one of the main challenges that trainers and coaches should address is to understand as much as possible whether what is being proposed is appropriate, in other words, whether the kinetic and kinematic characteristics of the stimuli proposed are appropriate to achieve the desired and necessary optimization. With this rationale, we should stay away from imitating procedures or "miracle" solutions since, regardless of the type of training used (strength training in this case), we must make sure it is the most suitable one at every given moment. We must not consider various strength training methods as better or worse, but rather as suitable or not for each moment and subject. This will force us to analyze them based on mechanics and allow us to make the most objective decisions possible, steering clear of "fads" or subjective preferences.

5 Take-Home Messages

1. Assess specific performance (applied force) based on mechanics, with objective methods endorsed by the scientific community; F–V–P profile in jumps and sprint, load-velocity ratio, force–time ratio.
2. Know the mechanical characteristics of the various methods of training and both the acute and chronic effects these may have.

3. Use the various strength training methods according to the enhancement sought (power, applied force at high or low velocity…) making sure the kinetics and kinematics of the method and selected exercise are appropriate for each specific situation.

References

Anderson CE, Sforzo GA, Sigg JA (2008) The effects of combining elastic and free weight resistance on strength and power in athletes. J Strength Cond Res 22(2):567–574

Aragón-Vargas LF, Melissa GM (1997) Kinesiological factors in vertical jump performance: differences among individuals. J Appl Biomech 13(1):24–44

Baker DG, Newton RU (2009) Effect of kinetically altering a repetition via the use of chain resistance on velocity during the bench press. J Strength Cond Res 23(7):1941–1946

Berning JM, Coker CA, Adams KJ (2004) Using chains for strength and conditioning. Strength Cond J 26(5):80–84

Bosco C, Luhtanen P, Komi V (1983) A simple method for measurement of mechanical power in jumping. Eur J Appl Physiol 17:1865–1871

Cardoso Marques MA, González-Badillo JJ (2006) In-season resistance training and detraining in professional team handball players. J Strength Cond Res 20(3):563–571

Cormie P, McGuigan MR, Newton RU (2010) Changes in the eccentric phase contribute to improved stretch-shorten cycle performance after training. Med Sci Sports Exerc 42(9):1731–1744

Cronin J, Sleivert G (2005) Challenges in understanding the influence of maximal power training on improving athletic performance. Sport Med 35(3):213–234

Ebben WP, Jensen RL (2002) Electromyographic and kinetic analysis of traditional, chain, and elastic band squats. J Strength Cond Res 16(4):547–550

Edman K (1992) Contractile performance of skeletal muscle fibers. Strength power sport

Foster C (2016) Applied exercise physiology: a serendipitous personal journey toward a place that didn't exist when the journey started. Fronteiras 5(3):172–187

Frost DM, Cronin J, Newton RU (2010) A biomechanical evaluation of resistance—fundamental concepts for training and sports performance. Sport Med 40(4):303–326

González-Badillo JJ (2000) Concepto y medida de la fuerza explosiva en el deporte. Posibles aplicaciones al entrenamiento. RED XIV(I):5–16

González-Badillo JJ, Gorostiaga E (1995) Fundamentos del entrenamiento de fuerza. Aplicación al alto rendimiento deportivo. InDE, Barcelona

González-Badillo JJ, Serna JR (2002) Bases de la programación del entrenamiento de fuerza, 1ª. InDE, Barcelona

González-Badillo JJ, Gorostiaga EM, Arellano R, Izquierdo M (2005) Moderate resistance training volume produces more favorable strength gains than high or low volumes during a short-term training cycle. J Strength Cond Res 19(3):689–697

Hubley C, Wells R (1983) Physiology to vertical jump performance. Eur J Appl Physiol 50:247–254

James RS, Navas CA, Herrel A (2007) How important are skeletal muscle mechanics in setting limits on jumping performance? J Exp Biol 210(6):923–933

Jaric S, Markovic G (2009) Leg muscles design: the maximum dynamic output hypothesis. Med Sci Sports Exerc 41(4):780–787

Jiménez-Reyes P, Samozino P, Cuadrado-Peñafiel V, Conceição F, González-Badillo JJ, Morin JB (2014) Effect of countermovement on power–force–velocity profile. Eur J Appl Physiol 114 (11):2281–2288

Jiménez-Reyes P, Samozino P, Brughelli M, Morin JB (2017) Effectiveness of an individualized training based on force-velocity profiling during jumping. Front Physiol 7

Jiménez-Reyes P, Garcia-Ramos A, Párraga-Montilla JA, Morcillo-Losa JA, Cuadrado-Peñafiel V, Castaño-Zambudio A et al (2020) Seasonal changes in the sprint acceleration force-velocity profile of elite male soccer players. J Strength Cond Res

Joy JM, Lowery RP, Oliveira De Souza E, Wilson JM (2016) Elastic bands as a component of periodized resistance training. J Strength Cond Res 30(8):2100–2106

Kawamori NHG (2004) The optimal training load for the development of muscular power. J Strength Cond Res 18(3):675–684

Komi PV (1994) Strength and power in sport. Med Sci Sports Exerc 26(11):1422

Lieber RL, Ward SR (2011) Skeletal muscle design to meet functional demands. Philos Trans R Soc B Biol Sci 366(1570):1466–1476

Maffiuletti NA, Aagaard P, Blazevich AJ, Folland J, Tillin N, Duchateau J (2016) Rate of force development: physiological and methodological considerations. Eur J Appl Physiol 116 (6):1091–1116

Maroto-Izquierdo S, García-López D, Fernandez-Gonzalo R, Moreira OC, González-Gallego J, de Paz JA (2017) Skeletal muscle functional and structural adaptations after eccentric overload flywheel resistance training: a systematic review and meta-analysis. J Sci Med Sport 20 (10):943–951

Marques MC, van den Tilaar R, Vescovi JD, Gonzalez-Badillo JJ (2007) Relationship between throwing velocity, muscle power, and bar velocity during bench press in elite handball players. Int J Sports Physiol Perform 2(4):414–422

McMaster DT, Cronin J, McGuigan M (2009) Forms of variable resistance training. Strength Cond J 31(1):50–64

Morin JB, Samozino P (2016) Interpreting power-force-velocity profiles for individualized and specific training. Int J Sports Physiol Perform 11(2):267–272

Norrbrand L, Fluckey JD, Pozzo M, Tesch PA (2008) Resistance training using eccentric overload induces early adaptations in skeletal muscle size. Eur J Appl Physiol 102(3):271–281

Núñez FJ, Galiano C, Muñoz-López A, Floria P (2020) Is possible an eccentric overload in a rotary inertia device? Comparison of force profile in a cylinder-shaped and a cone-shaped axis devices. J Sports Sci 38(14):1624–1628

Page P, Ellenbecker T (2005) Kinetics H (ed) Strength band training, 3rd ed. Champain, IL, 206 p

Petré H, Wernstål F, Mattsson CM (2018) Effects of flywheel training on strength-related variables: a Meta-analysis. Sport Med Open 4(1):55

Rassier DE, MacIntosh BR, Herzog W (1999) Length dependence of active force production in skeletal muscle. J Appl Physiol 86(5):1445–1457

Rivière M, Louit L, Strokosch A, Seitz LB (2017) Variable resistance training promotes greater strength and power adaptations than traditional resistance training in elite youth rugby league players. J Strength Cond Res 31(4):947–955

Samozino P, Morin JB, Hintzy F, Belli A (2008) A simple method for measuring force, velocity and power output during squat jump. J Biomech 41(14):2940–2945

Samozino P, Rejc E, Di Prampero PE, Belli A, Morin JB (2012) Optimal force-velocity profile in ballistic movements-altius: citius or fortius? Med Sci Sports Exerc 44(2):313–322

Sanchez NFJ, Sáez de Villarreal E, De Villarreal ES (2017) Does flywheel paradigm training improve muscle volume and force? A meta-analysis. J Strength Cond Res 31(11):3177–3186

Shoepe TC, Ramirez DA, Rovetti RJ, Kohler DR, Almstedt HC (2011) The effects of 24 weeks of resistance training with simultaneous elastic and free weight loading on muscular performance of novice lifters. J Hum Kinet 29(1):93–106

Soriano MA, Jiménez-Reyes P, Rhea MR, Marín PJ (2015) The optimal load for maximal power production during lower-body resistance exercises: a meta-analysis. Sport Med 45(8):1191–1205

Spudić D, Smajla D, Šarabon N (2020) Validity and reliability of force–velocity outcome parameters in flywheel squats. J Biomech 107:109824

Tillin NA, Folland JP (2014) Maximal and explosive strength training elicit distinct neuromuscular adaptations, specific to the training stimulus. Eur J Appl Physiol 114(2):365–374
Vandewalle H, Péerès G, Monod H (1987) Standard anaerobic exercise tests. Sport Med Int J Appl Med Sci Sport Exerc 4(4):268–289
Worcester KS, Baker PA, Bollinger LM (2020) Effects of inertial load on sagittal plane kinematics of the lower extremity during flywheel-based squats. J Strength Cond Res 1
Zatsiorsky V, Kraemer W (2006) Science and practice of strength training. Google Books

Víctor Cuadrado Victor is Phd in sports science and Master in High Sports Performance by the Spanish Olympic Committee. He also hava a Master in Physical and Sports Performance from Pablo de Olavide University. He is currently lecturer at the Department of Physical Education, Sports and Human Motricity of the Autonomous University of Madrid and is the Director of the Master in Sports Sciences at the Autonomous University of Madrid. He also teachs in the Master in Physical Training and Sports Rehabilitation in Soccer and is Associate Professor in the Master in Research in Physical Education and Health. Victor has published more than twenty articles published in indexed journals (JCR, SJR ...) and contributed to national and international conferences.

Equipment and Training Devices

Javier Sanchez-Sanchez and Alejandro Rodríguez-Fernández

Abstract Recent classifications of resistance training methods classify them in gravitational, to move the resistance we will have to apply an acceleration greater than the force of the gravity (9.81 $m{\cdot}s^{-1}$), and non-gravitational methods that allow us to work in multiple directions, not being exclusively conditioned by gravity. Bodyweights training, also known as calisthenic exercise; Free-weights training, weight lifting through barbells, dumbbells, and kettlebells; inertial devices, how flywheel devices that an active stretch, while trying to brake an external resistance that exceeds the capacity of the muscle; and external resistance variable resistance, where resistance varies throughout range of motion are, among others, the main methods of resistance training currently. In this chapter, a contextualization of them will be carried out, exposing those essential aspects that it is necessary to know from a practical application and allowing the reader to understand the following chapters where each of them will be addressed in depth.

Keywords Free-weight training · Bodyweight training · Rotary inertial devices · Resistance variable training

1 Introduction

Resistance training has become fundamental content in conditioning programs for health and sports performance (Maestroni et al. 2020). Training programs in team and individual sports, whether performance dependent on power or endurance-based, should include strength-enhancing exercises (Suchomel et al. 2016; Trowell et al. 2020). In this sense, although some studies did not observe benefits in subjects with a higher level of strength (Cronin and Hansen 2005; Baker and Newton 2008), a large body of research has highlighted the importance of

J. Sanchez-Sanchez (✉)
Pontifical University of Salamanca, C/ Henry Collet, 52-70, 37007 Salamanca, Spain
e-mail: jsanchezsa@upsa.es

A. Rodríguez-Fernández
Faculty of Sport Sciences, University of Leon, Campus Vegazana, s/n, 24071 León, Spain

A. Muñoz-López et al. (eds.), *Resistance Training Methods*, Lecture Notes in Bioengineering, https://doi.org/10.1007/978-3-030-81989-7_4

muscle strength for success in competition (Comfort et al. 2012; Chelly et al. 2009; Sanchez-Sanchez et al. 2018; Haff and Nimphius 2012). Currently, a large number of coaches and researchers accept that the strongest athletes are those who best respond to the demands of the sport (Maestroni et al. 2020). Stronger athletes can better overcome the fatigue caused by training and competition (Suchomel et al. 2016), reducing the risk of injury (Lehance et al. 2009) and improving recovery, with fewer recurrences (Lauersen et al. 2018).

Although the level of strength is conditioned by genetic factors (Suchomel et al. 2018), to a greater extent, the level and type of resistance training conditions the structural and neuromuscular development of the athlete (Stone et al. 2002). To date, the scientific evidence does not establish which methods are ideal for improving these performance factors (Suchomel et al. 2018). Different review studies have mentioned some methods which could be more interesting, such as bodyweight resistance training, machine-based resistance training, weightlifting, plyometrics, eccentric training, potentiation complexes, unilateral or bilateral exercises, variable resistance, kettlebell training, and ballistic training (Suchomel et al. 2018). In this chapter, we will provide an introduction to gravitational methods that use free-weights and bodyweight and nongravitational methods such as rotating inertial devices and variable resistance training.

2 From Theory

2.1 Free-Weight Training

Among the different methods of resistance training, free-weight training is one of the most commonly used and the first to be developed among the population. Free-weight training refers to a load that moves freely in space (e.g. the barbell bench press), and that is not attached to another support structure like the smith machine (McQuilliam et al. 2020). The basic requirement for the development of muscle mass and strength is muscle resistance that is greater than usual (Weltman and Stamford 1982). The ability to generate high levels of muscular power is an important component of sport performance (Young 2006) and training to increase maximum strength augments the capacity to develop power (Plisk and Stone 2003). As movements in sport performance are multi-joint, incorporating multiple muscle groups may provide superior development of strength-power characteristics (Gentil et al. 2017). Thus, free-weight training has been widely used in athletic performance, injury recovery, and health-oriented programs. Different free-weight training methods have been proposed.

Maximal or heavy strength training: High levels of strength may influence sport-specific skills and increase jump height and sprint performance (Wing et al. 2020) and maximal strength training has been widely used in this context. This method is based on high-load RT (i.e., $\geq 80\%$ 1RM), utilizing 2 to 4 sets at low-repetition ranges (i.e., ≤ 6), while young athletes use a load of 80–89% 1RM (Lesinski et al. 2016).

Weightlifting: the emphasis is on movement speed at moderate to heavy loads, increasing motor-unit recruitment and rate of force development. The main lifts are clean, jerk, and snatch and their derivatives (i.e., hang snatch, hang clean, snatch pull, and clean pull) (Hori et al. 2005). Weightlifting movements have become commonplace within resistance training programs since they produce superior strength-power adaptations compared to traditional methods and exploit both the force and velocity aspects of power.

Peak-power training: low-load and high-velocity movements are beneficial for power, such as sprinting and jumping. As the peak power occurs at different intensities according to the exercise, different intensities are necessary (Cormie et al. 2007). Peak-power training improves strength and power via neural mechanisms but limits further improvements as strength remains underdeveloped (Haff and Nimphius 2012).

Complex training: since sports performance depends on a multitude of strength manifestations and different aspects of the force–velocity curve are involved, a combination of different training methods appears to be the most appropriate strategy. This training method is time-efficient and based on the combination of high, medium, and very light loads (Seitz et al. 2014).

The equipment used in free-weight training includes barbells, dumbbells and kettlebells. The usual exercises performed with this equipment are swings, goblet squats, and modified weightlifting exercises (Suchomel et al. 2018). Kettlebell training may produce strength improvements but its capacity to provide an overload stimulus to the lower extremities is limited, while more traditional training methods, such as weightlifting movements, may produce superior strength-power adaptations (Otto et al. 2012). However, kettlebells are interesting in a small space, for inexperienced athletes, and when barbells and dumbbells are unavailable.

Free-weight training presents advantages such as being less expensive, more versatile, better simulating the movements demanded by sports activity, and developing greater power. In addition, free-weight training may allow greater recruitment of muscle mass, since the machines provide a stable environment and free-weights require more stabilization and balance with greater activation in bench press (Schick et al. 2010) and squat exercises (Schwanbeck et al. 2009), although this aspect may be determined by the intensity of the load (McCaw and Friday 1994). This greater activation occurs in the muscles implicated in the stabilization of the movement, since activation of the trunk has been shown to be similar when performing a squat both in the smith machine and in free-weights (Schwanbeck et al. 2009). This greater activation in free-weight training causes an increase in anabolic response to training in men but not in women (Mitchell et al. 2013). Training with machines does not have the same mechanical stress or balance and stability needed for a free-weight training session which may add stress, resulting in acute free testosterone in men, although this is not accompanied by a further increase in muscle thickness and strength compared to machines (Schwanbeck et al. 2020). Resistance training requires a continuum from methods that enhance tissue capacity, greater coordination, and development of maximum strength, as well as producing a high power base according to the needs and characteristics of the athlete.

2.2 Bodyweight Training

Bodyweight training, also known as calisthenic exercise, is one of the earliest developed resistance training methods, and currently has many followers as it allows the individual to work in a 3-dimensional or multiplanar environment to overcome the force of gravity (Harrison 2010). Bodyweight training generally includes (i) push-ups, pull-ups, chin-ups and dips for upper-body; (ii) squats, lunges, skips and bounds for lower-body; (iii) squat thrusts, jumping jacks, burpees, mountain climbers, body builders, body killers and jump rope for total body.

Bodyweight training has a number of advantages over machine and free-weight training. The majority of bodyweight training exercises are closed-chain exercises (i.e., the distal aspect of the extremity is fixed), which are often more functional movements, to improve relative strength. This method is accessible and versatile as the exercises only require the body. On the other hand, the provision of overload stimulus is limited and makes improvements in maximal strength difficult. Adjustable training variables in bodyweight training include total repetitions, multiple sets, repetitions for time, various speeds and tempos, varying the base of support (i.e., 1 leg versus 2), and varying the initial position (i.e., incline push up).

Previous studies did not show increases in strength after bodyweight resistance training (Kotarsky et al. 2018; Campney and Wehr 1965; Tsourlou et al. 2003), however, the increase in intensity in these studies was based on an increase in the number of repetitions, so training-induced adaptations were oriented more towards endurance rather than resistance. Applying an increase in difficulty can be used as a strategy to increase intensity, since Kotarsky et al. (2018), through push-up progression based-training obtained increases in strength and muscle thickness comparable to traditional bench press training after short-term training. Similarly, push-up training induced similar increases in muscle thickness and strength to the 40% 1RM bench press free-weight training (Kikuchi and Nakazato 2017). In addition, the inclusion of devices such as a system of straps and handles for performed suspension exercise (Cayot et al. 2017), weighted vests or dip belts (Harrison 2010), elastic bands (Calatayud et al. 2015), and low-intensity exercise with slow movement (Tanimoto and Ishii 2006), or the combination of different exercises in circuit training (Klika and Jordan 2013) are commonly practiced.

2.3 Rotary Inertial Devices

Eccentric training (ET) is a strength training method based on active stretching of the muscle fiber, to decelerate an external resistance (Isner-Horobeti et al. 2013). Compared with other contraction modalities, such as concentric or isometric, this type of muscular action (i) may generate greater maximal force, (ii) have a lower metabolic cost for the same work produced, (iii) tend to be more effective to stimulate gains in muscle strength, (iv) preferentially activate type II fibers, and (v) lead to distinct regional hypertrophy and muscle architectural remodeling (Douglas et al. 2017). ET involves the use of isokinetic machines (Franchi and Maffiuletti 2019), free-weights with supra-maximal loads (Suchomel et al. 2018), or complex devices to accentuate the eccentric phase (Wagle et al. 2017), and rotational devices have recently become very popular (Franchi and Maffiuletti 2019; Beato and Iacono 2020). The first evidence supporting the efficacy of ET as a conditioning method only dates back to the early 1990s (Colliander and Tesch 1990), when it was used by astronauts, during long-duration missions, to avoid neuromuscular atrophy derived from the absence of gravitational stimuli (Berg and Tesch 1994).

Rotational device equipment is based on a flywheel or conic pulley which is wound by a strap or rope, respectively (Franchi and Maffiuletti 2019). When compared to gravitational devices, these devices offer the possibility to create greater eccentric loading than classical weight-based resistance training (Franchi and Maffiuletti 2019), without the athlete using heavy loads or complex implementation processes (Raya-González et al. 2021). In addition, with the flywheel and, especially, the conic pulley the athlete is able to perform very specific movements in multiple conditions (Gonzalo-Skok et al. 2017), and at high speed (Beato and Iacono 2020). Training with these devices causes (i) a combination of both concentric and overloaded eccentric contractions, (ii) higher force and power production combined with lower energy expenditure, (iii) preferential recruitment of high threshold motor unit and greater cortical activity, (iv) post-activation potentiation (PAP), (v) chronic adaptations in muscular strength, power or hypertrophy and (vi) improve athletic performances (Beato and Iacono 2020).

The athlete pulls on a strap to rotate the device to a limit that causes rewinding (i.e., rotation in the opposite direction), causing a braking movement in the athlete (i.e., eccentric load) (Norrbrand et al. 2008). The speed of the concentric phase determines the value of the eccentric phase (Nuñez and Sáez de Villarreal 2017). In addition, the eccentric load can be increased when the inertia is braked in the final phase of the movement. (Nuñez and Sáez de Villarreal 2017). This phenomenon is called eccentric overload (Maroto-Izquierdo et al. 2017). Although eccentric overload has been described in multiple publications, it is currently a theoretical concept with limited practical evidence (Núñez et al. 2020).

The main drawbacks of this modality are represented by the fact that the eccentric phase is largely reliant on the concentric phase, and exercise intensity is not easy to quantify (Franchi and Maffiuletti 2019). On the other hand this training method has some limitations (i) a familiarization process required, (ii) lack of standard procedures for exercise loading prescription and monitoring, (iii) limited evidence for elite sport athletes, (iv) no clear superiority compared to traditional resistance training methods for both acute and chronic effects, (v) limited use in clinical settings due to the maximal concentric effort required to generate overload as well as proper technique of execution (Beato and Iacono 2020).

2.4 *Variable Resistance Training*

Neuromuscular training is based on the use of three types of resistance: constant external resistance, accommodating resistance, and variable resistance (McMaster et al. 2009). Some of the current forms of variable resistance equipment include cams and levers, rubber-based resistance, and chains. These materials enable the resistance to acquire different values and, therefore, cause variable stress levels (McMaster et al. 2009; Behm 1988).

3 From Practice

3.1 *Free-Weight Training*

Low-load, high-velocity training may be more beneficial for younger than for more mature athletes. For physically mature athletes, however, the incorporation of high-intensity strength training is likely required to elicit greater improvements in performance.

Bench press and squat exercises are commonly performed by athletes and sports practitioners and are carried out using both free-weights and on a smith machine. However, these exercises present certain aspects that must be taken into account from a perspective of practical application, (i) squats lead to higher biceps femoris and gastrocnemius activity during free-weights, attributed to the increased role that knee flexors play in stabilizing and supporting the ankle, knee and hip joints; (ii) bench press leads to higher activation of the medial deltoid during free-weights, caused by the instability that necessitates a greater response. On the other hand, for the anterior deltoid and pectoralis major, there were no differences in activation between free-weights and machines.

3.2 *Bodyweight Training*

Regular use of bodyweight resistance training can compromise the development of maximum strength, since increasing the number of repetitions to increase the training stimulus can lead to physiological adaptations related to resistance. However, young children, novices, older adults, or athletes returning to sport and athletes who want to increase explosive performance can benefit from this method.

Bodyweight training, such as push-ups, performed consistently over 4–6 weeks can significantly increase muscle strength (Sperlich et al. 2018) and thickness (Kotarsky et al. 2018).

Push-ups and their different variations (i.e., different hand positions, inclined or loaded) are undoubtedly one of the most commonly used bodyweight resistance training exercises and have shown similar kinematics to the free-weight bench press. For external loads with vest weight of 10, 20 and 30 kg versus external loads in bench press at 50–80% 1RM, the muscle activation was the same, indicating that push-ups with weight vests can be used interchangeably with the bench press to enhance upper body strength (van den Tillaar 2019). In addition, push-ups showed greater activity in the anterior core muscles, regardless of hand positions (Gottschall et al. 2018), (i) the pectoralis major, triceps brachii, and deltoid anterior activity during knee push-ups was 15, 14, and 30% less compared to toe push-ups (Sandhu et al. 2008; Herrington et al. 2015); (ii) the pectoralis major and triceps brachii was 24 and 69% greater in narrow push-ups than wide push-ups (Gottschall et al. 2018); (iii) knee and wide push-ups would be an ideal adjustment for novice participants.

3.3 Rotary Inertial Devices

The use of rotational devices in training programs has increased because it allows power training to be performed, prioritizing the eccentric phase of the movement (Raya-González et al., in press). Although the number of studies that aim to describe the response to these training programs has increased, unfortunately it has not been possible to establish application protocols, especially for high-level athletes (Beato and Iacono 2020). Based on recent evidence, the following can be recommended (i) the use of inertia discs of 0.05–0.11 kg/m^2, which move at maximum speed, for the development of force (i.e., high inertia) or power (i.e., low inertias) (Martinez-Aranda and Fernandez-Gonzalo 2016); (ii) 1–2 sessions/week, for 5–10 weeks to improve sports performance (Coratella et al. 2019).

Rotational devices can be used to generate PAP (Beato and Iacono 2020) with the following conditions, (i) the use of inertia 0.029–0.110 kg/m^2, in movements at maximum speed, within specific exercises (Cuenca-Fernández et al. 2015); (ii) multiple sets of exercises; (iii) a rest interval of at least 3 min depending on the athlete.

Rotational inertia devices are considered a good alternative to reduce the rate and severity of injuries during sports practice, since they improve the ability of the muscle to generate tension during the stretching phase (Norrbrand et al. 2008) and during sport-specific movements performed at high speed (Raya-González et al. 2021).

3.4 Variable Resistance Training

The cam is a material designed to change the external moment arm or to approximate the body's changing moment arm during the lift (McMaster et al. 2009), thus, forcing the muscles to exert near maximal effort throughout the range of motion. Cam and lever equipment is suitable for beginner and weak resistance trainers because it follows a fixed movement path and requires less skill, decreased intermuscular coordination, and is less likely to cause injury compared with other modes of resistance, as it is easier to maintain control of the load (Foran 1985). On the other hand, the cam is common in rehabilitation settings where multi-joint exercise movements are used in combination for the rehabilitation of injuries (Fleming et al. 2005).

Rubber bands and elastics are a portable, inexpensive, and easy-to-use material (McMaster et al. 2009), which enables different degrees of tension to be applied during movement (Wallace et al. 2006). Although this allows a great diversity of stimuli to be offered, it also makes it difficult to control the dose. In addition to the hardness of the material, the different levels of resistance depend on the application of force by the athlete and the degree of stretch to which the band is subjected at the time of the movement (Behm 1988). In general, the resistance imposed by the elastic band is low at the beginning of the movement, enabling a resisted execution at high speed; however, as the range of motion increases, resistance increases, forcing the athlete to strain to maintain power in higher overload ranges (Behm 1988). These materials can also be used to overload free-weight exercises (i.e., back squat and bench press), especially in the final range of motion. (Wallace et al. 2006). In this type of exercise, it is recommended that the resistance included in the elastic band represents 20–35% of the load and the free-weight 65–80%, with respect to the maximum repetition (Stevenson et al. 2010).

Chains allow different levels of resistance to be applied during free-weight movements executed in the vertical plane (McMaster et al. 2009). On the one hand, the resistance imposed by the chains depends on the structure, density, length, and diameter of the material. On the other hand, the resistance depends on the movement itself, since as the barbell is raised the chains come off the ground, increasing the weight to overcome, while when the barbell is lowered the weight decreases, since the chains are fixed to the ground (Berning et al. 2004). It is interesting to see the possibility of increasing resistance in the final range of movement as a consequence of the total release of the chain, which enables regulation of the incidence of maximum loads during exercise (Ebben and Jensen 2002). Under these conditions, the agonist muscles are activated at different levels, allowing multiple improvements in the factors involved in strength performance: power, acceleration, and motor control (Berning et al. 2004). Regarding motor control, although some recommendations establish that the chain must be anchored to the ground to avoid oscillations and thus avoid damage, movements in the horizontal plane will stimulate motor control and thus the activation of synergists and stabilizers (McMaster et al. 2009). Finally, properly designed training involves the use of resistance associated with the chain of 10–15% and 85–90% of free-weight, with respect to the maximum repetition.

4 Filling Gaps

For sports such as soccer, handball, and futsal or basketball, tennis, and volleyball which include movements that require strength and stability of the lower and upper extremities respectively, the utilization of free-weight training may be an interesting resistance training method, due to the potential for more sports-specific muscular development and for practitioners who have not developed the adaptation necessary for adequate stabilization in the smith machine. If the aim of free-weight resistance training is to increase strength or muscle thickness, increasing the number of repetitions is not the best variable to manipulate, instead, increasing the load, through, for example, difficulty, incline, or performing the exercise with one limb, is more appropriate. Training based on eccentric contractions has shown positive effects both on performance and in reducing the risk of injury but requires cautious application with respect to execution and recovery.

5 Take-Home Messages and Practical Resources

1. Different methods such free-weight, bodyweight, rotary inertial devices or variable resistance training have been widely used in athletic performance, injury recovery, and health-oriented programs.
2. Free-weight training refers to a load that moves freely in space, and the basic requirement for the development of muscle mass and strength is muscle resistance that is greater than usual.
3. Bodyweight training allows the individual to work in multiplanar environment to overcome the force of gravity. This method is accessible and versatile as the exercises only require the body but the provision of overload stimulus is limited and makes improvements in maximal strength difficult.
4. The rotational devices allows power training to be performed, prioritizing the eccentric phase of the movement. This devices can can be used to generate PAP and they are considered a good alternative to reduce the rate and severity of injuries during sports practice.
5. Some of the current forms of variable resistance equipment include cams and levers, rubber-based resistance, and chains. In general, These materials can be used to overload free-weight exercises, allows great diversity of stimuli to be offered, but it also makes it difficult to control the dose.

References

Baker DG, Newton RU (2008) Comparison of lower body strength, power, acceleration, speed, agility, and sprint momentum to describe and compare playing rank among professional rugby league players. J Strength Cond Res 22(1):153–158
Beato M, Iacono AD (2020) Implementing flywheel (isoinertial) exercise in strength training: current evidence, practical recommendations, and future directions. Front Physiol 11:1–6
Behm DG (1988) Surgical tubing for sport and velocity specific training. Natl Strength Cond Assoc. 10:66–67
Berg HE, Tesch A (1994) A gravity-independent ergometer to be used for resistance training in space. Aviat Space Environ Med 65(8):752–756
Berning JM, Coker CA, Adams KJ (2004) Using chains for strength and conditioning. Strength Cond J 26(5):80–84
Calatayud J, Borreani S, Colado JC, Martin F, Tella V, Andersen LL (2015) Bench press and push-up at comparable levels of muscle activity results in similar strength gains. J Strength Cond Res 29(1):246–253
Campney HK, Wehr RW (1965) Effects of calisthenics on selected components of physical fitness. Res Q 36(4):393–402
Cayot TE, Lauver JD, Scheuermann BW (2017) The acute effects of bodyweight suspension exercise on muscle activation and muscular fatigue. Eur J Sport Sci 17(6):681–689
Chelly MS, Fathloun M, Cherif N, Amar M Ben, Tabka Z, Van Praagh E (2009) Effects of a back squat training program on leg power, jump, and sprint performances in junior soccer players. J Strength Cond Res 23(8):2241–2249
Colliander EB, Tesch PA (1990) Effects of eccentric and concentric muscle actions in resistance training. Acta Physiol Scand 140(1):31–39
Comfort P, Haigh A, Matthews MJ (2012) Are changes in maximal squat strength during preseason training reflected in changes in sprint performance in rugby league players? J Strength Cond Res 26(3):772–776
Coratella G, Beato M, Cè E, Scurati R, Milanese C, Schena F et al (2019) Effects of in-season enhanced negative work-based versus traditional weight training on change of direction and hamstrings-to-quadriceps ratio in soccer players. Biol Sport 36(3):241–248
Cormie P, Mccaulley GO, Triplett NT, Mcbride JM (2007) Optimal loading for maximal power output during lower-body resistance exercises. Med Sci Sports Exerc 39(2):340–349
Cronin JB, Hansen KT (2005) Strength and power predictors of sports speed. J Strength Cond Res 19(2):349–357
Cuenca-Fernández F, López-Contreras G, Arellano R (2015) Effect on swimming start performance of two types of activation protocols: Lunge and YoYo squat. J Strength Cond Res 29(3):647–655
Douglas J, Pearson S, Ross A, McGuigan M (2017) Eccentric exercise: physiological characteristics and acute responses. Sport Med 47(4):663–675
Ebben WP, Jensen RL (2002) Electromyographic and kinetic analysis of traditional, chain, and elastic band squats. J Strength Cond Res 16(4):547–550
Fleming BC, Oksendahl H, Beynnon BD (2005) Open-or closed-kinetic chain exercises after anterior cruciate ligament reconstruction? Exerc Sport Sci Rev 33(3):134–140
Foran B (1985) Facility considerations: Advantages and disadvantages of isokinetics, variable resistance and free weights. Strength Cond J. 7(1):24–25
Franchi MV, Maffiuletti NA (2019) Distinct modalities of eccentric exercise: different recipes, not the same dish. J Appl Physiol 127(3):881–883
Gentil P, Fisher J, Steele J (2017) A Review of the acute effects and long-term adaptations of single- and multi-Joint exercises during resistance training. Sport Med 47(5):843–855
Gonzalo-Skok O, Tous-Fajardo J, Valero-Campo C, Berzosa C, Bataller AV, Arjol-Serrano JL et al (2017) Eccentric-overload training in team-sport functional performance: constant

bilateral vertical versus variable unilateral multidirectional movements. Int J Sports Physiol Perform 12(7):951–958
Gottschall J, Hastings B, Becker Z (2018) Muscle activity patterns do not differ between push-up and bench press exercises. J Appl Biomech 24(6):442–447
Haff GG, Nimphius S (2012) Training principles for power. Strength Cond J 34(6):2–12
Harrison JS (2010) Bodyweight training: a return to basics. Strength Cond J 32(2):52–55
Herrington L, Waterman R, Smith L (2015) Electromyographic analysis of shoulder muscles during press-up variations and progressions. J Electromyogr Kinesiol 25(1):100–106
Hori N, Newton RU, Nosaka K, Stone MH (2005) Weightlifting exercises enhance athletic performance that requires high-load speed strength. Strength Cond J 27(4):50–55
Isner-Horobeti M-E, Dufour SP, Vautravers P, Geny B, Coudeyre E, Richard R (2013) Eccentric exercise training: modalities, applications and perspectives. Sports Med 43(6):483–512
Kikuchi N, Nakazato K (2017) Low-load bench press and push-up induce similar muscle hypertrophy and strength gain. J Exerc Sci Fit 15(1):37–42
Klika B, Jordan C (2013) High-intensity circuit training using body weight. Acsms Health Fit J 17 (3):8–13
Kotarsky CJ, Christensen BK, Miller JS, Hackney KJ (2018) Effect of progressive calisthenic push-up training on muscle strength and thickness. J Strength Cond Res 32(3):651–659
Lauersen JB, Andersen TE, Andersen LB (2018) Strength training as superior, dose-dependent and safe prevention of acute and overuse sports injuries: a systematic review, qualitative analysis and meta-analysis. Br J Sports Med 52(24):1557–1563
Lehance C, Binet J, Bury T, Croisier JL (2009) Muscular strength, functional performances and injury risk in professional and junior elite soccer players. Scand J Med Sci Sports 19(2):243–251
Lesinski M, Prieske O, Granacher U (2016) Effects and dose-response relationships of resistance training on physical performance in youth athletes: a systematic review and meta-analysis. Br J Sports Med 50(13):781–795
Maestroni L, Read P, Bishop C, Papadopoulos K, Suchomel TJ, Comfort P et al (2020) The benefits of strength training on musculoskeletal system health: practical applications for interdisciplinary care. Sport Med 50(1):1431–1450
Maroto-Izquierdo S, García-López D, Fernandez-Gonzalo R, Moreira OC, González-Gallego J, de Paz JA (2017) Skeletal muscle functional and structural adaptations after eccentric overload flywheel resistance training: a systematic review and meta-analysis. J Sci Med Sport 20 (10):943–951
Martinez-Aranda LM, Fernandez-Gonzalo R (2016) Effects of inertial setting on power, force, work and eccentric overload during flywheel resistance exercise in women and men. J strength Cond Res 31(6):1653–1661
McCaw S, Friday J (1994) Acomparison of muscle activity between a free weight and machine bench press. J Strength Cond Res 8(4):259–264
McMaster DT, Cronin J, McGuigan M (2009) Forms of variable resistance training. Strength Cond J. 31(1):50–64
McQuilliam SJ, Clark DR, Erskine RM, Brownlee TE (2020) Free-weight resistance training in youth athletes: a narrative review. Sport Med 50(9):1567–1580
Mitchell CJ, Churchward-Venne TA, Bellamy L, Parise G, Baker SK, Phillips SM (2013) Muscular and systemic correlates of resistance training-induced muscle hypertrophy. PLoS ONE 8(10):1–10
Norrbrand L, Fluckey JD, Pozzo M, Tesch PA (2008) Resistance training using eccentric overload induces early adaptations in skeletal muscle size. Eur J Appl Physiol 102(3):271–281
Nuñez FJ, Sáez de Villarreal E (2017) Does flywheel paradigm training improve muscle volume and force? A meta-alaysis. J Strength Cond Res 31(11):3177–3186
Núñez FJ, Galiano C, Muñoz-López A, Floria P (2020) Is possible an eccentric overload in a rotary inertia device? Comparison of force profile in a cylinder-shaped and a cone-shaped axis devices. J Sports Sci 38(14):1624–1628

Otto WH, Coburn JW, Brown LE, Spiering BA (2012) Effects of weightlifting vs. kettlebell training on vertical jump, strength, and body composition. J strength Cond Res 26(5):1199–1202

Plisk SS, Stone MH (2003) Periodization strategies. Strength Cond J. 25(6):19–37

Raya-González J, Keijzer KL, Bishop C, Beato M. Effects of flywheel training on strength-related variables in female populations: a systematic review. In press

Raya-González J, Castillo D, Beato M (2021) The flywheel paradigm in team sports. Strength Cond J 43(1):12–22

Sanchez-Sanchez J, Rodriguez A, Petisco C, Ramirez-Campillo R, Martínez C, Nakamura FY (2018) Effects of different post-activation potentiation warm-ups on repeated sprint ability in soccer players from different competitive levels. J Hum Kinet 61(1):189–197

Sandhu JS, Mahajan S, Shenoy S (2008) An electromyographic analysis of shoulder muscle activation during push-up variations on stable and labile surfaces. Int J Shoulder Surg. 2(2):30–35

Schick EE, Coburn JW, Brown LE, Judelson DA, Khamoui AV, Tran TT et al (2010) A comparison of muscle activation between a smith machine and free weight bench press. J Strength Cond Res 24(3):779–784

Schwanbeck S, Chilibeck PD, Binsted G (2009) A comparison of free weight squat to smith machine squat using electro-myography. J Strength Cond Res 23(9):2588–2591

Schwanbeck SR, Cornish SM, Barss T, Chilibeck PD (2020) Effects of training with free weights versus machines on muscle mass, strength, free testosterone, and free cortisol levels. J Strength Cond Res 34(7):1851–1859

Seitz LB, Reyes A, Tran TT, de Villarreal ES, Haff GG (2014) Increases in lower-body strength transfer positively to sprint performance: a systematic review with meta-analysis. Sport Med 44 (12):1693–1702

Sperlich B, Hahn LS, Edel A, Behr T, Helmprobst J, Leppich R et al (2018) A 4-week intervention involving mobile-based daily 6-minute micro-sessions of functional high-intensity circuit training improves strength and quality of life, but not cardio-respiratory fitness of young untrained adults. Front Physiol 9:1–10

Stevenson MW, Warpeha JM, Dietz CC, Giveans RM, Erdman AG (2010) Acute effects of elastic bands during the free-weight barbell back squat exercise on velocity, power, and force production. J Strength Cond Res 24(11):2944–2954

Stone MH, Moir G, Glaister M, Sanders R (2002) How much strength is necessary? Phys Ther Sport 3(2):88–96

Suchomel TJ, Nimphius S, Stone MH (2016) The importance of muscular strength in athletic performance. Sport Med 46(10):1419–1449

Suchomel TJ, Nimphius S, Bellon CR, Stone MH (2018) The importance of muscular strength: training considerations. Sport Med 48(4):765–785

Tanimoto M, Ishii N (2006) Effects of low-intensity resistance exercise with slow movement and tonic force generation on muscular function in young men. J Appl Physiol 100(4):1150–1157

Trowell D, Vicenzino B, Saunders N, Fox A, Bonacci J (2020) Effect of strength training on biomechanical and neuromuscular variables in distance runners: a systematic review and meta-analysis. Sport Med 50(1):133–150

Tsourlou T, Gerodimos V, Kellis E, Stavropoulos N, Kellis S (2003) The effects of a calisthenics and a light strength training program on lower limb muscle strength and body composition in mature women. J Strength Cond Res 17(3):590–598

van den Tillaar R (2019) Comparison of kinematics and muscle activation between push-up and bench press. Sport Med 3(3):74–81

Wagle JP, Taber CB, Cunanan AJ, Bingham GE, Carroll KM, DeWeese BH et al (2017) Accentuated eccentric loading for training and performance: a review. Sport Med 47(12):2473–2495

Wallace BJ, Winchester JB, McGuigan MR (2006) Effects of elastic bands on force and power characteristics during the back squat exercise. J Strength Cond Res 20(2):268–272

Weltman A, Stamford B (1982) Strength training: free weights vs machines. Phys Sportsmed 10 (11):197–197

Wing CE, Turner AN, Bishop CJ (2020) Importance of strength and power on key performance indicators in elite youth soccer. J Strength Cond Res 34(7):2006–2014

Young WB (2006) Transfer of strength and power training to sports performance. Int J Sport Physiol Perform. 1(2):74–83

Javier Sanchez-Sanchez is a full professor at the University Pontificia de Salamanca (UPSA, Spain). He is a professor of the training and sports performance area and the director of the M. Sc. in Training and Rehabilitation in Soccer in UPSA and the main investigator of the Programming and Testing in performance Research Group. He a is author and reviewer of research related to the performance analysis, training load quantification, and team sports periodization. He has also worked as football fitness coach.

Alejandro Rodríguez-Fernández is a profesor at the University of León (Spain) and has Ph.D. in Sports Sciences and M.Sc. in Research and Innovation in Sports Sciences. He is an author and reviewer of research published in indexed JCR journal, related to the training and detraining effects, in addition to the training load control in team sports. He has also worked as football fitness coach.

Developing and Building Training Paradigms

Resistance Training for the Maximisation of the Vertical Force Production: Jumps

Eduardo Sáez de Villarreal and Rodrigo Ramírez-Campillo

Abstract In this chapter, the authors explore jump training exercises as a mean to maximise vertical force production and related physical fitness traits. Jump training may enhance muscular force, the rate of force development, muscular power, muscle contraction velocity, cross-sectional area, muscle stiffness, among other biological and biomechanical factors associated with enhanced physical function and athletic performance. Jump training exercises are characterised by the stretch–shortening cycle of the muscle–tendon complex, usually involving a pre-activation, stretching, and a shortening phase. Athletes have used jumps as a training method at least in the last 3000 years. From a scientific perspective, the number of scientific publications increased tremendously in recent years, with a 25-fold increase between 2000 and 2017. Scientific evidence supports the role of jump training for the improvement of physical performance in male and female athletes, from pre-pubertal to adult and senior age. However, evidence also supports the role of modified jump training exercises for several health-related outcomes (e.g., fat mass; muscle hypertrophy; bone density). In this chapter, the reader will find a summary of current scientific evidence regarding the biological foundations for jump training exercises, the scientifically proven methodological principles and practical guidelines regarding the programming of jump training exercises.

Keywords Human physical conditioning · Plyometric exercises · Athletic performance · Sports · Resistance training · Stretch reflex · Physical education and training · Exercise therapy

---

E. Sáez de Villarreal (✉)
Physical Performance and Sports Research Center, Universidad Pablo de Olavide, Seville, Spain
e-mail: esaesae@upo.es

R. Ramírez-Campillo
Department of Physical Activity Sciences, Universidad de Los Lagos, Santiago, Chile

A. Muñoz-López et al. (eds.), *Resistance Training Methods*, Lecture Notes in Bioengineering, https://doi.org/10.1007/978-3-030-81989-7_5

1 Introduction

Vertical force production (VFP) is widely recognised as a critical determinant of performance in a series of sport-specific activities, such as sprinting, jumping, and changing direction and is one of the most important variables required for success in many individuals and team sports. Therefore, practitioners are constantly seeking better and more effective strategies to improve VFP. Although many training routines have been proposed for athletes, plyometric exercises with a significant vertical and horizontal force production component have been shown to be more effective than other training strategies such as resistance training (Saez de Villarreal et al. 2009). Plyometric training (PT) enhances muscular force, the rate of force development, muscular power, muscle contraction velocity, cross-sectional area, muscle stiffness allowing greater storage and release of elastic energy.

Plyometrics refers to exercises that are designed to enhance muscle, mainly through the use of jump training. Plyometric exercises constitute a natural part of most sport movements because they involve jumping, hopping, and skipping (i.e., such as high jumping, throwing, or kicking) (Anderst et al. 1994; Asmussen and Bonde-Petersen 1974; Bauer et al. 1990). Plyometric exercises come in various forms depending on the purposes of a training program. Typical plyometric exercises include the countermovement jump (CMJ), the drop jump (DJ), and the squat jump (SJ). These exercises either can be combined within a training program or can be applied independently. Furthermore, plyometrics can be performed at various intensity levels, ranging from low-intensity double-leg hops to high unilateral-intensity drills. As far as the lower body is concerned, plyometrics includes the performance of various types of body-weight jumping exercises, such as DJ, CMJ, alternate-leg bounding, hopping, and other stretch–shortening cycle (SSC) jumping exercises (Bobbert 1990a; Bobbert et al. 1986; Bosco et al. 1982a; Cavagna et al. 1968). These exercises are characterised by SSC actions; that is, they start with a rapid stretch of a muscle (eccentric phase) and are followed by a rapid shortening of the same muscle (concentric phase) (Bosco and Komi 1979; Bosco et al. 1981, 1982b; Komi and Bosco 1978).

2 From Theory

2.1 *The Development of Plyometric Training: What is Plyometrics?*

For many years, coaches and athletes have sought to improve muscular power and performance. Throughout the last years and no doubt long before, jumping, bounding and hopping exercises have been used in various ways to enhance athletic performance. In recent years, this distinct method of training for power or explosiveness has been termed plyometrics. Whatever the origins of the word are, the term is used to describe the method of training that seeks to enhance the explosive reaction of the individual through powerful muscular contractions preceded by rapid eccentric contractions (Bosco et al. 1982a; Cavagna et al. 1968; Komi and Bosco 1978).

Plyometric training is known to be an intense form of exercise that requires maximal efforts to create the physiological change associated with athletic performance. This system became popular in the late 60s to the early 70s and was credited with being responsible for much of the East European success in athletics during that time. Since then, the use of plyometric training has evolved into a mainstay of the training and development programs of virtually all-sporting events. With this transition came many questions, including the age, gender and strength levels of the athletes who would benefit from this form of training (Verhoshanski 1966; Markovic 2007).

Plyometric type exercises have been used successfully by many athletes as a method of training to enhance power. In order to realise the potential benefits of plyometric training, the SSC must be invoked. This requires careful attention to the technique used during the drill or exercise. The rate of stretch rather than the magnitude of stretch is of primary importance in plyometric training. In addition, the coupling time or ground contact time must be as short as possible. The fundamental reason to train with plyometrics is to reduce the ground contact time that an athlete spends when running or jumping. This time is reduced as the athlete matures, gets stronger and practices the skills of their game (Verhoshanski 1966; Markovic 2007).

2.2 General Effects of Plyometric

- Improvement of neuromuscular processes (Komi 1992).
- Special effect on the inhibitor mechanisms and facilitators of muscle contraction (Bobbert 1990a).
- No improvement in maximal strength (in highly trained subjects), but greatest application in other variables (power) (Komi 1992).
- Possible improvement of elastic energy storage capacity by positive effect on nerve mechanisms (Komi 1992).
- Greater energy is absorbed only if the strength is greater (Bobbert 1990a).
- Improvement of mechanical efficiency (ratio work/energy) (Komi 1992).
- Improvement of the degree of tolerance to high stretching loads (Bosco and Komi 1979; Komi 1992).

2.3 Stretch–Shortening Cycle (SSC)

Although concentric contractions provide the propulsion force needed for movements such as running, jumping, throwing or lifting loads, a common strategy of human movement is to combine eccentric and concentric muscle contractions into a sequence known as a stretch–shortening muscle cycle, (SSC). The predominance of the SSC in numerous sporting gestures can be attributed to its positive effects on maximum muscle performance, increasing mechanical efficiency and attenuation of impact forces on the human body (Komi 1992; Steben and Steben 1981).

The most important feature of the SSC is its ability to achieve maximum muscle performance (Cavagna et al. 1968). Cavagna et al. (1968) found that this effect was due to the fact that in the SSC, the eccentric muscle contraction (muscle elongation) prior to concentric movement, enabled greater muscle strength at the beginning of the concentric contraction, compared to that obtained in a movement that only involved a concentric contraction. Later, Ettema et al. (1990) considered that the effects of pre-muscle stretching on positive muscle work (concentric contraction) on the tendon–muscle complex could be explained by three main mechanisms: the release of additional elastic energy, the interaction between the elongation of tendon structures and muscle fibres, and finally the potentiation of the contractile elements.

According to these authors (Cavagna et al. 1968; Ettema et al. 1990), additional elastic energy would be released by the elastic return of the muscle after stretching. The contribution relating to the total muscle work of this mechanism would depend on the specific conditions of the contraction. In this way, under some conditions, this contribution related to muscle work could be due only to the increase in work carried out by the tendon–muscle complex, while in others, the other two mechanisms mentioned should be used. Thus, when performing a muscle contraction involving a previous stretch, the initial increase in strength in the concentric phase has side effects on tendon structures and muscle fibres activated during contraction (Cavagna et al. 1968; Ettema et al. 1990).

For example, when a muscle strength causes an increase in the length of the tendon tissues involved in contraction, at the same time the shortening of the muscle fibres of that muscle-tendinous unit occurs and the additional elastic energy that produces the elastic return of tendon tissues is released and this in turn induces the increase in the ratio strength-speed of the muscle (Ettema et al. 1990).

The last mechanism that explains the positive effects of pre-muscle stretching on positive muscle work in the tendon–muscle complex is the potentiation of the contractile material. In this regard, Ettema et al. (1990) considered that in cases where in pre-isometric and pre-stretching actions the concentric phase was performed isotonically (at constant speed, after a very rapid initial decrease), the extra release of elastic energy could not be the factor justifying the additional work done by the muscular complex. In this way, they considered that the isotonic shortening of the muscle could be explained exclusively by differences in the length of muscle fibres between pre-stretch and pre-isometric actions (Ettema et al. 1990).

This means that the force–velocity curve of the SSC reaches very high velocity for a given load. In any case, different authors believe that the potentiation does not appear to be related to the elastic properties of muscle (Cavagna et al. 1968; Ettema et al. 1990; Ferragut and Lopez-Calbet 1998). We can summarise that the elastic potential of muscle depends on the elastic shortening of tendon tissues and the interaction between fibre length and shortening speed (Komi and Bosco 1978), and there are several factors that influence the effectiveness of an SSC (Komi and Bosco 1978) such as the intensity (speed) with which the stretch occurs, the magnitude of the increase in muscle stiffness at the beginning of the stretch (reflex mechanism), the coupling time between the stretch phase and the contraction phase, as well as the amplitude of the stretch (determines the area of the length-stress curve in which concentric muscle contraction will occur). In general, wide stretches are usually slower than low-amplitude stretches.

Therefore, the SSC would consist of three phases:

1. *Pre-activation* has the function of providing sufficient stiffness to the muscle to oppose stretching. From the moment the myoelectric activity on baseline levels increases to the time of contact with the ground at this stage, the upper centres of the central nervous system adjust the degree of pre-activation and muscle stiffness based on the magnitude of the intended stretch (higher drop height, higher pre-activation and therefore greater stiffness). The lower the pre-contact stiffness, the lower the subsequent reactive motion capacity.
2. *Stretching* causes the accumulation of elastic potential energy and/or a response of muscle spindles that determines a more intense contraction of extrafusal fibres (muscle fibres outside the muscle spindle). In the initial phase of stretching, the myotatic reflex can contribute to increased muscle stiffness and therefore increase the ability to accumulate potential elastic energy, during the stretching of a muscle that is in a more or less intense state of contraction. Stretching under these conditions allows the accumulation of elastic potential energy in both the cross bridges, tendons and other elastic elements of connective tissue.

    a. From contact with the ground to the completion of muscle elongation, this phase detects widening spikes in muscle electrical activity, due in part to opposition of muscle spindles to stretching (voluntary response) and the myotatic reflex (reflected response), which facilitates the activation of muscles undergoing stretching (Kilani et al. 1989). But the myotatic reflex is not the only reflex-type response. In the face of important stretches (when the drop height is very high) Golgi's tendon reflex is activated, which opposes the action of the myostatic reflex, protecting muscle integrity.
    b. It is also considered that the contract apparatus alone is able to generate more force when it has previously been stretched quickly and the time between the eccentric and concentric phase is minimal. This is what has been called the "potentiation effect", although it is not fully understood (López-Calbet et al. 1995). It is likely to be because of the special characteristics of the myosine heads and their behaviour when establishing crossed bridges.

3. *Shortening:* at this phase the forces provided by the shortening of the sarcomeres plus the force provided by the elastic return will be added, where the previously accumulated elastic energy is harnessed. To optimally use this energy, it is necessary that the concentric phase immediately happens in time to the eccentric phase. If this does not occur, the accumulated elastic energy dissipates in the form of heat. Mouche (2001) indicates that the transition phase should not last more than 200 ms in order to reduce the dissipation of elastic energy.

2.4 Determining Factors of Plyometric Performance

The force with which the muscle contracts at the time of the jump and the rate at which the muscle is able to generate tension during the jump are determining factors for plyometric performance. Both the contraction force and the speed at which stress is generated depends in turn on other factors, such as the speed at which motor units involved in the movement are recruited and activated, as well as the number of motor units recruited and their discharge frequency (stimulation dynamics); time spent attaining a state of maximum muscle stimulation, or time spent coupling between stimulation and contraction (excitation dynamics) and in part the interaction between the contractile and elastic elements (contraction dynamics). Once muscle contraction is initiated, a part of the force is invested in tightening the elastic elements in set, which work by dampening the transfer of force to the bone levers. However, if muscle contraction starts in conditions where the elastic elements are already tensioned, the transfer of force to the bone levers is faster (Saez de Villarreal et al. 2009; Bobbert et al. 1996).

Another determining factor is the efficiency of motor control for the orders generated to produce the jump or simple motor coordination. At least two levels of motor control are especially important in terms of jump efficiency. One being, agonist-agonist, the other agonist-antagonistic coordination, both intramuscular and intermuscular. For the muscles involved in the jump to act more effectively, massive recruitment of the agonist muscles, conveniently sequenced over time, is necessary. Simultaneously, antagonistic activity should be minimised to ensure joint stability and coactivation. Then, the resulting force vector must be properly directed, so that its vertical component is maximised in the centre of mass. The direction of the vertical component of the resulting force vector depends primarily on the activity of the biarticular muscles (Ingen Schenau et al. 1987, 1992; Jacobs and Ingen Schenau 1992).

Table 1 Intensity guide for plyometric training

Type of plyometrics move	Examples	Low
Standing-based jumps performed on the spot	Tuck-jumps Spilt jumps Squat jumps	Low-medium
Jumps from standing	Standing long jump Standing hop Standing jump for height	Medium
Multiple jumps from standing	5 consecutive bounds 2 × 6 bunny jumps Double-footed jumps over 4 hurdles Double-footed jumps up steps	High
Multiple jumps with run-up	3 × 2 hops and jump into sand pit with 11 stride approach 2 × 10 bounds with a 7-stride run-up	High Very high Very high
Depth jumping *Recommended drop height 40–100* cm. *the greater the height the greater the strength component, the lower the height the greater the speed*	2 × 6 jumps—down and up Run and hop off low box onto one leg landing followed by three subsequent hops Bounding up hill	High High Very high
Eccentric drop and hold drills	Hop and hold 5 time Bound/hop/bound/hop and hold over 30 m *(To perform the above two, the athlete literally stops on each landing before springing into the next move when required)* Drop and hold from height above 1 m	Low
Pre-season/early conditioning phase	Moves such as split squats, jump squats and straight leg jumps High reps short recovery (low intensity)	
Main power conditioning phase	Athletes who are sufficiently skilled should use drills from the medium intensity However do not neglect lower leg drills like 'pogos' up and down on the spot	
Pre-competition phase	Concentrate on quality plyometrics drills High intensity, however, do not allow fatigue to impair performance	
Competition phase	During this phase of training athletes should continue to perform high-quality plyometric drills in low numbers, well away (Bosco et al. 1981, 1982a; Cavagna et al. 1968; Bosco and Komi 1979) from important competitions	
Volume and intensity guidelines		
Beginner	Intermediate	Advanced
60–100-foot contacts of low intensity	100–150-foot contacts of low intensity or 100 moderate-intensity	150–200-foot contacts of low to moderate intensity

3 From Practice

Through evolution, several animal species, including humans, have evolved the ability to jump. Indeed, jumping actions are common with, among others, mammals, reptiles, and insects. For humans, jumping allows them to perform a wide range of activities, from dancing to moon exploration. From practice, jumps have been used by athletes as a training method in preparation for competitions for the last 3000 years at least (Minetti and Ardigo 2002).

Jump training exercises involve multi-joint drills and large muscle groups (e.g., quadriceps). Depending on the type of jump, these may involve an SSC, with a considerable voluntary effort (i.e., near maximal or maximal) during the concentric portion of the jump (i.e., against the force of gravity). In addition, some jumps may also involve considerable eccentric forces upon landing, as high as 12 times body mass (Bobbert 1990b). In this sense, highly trained athletes use some jump training exercises most commonly, such as bounce drop jumps from relatively high drop heights (e.g., plyometrics). However, jump training exercises can be adapted safely and effectively for, among other, older adults, children with cerebral palsy and Down syndrome and injured athletes. (Ramirez-Campillo et al. 2020). Adaptation strategies may involve the performance of the concentric-only portion of the jump, assistive devices (e.g., suspension training), submaximal jumps, among other. In this sense, jump training may involve a wide range of jump training exercises, selected according to the participant's characteristics and goals.

Owing to the broad range of settings in which jump training can be applied, it is not surprising that the number of publications on jump training has increased tremendously over recent years (Ramirez-Campillo et al. 2018a), with a 25-fold increase between 2000 and 2017. Moreover, jump training seems to be equally (or even more) effective compared to other training methods (e.g., traditional resistance training) for the improvement of several outcomes (Ramirez-Campillo et al. 2018a, 2020). In addition, the implementation of jump training may be inexpensive compared to other resistance training methods, requiring little or no equipment, usually involving drills with the body weight used as resistance. Additionally, jump training may be conducted in a relatively small physical space, which may be an important advantage during certain scenarios (e.g., encountering pandemic restrictions) where athletes may be forced to train at their homes (Gentil et al. 2020). Moreover, jump training may be considered more fun compared to other training methods (e.g., flexibility, endurance), particularly among younger athletes (Ward et al. 2007). Further, jump training may reduce the risk of injury (Ter Stege et al. 2014). However, rather than an independent entity, jump training usually should be a component of an integrated approach to training, which targets multiple physical fitness and health attributes and aligns with the goals of long-term physical development strategies.

3.1 Application of Plyometric-Jump Training in Sports

Among youth sports, for both male and female, from pre-pubertal (e.g., <8 years old) to post-pubertal age, jump training exercises have demonstrated beneficial effects (e.g., physical fitness; physiological and biomechanical outcomes) on athletes from different sports. Such sports include among others, soccer, basketball, handball, volleyball, tennis, hockey, sprinters, combat sports, and artistic gymnasts. The beneficial effects derived from jump training exercises have been reported usually without adverse effects. Further, compared to adults, youth seem to experience reduced symptoms of exercise-induced muscle damage and a faster recovery after jump training exercises (Marginson et al. 2005). Among adult athletes, male and female, jump training programs have demonstrated beneficial effects on athletes from different sports, including the previously mentioned young athletes and also for; swimmers, water polo players, endurance runners, ice hockey players, rugby players, golf players and netball players among others (Ramirez-Campillo et al. 2018a, 2020).

Improvements in several physical fitness outcomes after jump training programs have been reported among athletes. Scientific literature reports improvements include muscle power, jumping (e.g. vertical, horizontal), linear sprinting (i.e. from 5-m up to 200-m), agility and change-of-direction sprint, repeated sprinting ability with and without change of direction, short-term endurance (e.g., up to 60s), long-term endurance (e.g., Yo-Yo test; 3-km running time trial), reduced contact times while running, better running economy, maximal strength (e.g., dynamic; isometric), dynamic and static balance, sport-specific performance (e.g. soccer ball kicking speed), range of motion and coordination, among others (Ramirez-Campillo et al. 2018a, 2020).

3.2 Applications of Plyometric-Jump Training for Physical Fitness and Health

Although commonly associated to athletes and sport competition, jump training exercises also have demonstrated significant favourable effects on several health-related outcomes. Such improvements include glucose metabolism markers (e.g., fasting glycaemia and insulin), fat mass reduction (Racil et al. 2016), skeletal muscle hypertrophy (Grgic et al. 2020), bone health improvement (Markovic and Mikulic 2010), and acute hypotensive effect (Ramirez-Campillo et al. 2016a). In participants with cerebral palsy and Down syndrome, improvements have been noted in neuromuscular control and body composition (González-Agüero et al. 2012; Elnaggar 2020). During prolonged bed rest, adapted jump training exercises preserved muscle mass and muscle power (Kramer et al. 2018). In older adults, adapted jump training exercises allow improvements in balance, rate of force development, maximal strength, muscle power and EMG (Moran et al. 2018).

In addition, adapted jump training exercises may reduce injury risk (Hewett et al. 1999), through reduction in factors associated to injury, such as reduced knee abduction–adduction, improved balance, better neuromuscular control (e.g., landing technique), and reduced strength asymmetries between knee extensors/flexors. Moreover, in case of injury, jump training exercises can be adapted and incorporated during rehabilitation programs (e.g., neuromuscular training) (Hewett et al. 2005).

3.3 Factors Associated with Plyometric-Jump Training Effectivity

Participant's characteristics (e.g., jump training exercise technique proficiency; type of sport; training age; biological maturity; sex; participant nutritional/ supplementation habits) are relevant factors to consider for jump training exercise prescription (Ramirez-Campillo et al. 2018a, 2020; Moran et al. 2017; Ramírez-Campillo et al. 2016b), particularly considering the inter-individual variability to jump training programs (Ramirez-Campillo et al. 2018b). In addition, an adequate prescription of jump training exercises should consider the total duration of the training program (e.g., weeks), total volume (e.g., number of jump repetitions; foot contacts) (Ramírez-Campillo et al. 2013), and volume progression rate (e.g., weekly) (Asadi et al. 2017), in addition to potential taper strategies (Ramirez-Campillo et al. 2021). Moreover, intensity markers such as reactive strength index, jump height, movement velocity, force–velocity profile, rating of perceived exertion, among other potential jump training intensity markers, should be considered during jump training exercise prescription (Ramirez-Campillo et al. 2018c, 2019). Further, the type of jump training exercise (e.g., bilateral; unilateral; vertical; horizontal; loaded; unloaded; combined) (Moran et al. 2020, 2021; Ramirez-Campillo et al. 2015, 2018d), the jump training exercise order randomisation between training sessions, the specificity of jump training exercises, sequencing (i.e., before vs. after regular sport practice), and external load (e.g., heavy vs. light) should also be considered (Ramirez-Campillo et al. 2018a, 2020). Furthermore, the type of surface (e.g., wood; grass), recovery duration (e.g., inter-session; inter-jump; inter-set; inter-repetition) and type of recovery (e.g., cluster set vs traditional set; active vs passive) may also affect the outcomes of a jump training program (Ramirez-Campillo et al. 2018a, 2020). Additionally, the combination of jump training with other training methods, such as heavy resistance training (e.g., complex training) may be effective (Thapa et al. 2021), potentially due to post-activation potentiation or post-activation performance enhancement mechanisms, with the advantage of combining training methods in a relatively short time period.

4 Filling Gaps

To reduce the risk of injury and facilitate the strength gains that plyometrics can give, the athlete must first establish a speed and resistance training base. Beginning plyometrics too early in the conditioning cycle, or with the inexperienced athlete, can be disastrous. Several criteria need to be met before instituting a plyometric training program. These criteria are:

4.1 Physical Maturity of the Athlete

The age of the athlete or the number of years that they have participated does not measure their physical maturity. The National Strength and Conditioning Association (NSCA) recommends that the strength level for the hips and legs be based on the ability to squat 1.5–2.5 the athlete's body weight. This should be considered the minimum standard for shock—and high-intensity plyometrics. The upper body levels, according to the NSCA, should be based on the ability to do five continuous clap push-ups. Larger athletes (weight >110 kg) should be able to bench press their body weight, while smaller athletes (<75 kg) should be able to bench press 1.5 times their weight, and athletes of intermediate body weight (75–110 kg) should use graduations of these guidelines.

4.2 Coachability

Coachability refers to the athlete being able to respond in a positive fashion to instructions and criticism. If not, plyometric training should be delayed preventing injury, overtraining, or undertraining. If the athlete does not respond to coaching direction, they will not perform the movements properly. This can result in poor training results or injury.

4.3 Demands of the Sport

The demands of the sport must be considered when designing the plyometric program. Determine if the sport movements are mostly linear, vertical, lateral, or a combination of these movements. For example, volleyball players require vertical and lateral movement, while long jumpers emphasise horizontal movement. The intensity and volume should also be considered in the program design. During a training phase a shot-putter may use low volume and high intensity while the 400-m hurdler may use moderate volume and intensity.

4.4 Fitness Level

The strength and conditioning level of the athlete must be considered prior to performing plyometrics. If the athlete does not possess sufficient muscular strength or sufficient fitness levels, injury or overtraining may result.

4.5 Other Factors

Several factors should be considered when the decision has been made to begin plyometric training. These should include the sport specific exercises desired, proper footwear, surface types, proper equipment needs, and training area. Other areas of consideration are the frequency, volume, intensity, progression, recovery, and the direction of motion recommendations for the exercises.

Safety includes many areas, including proper footwear, resilient surface, proper equipment, and training area size. Footwear should provide sufficient ankle and arch support to prevent injury. Running shoes should be avoided due to their narrow sole and poor upper support; cross-training shoes are the best for plyometrics.

To prevent injuries, the landing surface should possess good shock-absorbing properties. The best surface is a grass field. A good alternative would be wrestling mats. Wood, tile, concrete, and carpet should be avoided due to their poor shock-absorbing properties. The boxes used for jumps should be sturdy, have a nonslip top, and rounded edges. The size of the training area depends on the type of exercises being used. Long-response drills may require a straightaway of 100 m. Bounding drills require at least 30 m of straightaway. For box jumps, adequate ceiling height must be provided.

Frequency, volume, intensity, progression, and recovery all refer to the training session itself. Frequency is the number of workouts per week. Volume is the number of foot contacts per workout. Intensity refers to the amount of stress placed on the muscle during the workout. Progression is the change from low-intensity to medium-intensity and to high-intensity levels as the athlete progresses. Recovery is the rest that is allowed between the individual sets of the drills The focus must always be on quality and not quantity. Table 1 shows some aspects of the plyometric workout.

5 Take-Home Messages

1. *Volume of plyometric training*. There is little to substantiate the exact requirements for calculating appropriate volumes for athletes. The following represent "guidelines" that should be observed when prescribing exercise for the athletes.

a. *Consider the athlete.* If they are young and inexperienced, remember that there will be a learning curve associated with any exercise drill. Several sessions of training should be utilised to teach appropriate execution of the drill. The learning curve is very rapid for this age group, and proper execution is much more important than the number of repetitions achieved.
b. *Observe the execution of the drill.* Fatigue impairs drill execution and learning. When execution technique is downgraded below an acceptable level, the drill should be stopped. It is far more important to see a drill performed correctly than to perform repetition for the sake of repetition.
c. *Focus and concentration is limited.* The younger the athlete, the more likely they are to mentally wander during the course of a training session. It is best to perform fewer drills correctly than more drills incorrectly. Pre-planning is essential, the coach should consider which biomechanical traits (vertical jump, linear jump, change-of-direction) they would like to develop prior to beginning a training session. This allows for selection of exercises prior to practice and sets a plan of training into motion.

2. *Intensity of plyometric training.* Intensity of plyometric exercises is determined by "effort of execution". Jumping drills that require maximal distance or height to be achieved is going to be higher in intensity that efforts such as put forth during the footwork drills previously described. A 20 cm box as the maximal height utilised in the performance of jumping drills consisting of moving on and off the box should be sufficient to achieve statistically significant results in vertical jump improvement. Apparently, athletes can still physically benefit from a drill using relatively lower heights.
   Maximal efforts should be utilised once the youngster has mastered the execution of the drill. Once the learning phase has been accomplished the athlete is much more likely to direct their efforts at force development along appropriate lines. The result will be a movement that has the subjective quality of "ease of execution", "flowing movement" and/or "powerful effort".
3. *Frequency of plyometric training.* Traditional thinking on plyometric training discusses the need to perform maximal effort days twice within a training week. This tends to allow for a recovery period of 48–72 h between training days. When working with youth athletes, it is inadvisable that truly maximal days of training occur until they have accomplished all the learning, execution and adaptation necessary to perform maximal effort exercise. With this in mind, it is acceptable to have a higher frequency pattern for plyometric training days. Three days a week are perfectly acceptable, given that there are not competition days at the end of each weekly cycle of training. If there are competition days included within the week or on the weekend, then the frequency of plyometric training should be reduced to twice a week. If an active, formal warm-up is being conducted as part of each workout it is also acceptable to include 4–5 plyometric exercises within the context of the warm-up routine, rather than having a formal workout for a particular day. This gives the coach the opportunity to expose all of the athletes in a group to the same drills. It is also an

opportunity to prescribe the number of repetitions, or time allotted to the performance of each drill. This sort of planning helps in the administration of plyometric training programs.

4. *Recovery in plyometric training*. The effects of fatigue on plyometric training are an important aspect to take into account. Fatigue is most likely to occur when the athlete is asked to perform exercises that do allow for full recovery between execution of repetitions. Without full recovery, the muscular and nervous system does not have the opportunity to rid the systems of the effects of fatigue metabolites and will result in a decrement in performance. This decrement will serve to frustrate the athlete and the coach because of an inability to achieve desired efforts as measured by speed, distances, or heights. Learning new skills will be impeded for the same reasons. The level of metabolism that should be utilised when performing these types of workouts is the ATP-PC and anaerobic glycolytic systems. These metabolic systems require brief, intense work periods, followed by long (5–30 s/work) active recovery periods. Jogging, walking, moving about are all acceptable forms of active recovery between repetitions of effort. It is well documented that active recovery is a more effective method of clearing the systems in preparation for the next work bout.
5. *Progression in plyometric training*. Progression in learning of plyometric exercises includes, but is not limited to, drills that are increasingly more complex. As an example, straight depth jumps as compared to depth jumps with 180° of body rotation. It is imperative that the coach be able to biomechanically determine the required traits of their particular sport. An example would be the netball coach who decides that vertical jump is a priority and that the best way to improve that skill is to train using exercises that are specific to the particular tasks of shooting and rebounding. Progression can take the form of increasing the range of motion that a particular task requires. For example, the average angle at the knee during take-off in many sport events is approximately 140° (i.e., "short-amplitude" jump). A progression might implicate "long-amplitude" jumps with a 90-degree knee angle. Each jump technique has a particular purpose and places a different demand on the athletes' body. Long amplitude jumping activities are most valuable for those athletes involved in Olympic weightlifting, free-style and Greco-Roman wrestling, and rugby. Another form of progression includes increases in the intensity of exercise. For example, using hurdles of increased height serves to challenge the limits of each athlete. These changes in the height of hurdles or the distances covered represent advancement in the effort and complexity of task. The coach must be able to recognise the individual needs of the athlete and to design appropriate challenges that teach the athlete to respond quickly to the ground and reduce the amortization phase. This should be done without making the task excessively difficult. In other words, the jump drills should allow the athlete to accomplish adequate ground contact times.

References

Anderst WJ, Eksten F, Koceja DM (1994) Effects of plyometric and explosive resistance training on lower body power. Med Sci Sports Exerc 26:S31

Asadi A, Ramirez-Campillo R, Meylan C, Nakamura FY, Cañas-Jamett R, Izquierdo M (2017) Effects of volume-based overload plyometric training on maximal-intensity exercise adaptations in young basketball players. J Sports Med Phys Fitness 57(12):1557–1563

Asmussen E, Bonde-Petersen F (1974) Storage of elastic energy in skeletal muscle in man. Acta Physiol Scand 92:385–392

Bauer T, Thayer RE, Baras G (1990) Comparison of training modalities for power development in the lower extremity. J Appl Sport Sci Res. 4:115–121

Bobbert MF (1990a) Drop jumping as a training method for jumping ability. Sports Med 9:7–22

Bobbert MF (1990b) Drop jumping as a training method for jumping ability. Sports Med 9(1):7–22

Bobbert MJ, Mackay M, Schinkelshoek D, Huijing P, Van Ingen Schenau GJ (1986) Biomechanical analysis of drop and countermovement jumps. Eur J Appl Physiol 54:566–573

Bobbert MF, Gerritsen KGM, Litjsens CA, Van Soest AJ (1996) Why is countermovement jump height greater than squat jump height? Med Sci Sports Exerc 28:1402–1412

Bosco C, Komi PV (1979) Potentiation of the mechanical behaviour of the human skeletal muscle through pre-stretching. Acta Physiol Scand 106:467–472

Bosco C, Komi PV, Ito A (1981) Prestretch potentiation of human skeletal muscle during ballistic movement. Acta Physiol Scand 111:135–140

Bosco C, Vitasolo JT, Komi PV, Luhtanen P (1982a) Combined effect of elastic energy and myoelectrical potentiation during stretch shortening cycle exercise. Acta Physiol Scand 114:557–565

Bosco C, Tihanyi J, Komi P, Fekete G, Apor P (1982b) Store and recoil of elastic energy in slow and fast types of human skeletal muscles. Acta Physiol Scand 116:343–349

Cavagna GA, Dusman B, Margaria R (1968) Positive work done by a previously stretched muscle. J Appl Physiol 24:21–32

Elnaggar RK (2020) Effects of plyometric exercises on muscle-activation strategies and response-capacity to balance threats in children with hemiplegic cerebral palsy. Physiother Theory Pract:1–9

Ettema GJC, Van Soest J, Huijing PA (1990) The role of series elastic structures in pre-stretch induced work enhancement during isotonic and isokinetic contractions. J Exp Biol 154:121–136

Ferragut F, Lopez-Calbet JA (1998) Mecanismos responsables de la potenciación de la contracción muscular concéntrica en el curso del ciclo estiramiento-acortamiento. Revista De Entrenamiento Deportivo 12:5–10

Gentil P, Ramirez-Campillo R, Souza D (2020) Resistance training in face of the coronavirus outbreak: time to think outside the box. Front Physiol 11(859)

González-Agüero A, Vicente-Rodríguez G, Gómez-Cabello A et al (2012) A 21-week bone deposition promoting exercise programme increases bone mass in young people with down syndrome. Dev Med Child Neurol 54:552–556

Grgic J, Schoenfeld B, Mikulic P (2020) Effects of plyometric versus resistance training on skeletal muscle hypertrophy: a review. J Sport Health Sci S2095-2546(20)30076-4

Hewett TE, Lindenfeld TN, Riccobene JV, Noyes FR (1999) The effect of neuromuscular training on the incidence of knee injury in female athletes—a prospective study. Am J Sports Med 27 (6):699–706

Hewett TE, Myer GD, Ford KR (2005) Reducing knee and anterior cruciate ligament injuries among female athletes: a systematic review of neuromuscular training interventions. J Knee Surg 18(1):82–88

Jacobs R, Van Ingen Schenau GJ (1992) Control of external force in leg extensions in humans. J Physiol 457:611–626

Kilani HA, Palmer SS, Adrian MJ, Gapsis JJ (1989) Block of the stretch reflex of vastus lateralis during vertical jumps. Hum Mov Sci 8(3):247–269

Komi PV (1992) Stretch-shortening cycle. In: Komi PV (ed) Strength and power in sport. Oxford, Blackwell Scientific Publications, pp 169–179

Komi PV, Bosco C (1978) Utilisation of stored elastic energy in leg extensor muscle by men and women. Med Sci Sports Exerc 10:261–265

Kramer A, Kümmel J, Gollhofer A, Armbrecht G, Ritzmann R, Belavy D, Felsenberg D, Gruber M (2018) Plyometrics can preserve peak power during 2 months of physical inactivity: an RCT including a one-year follow-up. Front Physiol 9:633

López-Calbet JA, Arteaga R, Chavaren J, Dorado C (1995) Comportamiento mecánico del músculo durante el ciclo estiramientoacortamiento. Factores neuromusculares. Archivos de Medicina del Deporte 12(47):219–223

Marginson V, Rowlands AV, Gleeson NP, Eston RG (2005) Comparison of the symptoms of exercise-induced muscle damage after an initial and repeated bout of plyometric exercise in men and boys. J Appl Physiol 99(3):1174–1181

Markovic G (2007) Does plyometric training improve vertical jump height? A meta-analytical review. Br J Sports Med 41:349–355

Markovic G, Mikulic P (2010) Neuro-musculoskeletal and performance adaptations to lower-extremity plyometric training. Sports Med 40(10):859–895

Minetti AE, Ardigo LP (2002) Halteres used in ancient Olympic long jump. Nature 420 (6912):141–142

Moran JJ, Sandercock GR, Ramirez-Campillo R, Meylan CM, Collison JA, Parry DA (2017) Age-related variation in male youth athletes' countermovement jump after plyometric training: a meta-analysis of controlled trials. J Strength Cond Res 31(2):552–565

Moran J, Ramirez-Campillo R, Granacher U (2018) Effects of jumping exercise on muscular power in older adults: a meta-analysis. Sports Med 48(12):2843–2857

Moran J, Ramirez-Campillo R, Liew B, Chaabene H, Behm DG, García-Hermoso A, Izquierdo M, Granacher U (2020) Effects of bilateral and unilateral resistance training on horizontally orientated movement performance: a systematic review and meta-analysis. Sports Med 51 (2):225–242

Moran J, Ramirez-Campillo R, Liew B, Chaabene H, Behm DG, García-Hermoso A, Izquierdo M, Granacher U (2021) Effects of vertically and horizontally orientated plyometric training on physical performance: a meta-analytical comparison. Sports Med 51(1):65–79

Mouche M (2001) Evaluación de la potencia anaeróbica con Ergojump. Lecturas Educación Física y Deportes. Revista Dig 30

Racil G, Zouhal H, Elmontassar W, Ben Abderrahmane A, De Sousa MV, Chamari K, Amri M, Coquart JB (2016) Plyometric exercise combined with high-intensity interval training improves metabolic abnormalities in young obese females more so than interval training alone. Appl Physiol Nutr Metab 41(1):103–109

Ramírez-Campillo R, Andrade DC, Izquierdo M (2013) Effects of plyometric training volume and training surface on explosive strength. J Strength Cond Res 27(10):2714–2722

Ramirez-Campillo R, Burgos CH, Henríquez-Olguín C, Andrade DC, Martínez C, Álvarez C, Castro-Sepúlveda M, Marques MC, Izquierdo M (2015) Effect of unilateral, bilateral, and combined plyometric training on explosive and endurance performance of young soccer players. J Strength Cond Res 29(5):1317–1328

Ramirez-Campillo R, Abad-Colil F, Vera M, Andrade DC, Caniuqueo A, Martínez-Salazar C, Nakamura FY, Arazi H, Cerda-Kohler H, Izquierdo M, Alonso-Martínez AM (2016a) Men and women exhibit similar acute hypotensive responses after low, moderate, or high-intensity plyometric training. J Strength Cond Res 30(1):93–101

Ramírez-Campillo R, González-Jurado JA, Martínez C, Nakamura FY, Peñailillo L, Meylan CM, Caniuqueo A, Cañas-Jamet R, Moran J, Alonso-Martínez AM, Izquierdo M (2016b) Effects of plyometric training and creatine supplementation on maximal-intensity exercise and endurance in female soccer players. J Sci Med Sport 19(8):682–687

Ramirez-Campillo R, Alvarez C, Garcia-Hermoso A et al (2018a) Methodological characteristics and future directions for plyometric jump training research: a scoping review. Sports Med 48:1059–1081

Ramirez-Campillo R, Alvarez C, Gentil P, Moran J, García-Pinillos F, Alonso-Martínez AM, Izquierdo M (2018b) Inter-individual variability in responses to 7 weeks of plyometric jump training in male youth soccer players. Front Physiol 9:1156

Ramirez-Campillo R, Alvarez C, García-Pinillos F, Sanchez-Sanchez J, Yanci J, Castillo D, Loturco I, Chaabene H, Moran J, Izquierdo M (2018c) Optimal reactive strength index: is it an accurate variable to optimize plyometric training effects on measures of physical fitness in young soccer players? J Strength Cond Res 32(4):885–893

Ramirez-Campillo R, Sanchez-Sanchez J, Gonzalo-Skok O, Rodríguez-Fernandez A, Carretero M, Nakamura FY (2018d) Specific changes in young soccer player's fitness after traditional bilateral vs. unilateral combined strength and plyometric training. Front Physiol 9:265

Ramirez-Campillo R, Moran J, Drury B, Williams M, Keogh JW, Chaabene H, Granacher U (2019) Effects of equal volume but different plyometric jump training intensities on components of physical fitness in physically active young males. J Strength Cond Res

Ramirez-Campillo R, Moran J, Chaabene H et al (2020) Methodological characteristics and future directions for plyometric jump training research: a scoping review update. Scand J Med Sci Sports 30(6):983–997

Ramirez-Campillo R, Pereira LA, Andrade DC, Mendez-Rebolledo G, De La Fuente CI, Castro-Sepulveda M, Garcia-Pinillos F, Freitas TT, Loturco I (2021) Tapering strategies applied to plyometric jump training: a systematic review with meta-analysis of randomized-controlled trials. J Sports Med Phys Fitness 61(1):53–62

Saez de Villarreal E, Kellis E, Kraemer WJ, Izquierdo M (2009) Determining variables of plyometric training for improving vertical jump height performance: a meta-analysis. J Strength Cond Res 23(2):495–506

Steben RE, Steben AH (1981) The validity of the stretch-shortening cycle in selected jumping events. J Sports Med 21:28–37

Ter Stege MHP, Dallinga JM, Benjaminse A, Lemmink KAPM (2014) Effect of interventions on potential, modifiable risk factors for knee injury in team ball sports: a systematic review. Sports Med 44(10):1403–1426

Thapa R, Lum D, Moran J, Ramirez-Campillo R (2021) Effects of complex training on sprint, jump, and change of direction ability of soccer players: a systematic review and meta-analysis. Front Psychol 11:627869

Van Ingen Schenau GJ, Bobbert MF, Rozendal RH (1987) The unique action of biarticular muscles. J Anat 155:1–15

Van Ingen Schenau GJ, Boots PJM, De Groot G, Snackers RJ, Van Woensel WWLM (1992) The constrained control offorce and position in multi-joint movements. Neuroscience 46:197–207

Verhoshanski Y (1966) Perspectives in improvement of speed-strength preparation of jumpers. Track Field. 9:11–12

Ward P, Hodges N, Williams AM (2007) The road excellence in soccer: deliberate practice and the development of expertise. High Abil Stud 18(2):119–153

Eduardo Sáez de Villarreal has a B.P.E., a M.Sc. in High Performance in Sports, and a Ph.D. in Sport Sciences. He is a very passionate strength and conditioning coach that progressively developed into an applied sport scientist, with a main emphasis on strength, power, and velocity development in collective and individual sports. His work (>100 peer-reviewed articles published, >4000 citations; h index 31; h 10 index 45) focuses on intervention strategies and profiling assessments that may improve players' physical and technical potential, using a scientific approach whenever possible. Using his background in strength and conditioning, he has extensive experience with fitness and strength training, injury prevention, recovery, and youth development. Over the past 30 year, he has worked and consulted a bit all over the world and in various sports (soccer, waterpolo, AFL, rugby, ice hockey, individual sports), which allow him to merge theory, experience and pragmatism for a holistic and contextualized approach of high performance.

Rodrigo Ramírez-Campillo has a B.P.E., a M.Sc. in Exercise Physiology, and a Ph.D. in Health Sciences. He has published nearly 300 peer-reviewed articles (>4500 citations; h index 38; h10 index 106). His research focus is on sports sciences, sports medicine, and medical and health sciences. He has guided national and international doctoral students, as well as master and undergraduate students. In addition, he participates as a reviewer for grant-related agencies and journals. He is ranked worldwide as the most productive researcher in plyometric exercise and is ranked among the most productive researchers in areas such as soccer, exercise-based therapy, resistance training, and human physical conditioning.

Resistance Training for the Maximization of the Horizontal Force Production

Pedro Jiménez-Reyes and Pierre-Jean Vazel

Abstract This chapter presents a double perspective from theory to practice contributing to the scientific-based evidence what is behind the real examples of best sprinters over the world in order to understand how best coaches train in a daily basis and elite competition context. During *From theory* section we will introduce from a general perspective (i) the description of a sprint from the velocity–time curve describing the *different* phases and (ii) the muscular implications needed to each phase characterizing the forces during the sprint. Taking all this together will help readers to understand and have a better knowledge about the adequateness of sprint training methods. During *From practice* section we will focus on the main sprint training methods, trying to understand how coaches could use them considering the velocity–time curve and the forces during a sprint. This section aims to cover the main parameters to take into account when designing a sprint training program independently of sport with the possibility to orient the target to acceleration or maximal velocity as described in the previous section. This second part intends to cover the very general training principles sprint coaches do have in common, and to shed light on how they differ.

Keywords Sprint · Acceleration · Speed-strength continuum · Speed-endurance continuum · Sprint speed training · Heavy sled training · Long to short distance approach · Short to long distances approach

P. Jiménez-Reyes (✉)
Centre for Sport Studies, Rey Juan Carlos University, Madrid, Spain
e-mail: pedro.jimenezr@urjc.es

P.-J. Vazel
Athlétisme Metz Métropole, Metz, France

A. Muñoz-López et al. (eds.), *Resistance Training Methods*, Lecture Notes in Bioengineering, https://doi.org/10.1007/978-3-030-81989-7_6

1 Introduction

In recent years, sprint or acceleration training have been used as a specific part in resistance training and S&C programs. Although sprint training has developed a scientific-based evidence in a specific way in track and field, recently it has received a growing interest and attention in team-sports as a key role component of success. The forces component and contribution which explain the sprint nature of this skill dramatically changes the way the kinetics and kinematics can be applied during training.

Considering sprint as specific skill directly related to performance in an athletics context, aspects such as acceleration, maximal velocity and the capacity to sustain high running velocity should be considered for a better understanding about how effectively manage and organize training components. In the specific context of team-sports, sprint could be considered as a transferable sport skill with a positive and beneficial transfer to success in sports such as rugby or soccer where being the quickest over various distances is valuable since acceleration ability has been recognized with a capacity to change pace being relevant to evasive play rapidly.

Several kinds of training methodologies have been used previously for acceleration and sprint training based on general conceptions with no consideration on how forces are applied during sprinting to adapt in a more adequate and reasonable way training methods. To date, the most used training methods have been classified as specific and non-specific sprint training methods (Rumpf et al. 2016), showing different effects depending on the distance analyzed. About the specific sprint training methods, free sprinting and resisted and assisted have been used both in the literature and practice. Nevertheless, even considering the specificity of these methods, sometimes they have been implemented with no consideration the forces, presenting a high variability and/or different training adaptations. What is more, there is an increasing interest in using those sprint training methods to support S&C coaches in team-sports tasked with delivering improved speed as a sub-component of their role.

2 From Theory

During this section we will introduce from a general perspective (i) the description of a sprint from the velocity–time curve describing the different phases and (ii) the muscular implications needed to each phase characterizing the forces during the sprint. Taking all this together will help readers to understand and have a better knowledge about the adequateness of sprint training methods.

2.1 *Understanding the Sprint—Velocity–time Curve*

Sprint running, and more specifically sprint acceleration, is a key component and central to performance in many sports such as athletics, soccer and rugby (Morin et al. 2015a). Sprint, power output and forward acceleration are key physical determinants of athletic performance, recognized as major components of successful performance and being an essential part of a strength and conditioning specific training program in many sports and recreational physical activities.

Although maximal straight-line single-bout speed is the focus of many track events, and sprint running speed is also considered a relevant parameter for field-based team-sports (Simperingham et al. 2016), it is essential to emphasize that the ability to accelerate over short distances should be prioritized in many sport activities rather than maximal velocity, since maximal velocity is rarely achieved in these kinds of sports (Spencer et al. 2005; Morin et al. 2015b). Sprinting in field-based team-sports can vary from short (e.g., Futsal, Rugby union forwards) (Deutsch et al. 2007) to long (e.g., Australian rules football) (Veale et al. 2007) distances. Acceleration is a key factor in field-based team sports since players who accelerate more rapidly have an advantage due to the frequent occurrence of such accelerations (e.g. 5–20 m, 2–3 s) during games (Spencer et al. 2005; Schimpchen et al. 2016). Recent studies suggest that 68% of sprints in rugby (Gabbett 2012) and 90% of sprints in soccer (Vigne et al. 2010) are shorter than 20 m, and this is not enough distance for players to reach maximum velocity. In addition, such short linear sprints are used in decisive actions (Faude et al. 2012).

A typical sprint-track running is characterized by the velocity time-curve and can be divided into three phases, acceleration, constant velocity and deceleration (Mero et al. 1992). In the recent years, sprint running has been also divided into three phases such as early acceleration, acceleration and maximal velocity. During these phases sprint technique is modified, nevertheless two different phases are present during the whole sprint cycle, stance phase (support phase) and flight phase (swing phase). Maximum speed is relevant in track events and limited field-based team-sport contexts such as Australian football league (Veale et al. 2007), although acceleration is of relatively greater importance when covering only short distances at maximal effort as in many field-based team-sport usually happens, being the ability to accelerate a successful parameter (Simperingham et al. 2016). For instance, in team sports, a first step quickness has been considered as an important parameter for acceleration, defined as the first 0–5 m and included in the acceleration phase is characterized by high propulsion force (Sleivert and Taingahue 2004). Traditionally, several factors/parameters have been investigated concerning acceleration and sprint performance providing the basis for understanding the underpinnings on the ability to run fast. Some of these parameters have been: stride length and frequency, ground contact time and flight time, joint angle movements, ground reaction forces, stiffness and EMG activity patterns. Classically, running speed has been proposed as the product of stride rate or frequency and stride length, assuming that to increase velocity it should be needed to increase at least one, if not both, whilst a no large decrease in the other (Weyand et al. 2000; Hunter et al. 2004). Typically, maximum stride frequency is reached between 10 and 20 m and at this point stride length is about 75% of maximum value (reached during the maximum velocity phase).

2.2 *Muscular Implications and Forces During a Sprint*

Recent years, research by Morin and colleagues (Morin et al. 2010, 2011, 2012) has highlighted the forward orientation of GRF's as a further determining factor of performance specifically in the acceleration phase of the sprint, and that the vertical component of the GRF was not related to performance in that phase, but specially with maximal velocity phase as it had been previously described by Weyand et al. (2000).

The acceleration of the athlete's center of mass during sprint running is determined by body mass and 3 external forces acting on the body: (a) ground reaction force (GRF), (b) gravitational force, and (c) air or wind resistance (Samozino et al. 2016; Hunter et al. 2005). GRF can be divided into three components (anterio-posterior, vertical and medio-lateral) although typically the anterio-posterior (horizontal) and vertical components are the most studied and relevant for sprint performance (Hunter et al. 2005; Morin et al. 2011; Rabita et al. 2015).

Sprint performance implies large forward acceleration, which is directly depending on the ability to develop and apply high amounts of horizontal external force onto the ground at various speeds over the sprint acceleration (Morin et al. 2011). This is why the ability to produce GRF with a magnitude and timing unique to that individual phase of the sprint becomes paramount, changing from a high force at low speed in early acceleration phase to low force at high speed in maximum speed phase (Morin et al. 2012).

The most widespread studied muscles during sprinting have been the hip extensors (hamstrings and gluteus maximus), knee extensors (quadriceps), and plantar flexors (soleus and gastrocnemius). Generally, it is widely accepted that most of muscles activates at highest levels just before or at the beginning of ground contact (Mann et al. 1986; Morin et al. 2015a). It is mainly during this support phase—single moment when force can be applied onto the ground- where muscles responsible for hip, knee, and ankle movements play a specific role in acceleration performance, propelling efficiently the body forward.

When analysing an acceleration run from a purely biomechanical perspective, great differences can be observed among the three different phases previously described. These singularities provide us critical information for a better understanding of the underlying parameters responsible for this differentiation when talking about muscles role or patterns of action.

These variances become more evident during the very early steps, when body positioning when force is applied -mainly characterized by a greater forward lean of the trunk (Debaere et al. 2013; Nagahara et al. 2014a, b)—and available time for the application of force—i.e. approximately 190 ms versus ± 101–108 ms when maximum velocity phase is reached (Wild et al. 2011; Yu et al. 2016)—are away from the remaining phases of an accelerated run.

In this first stage, the main hip and knee extensors work alongside soleus and gastrocnemius in order to achieve a triple joint extension of the lower limb will provide forward propulsion to body mass. This phase may be mechanically characterized by an involvement of the knee and hip extensors alongside the calves in order to act as propels during the stance phase, which is distinguished by large contact times enabling the development and application of high level of forces onto the ground, also made possible by the low levels of displacement velocities. During the evolution of this phase of acceleration and considering early acceleration and acceleration itself is mostly characterized by a gradual decrease of body forward lean (Nagahara et al. 2014a, b), the achievement of maximal stride frequency and a marked increase in stride length given the continuous rise of running velocity (Nagahara et al. 2014a, b). This higher speed is also associated with shorter ground contact time and important consequences on the kinetic patterns, depending on hip extensors –i.e. concentric action of gluteus and eccentric on hamstrings (Morin et al. 2015a)—while knee extensors and calves progressively adopt a more focused role on the stabilization and transmission of forces as the speed of movement is increased (Mann et al. 1986; Schache et al. 2015).

Lastly, during maximum speed phase, plantar flexors (gastrocnemius, soleus) in conjunction with dorsiflexor (tibialis anterior) muscles have a major influence in how effectively the forces of different body segments are transferred to the ground. In fact, gastrocnemius-soleus-achilles-complex (GSAC) has shown to play a relevant role contributing to horizontal propulsion during contact phase by storing and releasing elastic energy to help the body forward projection.

3 From Practice

There is a great interest for many strength and conditioning coaches and practitioners about the best training practice for improving sprinting performance. Typically, several training methods have been widely used for improving sprinting performance such as sprinting (Rumpf et al. 2016), technical skills (Bushnell and Hunter 2007), maximal power (Delecluse et al. 1995; McBride et al. 2002), reactive strength (plyometric training) (De Villarreal et al. 2012), ballistic training (Sheppard et al. 2011; Cormie et al. 2010), and combinations of these methods (Harris et al. 2000; Ronnestad et al. 2008) although with inconsistent results in many of them.

In this section we will focus on the main sprint training methods, trying to understand how coaches could use them considering the velocity–time curve and the forces during a sprint.

Training programme design for any sport requires the organisation of multiple training elements considering the target to work or develop. Several factors such as coaching philosophy, training theory, and evidence-based knowledge will modulate the response and it will be an objective information for making decision to improve the training process. This section aims to cover the main parameters to take into account when designing a sprint training program independently of sport with the possibility to orient the target to acceleration or maximal velocity as described in previous section. It isn't intended to represent a catch-all analysis but rather is a selection of current views based on the knowledge and experiences of the authors.

Before starting it is worth to note that the main general principal of programming should be a strategy for individualization. Although during this section we will present some general concerns about how coaches could manage key parameters for training purposes, it is mandatory to highlight that coaches should adapt taking into account the specific context, training background, moment of season, needs, etc. Readers will observe this individualization and adaptation to elite context in real examples of world-class sprinters.

To simplify these sprint training methods, we will use the classification proposed by Rump et al. (2016) which is derived from a review of studies considering training methods as specific (free sprinting, resisted and assisted sprinting), non-specific (resistance training and plyometric training) and combined (a combination of specific and non-specific). Furthermore, we will present real examples of best sprinters over the world in order to understand how best coaches have used in a daily basis and elite competition context.

The rationale of coaches during years to develop and use different sprint training methods has been the adequateness to the velocity–time curve and the constraints related to muscular implications and applied force regarding the available time during each sprint phase. The basis to use the different sprint training methods is supported on the development of fundamental actions that allows to athletes to reproduce the specific conditions during each sprint phase to develop the acquisition of pertinent aspects of stability and adaptability in movement, improving specifically according to the nature of movement.

3.1 Sprint Practice

"No sport is so open to variety in mode of practice" This quote about sprint training by Michael C. Murphy in 1894 (Murphy 1894) (coach to John Owen, the first amateur athlete to run under even-10 s at 100 yards), could still apply more than a century later to describe how sprinters prepare for competitions. This second part intends to cover the very general training principles sprint coaches do have in common, and to shed light on how they differ. The training designed by American coach Lance Brauman for Tyson Gay (TG) in his preparation for 2009 will serve as illustration of current practice by some of the fastest men and women of all-time. That year, TG became the 2nd fastest sprinter ever at 100 m in 9.71, placing 2nd at 100 m world championships behind Usain Bolt who set the still existing world record in 9.58, and a few weeks later, improved his mark to 9.69. TG also won two individual world championship gold medals in 2007 at 100 m and 200 m, and under Brauman's coaching, Veronica Campbell-Brown, Tori Bowie, Noah Lyles and Shaunae Miller-Uibo also have won Olympic or world titles in the past 15 years.

3.2 *Sprint Coaching Philosophy*

The first published training methods were mainly focused on sprinting over the competitive distance at top speed, from one to three times per day, every day to every 3rd day (Walsh 1859; Westhall 1863; Wheeler 1868). Sprint races being understood to be a contest of both strength and endurance (Ozolin 1945), coaches would increase the variety of the workout content using two main ways: training along the speed-endurance range and along the speed-strength range.

According to the earliest training manuals and testimonies, speed-endurance continuum includes distance ranging from the first meters for start practice to 440 yards (1894), and even extending to long continuous runs up to 10 miles (16 km) for that aerobic condition was understood as a "conditioning" measure (Hayes 1992); the pace obviously ranges from jogging for long distances and half-speed to maximum speed for shorter distances.

Training on the speed-strength continuum includes assisted sprinting (declined track or pulling devices, (Ozolin 1949)), resisted sprinting (inclined track, towing devices, (Bartenyev 1965)), training with jumping, throwing (Wheeler 1868; Alabin 1976) or lifting (Worman 1894). The complexification of workout contents along the two axis, speed-strength and speed-endurance, led all coaches to engage into training methodology and follow some general principles (Ozolin 1949; Matveyev 1965; Harre 1971; Bompa 1983) in their approach:

- *Principle of continuous load demand.* higher competition performances result from long-term progression, avoiding interruptions and making sure of the necessary recovery to prevent excessive strain and injuries. First sprint training plans published in early XXth century consisted in 6–13 weeks before being able to compete. By the 1950s, most of the successful training plans extended through the Fall until late Summer. The training plan used by TG in 2009 before the world championships contained 42 weeks, and he set his lifetime 100 m best after 47 weeks.
- *Principle of variety of training*. While training of professional sprinters in the XIXth century consisted mainly in hygiene focus (health, diet, sleep) and training through competitions, varying training allows to reach higher sprint performance and also requires longer time to reach sport form. To improve running speed, coaches use different exercises which not only focus on speed, but also on speed-endurance and speed-strength spectrum: in the early 1950s, Soviet sprinters started to focus more on longer sprint distances and included jumps, throws and lifts as part of their training regimen (Korobkov and Fillin

1956), an approach still used today by the best sprinters. TG's 2009 program contained sprints from 10 to 400 m, covering the running intensity spectrum from about 50–100% of maximum effort), vertical and horizontal jumps, medicine ball throws, weightlifting exercises, ability and mobility exercises, all year round.

- *Principle of periodization of training:* with the increase of variety in training, it became necessary to divide the longer training preparation into periods. As early as 1916, Russian author B.A. Kotov proposed three periods: general, preparatory and special. The periodization used by successful pre-World War II American sprint coach Dean Cromwell (Charley Paddock, Frank Wykoff) also had a three-phase progression, with a different terminology: (1) out of season (Fall-Winter), pre-season (Winter-Spring) and in-season training (Spring–Summer) (Cromwell 1929, 1939). This 3-phase classification is still widely used as it allows organization of the training year according to the principles previously described, although how the demand of training is increased may differ (Francis 1992). Some periodization scheme variations use a preparatory period, which actually includes the general phase, a competitive season (same as in-season training) and a transition period, which usually occurs in the early Fall, and is active recovery after the last competition events (Petrovsky 1978). TG used a simple periodization, with training microcycles of 4 weeks, with workload being reduced on the 4th in order to promote recovery. A corollary principle described by Matveyev is the *cyclicity of the training process.* It allows the repetition of the three phases in a cyclic way, two times in the case of a double periodization. In 1964, Karin Balzer was the first to use a sophisticated succession of 4 cycles during her 46-weeks long preparation (Birkemeyer 1966), repeating blocks of slow-extensive, fast-extensive and intensive running workout progression before the 4 main competition periods: indoor, preparatory, Olympic selections and Olympic Games where she became the first ever East German champion winning the 80 m hurdles. In 2009, Tyson Gay used a single periodization, as he didn't compete indoors during winter season, and started in April with two 400 m races, followed a month later by a 200 m (19.58, lifetime best), before racing over 100 m from 25 June, in 9.75 (wind assisted). From then on he maintained sport form, as his 100 m results remained consistent through the summer: 9.77 (American record) on 10 July, 9.79 (wind assisted) on 31 July, and 9.71 during World Championships on 16 August. This competition strategy was coherent with his training plan, with a progressive increase of intensity and training through the meets, getting benefits from higher speed stimulus in a competitive context.
- *Principle of increased demand of training.* To ensure continuous progression, coaches follow the principle of a progressive increasing overload on either or both quantity (volume) and quality (intensity) of training. The running meterage tables of 100 m European Champion Irina Turova (Soviet Union) in 1952–1954 were the first published material (Turova 1955) to illustrate a progressive demand. Here, increasing the monthly volume of running was the prime focus during the 1953 season, and from season to season. However, in TG's case, the

focus was to increase the speed endurance intensity from about 70% to 100% from December to June, at the expense of volume (number of run repetitions and meterage) in both short and long sprint workouts. Going from long to short distance (LtS) approach for speed endurance along the training season, TG used short to long distances (StL) for acceleration and maximum speed work. Since the early XXth century, most of the programs were either LtS or StL with equal success. LtS program was used by women's 100 and 200 m world record holder Florence Griffith-Joyner (Kersee 1989) while StL scheme was popularized by Ben Johnson (9.79 in 1988, disqualified for doping, (Francis 1992)).

- However, East German Karin Balzer was the first to use a "concurrent" program in the early 1960s, with parallel use of LtS for speed-endurance and StL for acceleration and maximum speed training, both ends joining towards the competition specific distance (100 or 200 m), similar to TG. This scheme found physiological justifications in East Germany a decade later: increasing the pace (from jogging to running to sprinting) by decreasing the speed endurance running distances (from long continuous runs to repetitions of shorter bouts getting closer to the competition distance) during the training season allowed a progressive workload stress adaptation measured with blood lactate concentration (Lang 1975); the same observation was found in short distances, when increasing the pace and workload stress as measured by lactate readings, by gradually lengthening the short sprint training from 10 m up to the competition distance, as used by 400 m world record holder Marita Koch (Hess 1979).
- *Principle of individualization of training*. Large variations in sprint volume and intensity tolerance have been found in elite sprinter's training. During the 1950s, they were mostly due to different "schools" of philosophy: Polish would cover 3200 m of sprint in one week, compared to only 420 m for Soviets, for mostly the same competitive results (Fillin 1955). For reference, TG's average weekly volume ranged between 460 and 1120 m. During the 1960s, Polish would become the firsts to test the response to various loads by screening the 100 m results of their best sprinters, and establish training plans using high, mid or low intensity training depending on their profiles (Zabierzowski 1962). In 2009 Lance Brauman training group members used mostly the same training, with minor adjustments to meet competition and travel requirements, TG sometimes using slightly longer training distances. Most of the individualization part was found in the strength training (Table 1).

3.3 Training in the Speed-Endurance Continuum

From a few meters to long continuous runs, sprint training covers the entire spectrum of running distances and velocities, along a speed-endurance continuum.

60 m and 100 m are the shortest distances in the world championships (indoor) and Olympic games programs, respectively, and as such maximum speed training is

Table 1 East-German concurrent training scheme using a LtS progression in endurance workouts and StL for speed workouts with the aim to increase running intensity and workload stress along the training preparation phases (Hess 1979)

Endurance run >80% *(200–500 m)*					
	Endurance run 80–89% *(120–400 m)*				
		Endurance run 90–94% *(100–200 m)*			
			Maximum speed run *(30–80 m)*	Short SE +95% *(60–120 m)*	Long SE +95% *(80–150 m)*
		Sub-maximum speed run *(30–60 m)*			
	Maximum acceleration run *(10–50 m)*				
Sub-maximum acceleration run *(10–40 m)*					
General preparation	→		**Special preparation**	→	**Pre-competition phase**

necessary. However, they also require specific forms of endurance in order for the athlete to be prepared for three tasks: (1) to be able to maintain maximum running velocity which is reached after about 6 s for men and women, and reduce subsequent deceleration; (2) to acquire the physical condition to adapt and recover from the training load required to excel in sprint events; (3) to have the capacity to warm-up and compete several times at high intensity in a short time span (60 m heats, semi-finals and finals are often held during the same day).

3.3.1 Sprint Speed Training

Generally, in order to focus on acceleration and considering the velocity–time curve and that during this phase the athlete has a higher available time to develop force during contact time and the specific evolution of stride frequency and length, free sprinting over short distances has been widely used.

Taking into account the muscular implications during this phase and the importance to create maximal horizontal projection through extension of the leg and maintenance of body orientation are important physical requirements for acceleration, and whilst they represent whole limb extension qualities, appear to be particularly dependent on the function of posterior chain. As such, limb extension with a bias towards gluteal and hamstring work tends to dominate during this phase, and this specificity has been one of the main works to train during gym sessions for coaches.

The use of free sprinting for improving acceleration should be take into account these specific features and for this reason the total volume, intensity and recovery are specific to get the target to develop and train acceleration or maximal velocity phases.

100 m race can be divided in consecutive phases: reaction to the gun, acceleration phase, maximum velocity phase, velocity drop phase (Gundlach 1969). The training distances commonly used correspond to these phases:

- *Reaction time to the start signal* is the only part of the race that cannot be improved indefinitely. Indeed, a limit has been used in international competitions since 1972 Olympics and set as 0.100 s in the IAAF Rule book since 1991. However, the shorter the distance, the shorter the reaction time, and it has been measured as 0.08 in training tests in elite men and women sprinters over 30 m trials (Vazel 2018). In practice, reaction skills are not trained separately from the push-off action from the starting-blocks and first steps execution. For the 2009 season, TG used starting-blocks from mid-December 2008, 8 weeks after resuming training.
- *From 10 to 40 m to work on acceleration phase.* Elite sprinters such as TG who reach their maximum velocity farther into the race (55.23 m for his 9.71 race at 2009 world championships final, (Graubner and Nixford 2009)) include distance up to 60 m. Workout example: 1 set of 10 m, 20 m, 30 m, 40 m, 50 m. The distance mostly used as reference for acceleration test is the 30 m; TG's coach would only use hand timing, but common practice is to time short runs electronically and compare the results with intermediate times taken in 60 m or 100 m competitions (Tepper 1989).
- *From 30 to 80 m to work on maximum speed.* The most common workout format is the "flying run" where the sprinter accelerates along a 20–30 m run-up into a 20–30 m active fly zone. TG used it from January until Summer, once a week, after acceleration work, for 3–4 repetitions of 20–60 m fly zone distance. The running velocity reached in these runs are higher than what is recorded during full-out sprint efforts from the start (Gundlach 1968). The golden standard for the fly zone is 30 m, and the times are used as tests for maximum speed and compared with the correspondent section of the 60 or 100 m races in competition (Petrovsky 1973).

In theory, sprinters are expected to run at maximum effort all year round in order to elicit a specific adaptation. In practice it has not been found to be effective, as sprinters need to be sufficiently physically prepared before engaging into high training intensity (Petrovsky 1974). Therefore, they increase effort in these runs throughout the season, meaning that they run at sub-maximum intensity in the first months of the preparation. Before his Olympic title in September 1972, Valeriy Borzov started flying 30 s in November 1972 at 90% of his best training time. It increased up to 98% in February–March for the indoor season, then dropped down to 85% in late April at the beginning of its 2nd training period, and progressively improved to 93% in June up to 100% in July (Petrovsky 1973). Maximum sprint too early would haste sport form, which would not be possible to maintain until the most important race held during Summer. TG started maximum speed work with efforts over 90% from January as he was using a single periodization without indoor competitive season. Sub-maximum sprinting is also used in order to be able to work on running form and relaxation.

Acceleration and maximum speed work are performed over small volume. Using 60 m distances, the minimal training load that serves entry into the state of sport form is 5 × 60 m or 300 m at an intensity ranging from 80 to 100% (2010). TG's volume per workout ranged between 300 to 350 m. However, the upper limit range in elite sprinter's practice is about 800 m. For example, 720 m total volume for a flying 30 m workout by European Record holder Christine Arron (100 m 10.73, 6th fastest woman ever): 12 × 30 m (from 30 m run-up) using ladders sets of 2 m to 2m45 intervals in order to control the step length and the intensity. Rest intervals between runs for top speed work are usually over 5 min, and in experienced sprinters, duration is self-chosen. Reducing rest time, or increasing volume would elicit incomplete recovery and thus speed endurance adaptation, which is not desired in these workouts (Petrovsky 1973).

3.3.2 Sprint Speed Endurance Training

Competing at the shortest of the Olympic running distances is also a contest of endurance since top speed is reached at about 60% of the race time, thus the remaining 40% are about maintaining speed and preventing a large velocity drop. Two types of workouts are used for speed-endurance, *over-distance* with runs longer than the competition distance, and *under-distance* with runs ranging from 30 to 90 m (McDonald-Beiley 1953).

- *Over distances*: TG used distances up to 400 m, in a long-to-short approach: the mean distance used weekly in winter was 200–300 m, down to about 100 m in spring and summer. The weekly speed endurance volume went from 3000 m in winter, to 2500 m in spring and 1500 m in summer. The longest workout occurred in January: 400 m in 49 s (92% of his personal best at the distance), rest 6 min, 4 × 200 m in 28 s (70%) with rest 2 min and 4 × 150 m in 17 s (88%) with rest 4 min. The total volume was 1800 m.
- In April, a sample workout was 300 m in 36 s, 200 m 24 s, 2 × 100 m 12 s (all at about 88–90% of personal bests) with 6 min rest, followed by *under-distance* speed endurance of 2 × 2 x 80 m, for a total volume of 1020 m. Interestingly, during the same week, his rival Usain Bolt (who would set the 100 m world record in August with 9.58) was working on similar volume (1080 m) and intensity range of 88–90%: 6 × 180 m in about 20 s with 8 min rest. Getting closer to the competitions, TG reduced the volume, for example 3 × 150 m in 14.6 (97%) with 8 min rest. Women's 100 m and 200 m world record holder Florence Griffith-Joyner adopted a similar approach as TG and Usain Bolt, starting with even longer distances: 5 × 600 m at 80–85% pace in the Fall,

reducing the distances to 450 m in the Winter (Kersee 1989). The week before the 100 m world record in July, her three last speed-endurance workouts were as follow: Monday: 6 × 240 m (over 95% of her estimated personal best) walk across the field as rest, Wednesday: 3 × 160 m (close to 100%) walk across the field as rest, Friday (8 days before the WR): 320 m time trial in 37.86. These are extreme examples of volume at high intensity with short recovery, achieved by an athlete who was also the fastest American in the 4 × 400 m relay. By contrast, Christine Arron didn't go over 150 m the year of her European Record at 100 m, and a typical workout was 3 × 120 m at 100% intensity with 15 min rest.

- *Under distances*: repetitions of short distances with incomplete recovery are used in order to work endurance at a higher intensity than if long runs would have been employed. During the course of an all-out run at 300 m, TG would reach 95% of his maximum sprint speed during the first half of the race while during a 4 × 80 m workout, TG would reach the same speed but doing it multiple times for a similar volume. This method was first employed by West Germans Gert Metz (100 m European Record 10.0 in 1970) and Gerhard Wucherer (60 m World indoor bests in 1969): 3 sets of 3 reps of 70 m (total volume 630 m) with 30 s rest between reps and 10 min between sets. Italian Pietro Mennea (200 m world record holder 19.72 in 1979) did 4 × 6 × 60 m early in his career (1972), with 75 s and 5 min as rest; the mean time for these runs being at 5% from his personal best (Letzelter 2010). The volume of 1440 m at this intensity represents the upper limits of human speed endurance abilities. In February, March and April 2009, TG used a reduced volume of under distance speed endurance training, 2 × 2 × 60 m or 80 m with walk back as rest between reps and 5 min between sets; the volume was kept small as this was done at the end of *over-distance* runs (Table 2).

Endurance training for short sprinters is directly made of running workouts. Although it could be trained using other forms of exercises such as weight training (Jesse 1971; State 1955), these are mostly used to improve the strength component, along the speed-strength continuum.

Table 2 Overview of TG's running training plan made by Lance Brauman in preparation for the 100 m at the 2009 World Championships

	Acceleration	Speed phase	Under dist. end	Over dist. end
Nov–Dec	50 m uphill, 60 m technical runs			100–400 m @70–80%
Jan–Feb	10–40 m, 30 m sled/hill	20–40 m flys		100–400 m @80–90%
Mar–Apr	10–50 m, 30 m sled/hill	20–60 m flys	60–80 m	100–400 m @88–95%
May–Jun	10–50 m	30 flys, 80 m	80–90 m	100–250 m @88–100%
Jul–Aug	20–60 m	60–80 m	80–90 m	100–180 m @90–100%

3.3.3 Training in the Speed-Strength Continuum

The benefits of strength training have been demonstrated experimentally in the 1940s in a landmark thesis in Soviet Union (Ter Ovanesyan 1946) comparing performances of groups of sprinters who included it with the groups who did not; injury prevention was also cited as positive effect by sprint world record holders from the 1950s (McDonald-Beiley 1953; Cuthbbert 1966). However, the modalities of such training have since been subject to controversy regarding the negative effects on speed of excessive focus on strength and on which forms of workouts are the most appropriate for sprinters. Zatsiorskiy and Verkhoshanskiy were the first to classify commonly used exercises along a speed-strength continuum, depending on the external resistance applied during the muscular activity, adapting Hill's force–velocity relationship graphic (Zatsiorsky 1966; Verkhoshanskiy 1970). Sprinting would rank on the speed side, lifting weights on the opposite side as expression of absolute strength, and the bounding and throwing exercise variations would place in between those two ends.

3.3.4 Lifting Weights

Exercises with additional loads using barbells and weights are the most widely used to improve strength, especially maximum strength and hypertrophy. This was made popular by the success of Canadian Ben Johnson during the 1980s: from 80 kg in bench press for a body weight of 68 kg in 1979 at age 18, he improved to 3 × 165 kg in bench press and 6 × 250 in squats (for a 72 kg body weight) in 1986 the year he ran under 10 s at 100 m, and finally 2 × 185 kg and 6 × 272 kg for the same exercises respectively (80 kg body weight), in 1988 when he ran 9.79 before his doping disqualification (Francis 1992; Horrigan 1990; Francis 1990).

By contrast, the two fastest men ever (as for 31.12.2020), Usain Bolt and TG, have not been able to engage in intensive maximum weight training to such extent, due to back condition and health issue (Bolt 2013 season). TG started to lift weights consistently from age 23 in 2006, the year he ran under 10 s for the first time, working with loads up to 4 × 85 kg in bench press and same mass in squats. By 2009, he improved his lifting performance by 20 kg in both exercises. According to his coach, non-typical weight room is one of the major reasons that he struggled with force application early in races. However, more classical was TG's strength training periodization, divided in three periods: general strength work in Fall and Winter, followed by 8 weeks of strength development, and finally from May, speed-strength work with bar velocity as a focus. A similar approach was employed by Christine Arron, who was lifting 10 × 80 kg in squats in the first period, progressing to less repetitions up to 120 kg (front squats), and using lighter and faster exercises from Spring until major competitions. Large variations in maximum lifting personal bests are found in all-time best sprinters: among the all-time top 10 women at 100 m (10.75 or faster), the performance in squat ranges from 120 to 240 kg, in bench press and power clean, from 50 to 100 kg.

Maximum strength influence on sprint performance diminishes with the level of expertise of sprinters: maximum squat result can explain 40% of the 100 m results for a group of 11.9–13.2 s. performers, 18% for a 10.9–11.9 s. group, and only 2% for a 10.0–10.5 s. group (Letzelter 1986).

Furthermore, its relevance for sprinters has been questioned since the expression of maximum strength requires about 0.4 s to reach its peak and sprinters only have less than 0.1 s to apply it on the ground during step impulse; hence measuring the level of strength reached after 0.1 s during an exercise has been proposed as a useful test and work direction (Alabin 1976; Letzelter and Faubel 1973; Letzelter and Schilling 1974).

Lifting weights being theoretically classified on the opposite side of the speed-strength continuum regarding speed, sprinters have used a method consisting in moving lighter weights as fast as possible. The first athlete to have used this kind of form of velocity-based training was GDR's Gisela Birkemeyer in the 1959–1960 season (Birkemeyer 1966). She was timing (in seconds) her sets of strength exercises in order to control the intensity of the exercises, thus velocity and not load (kg). Nevertheless, in practice no consensus on efficiency of fast (light) over slow (heavy) weightlifting, both ends having been successfully used. Indeed, elite sprinters receive such a stimulus in term of intensity and workload volume form sprinting that the transfer from heavy weightlifting is made via the sprint training itself, when performed all year round, in a vertical integration (Francis 1992).

3.3.5 Jumping and Throwing

Ballistic exercise variations have been classified between speed and strength extreme ends of the speed-strength continuum. The problem of transfer of maximum strength into sprint performance has led coaches to find jumping and throwing exercises that are more specific to the sprinting motion and neuromuscular activities and adaptations, called *special exercises* (Bartenyev, 1965; Chevychalov and Gorozhanin 1965). Jumping exercises have been divided into horizontal & vertical categories, according to the orientation of force application: vertical for general strength and horizontal for special strength (Handreck 1965). Short jump training, such as simple jumps, has been found to be effective to improve short sprint distance performance, and long jump exercises, such as mutlibounds, to improve speed endurance (Verkhoshanskiy 1974).

More complex training plans in East Germany during the late 1970s have classified exercises according to their congruence with sprint race phases, depending on the contact times, correlation between jump and sprint performances, or stress intensity in the energetic supply (Jaenicke 1981): to improve sprint acceleration, the training catalogue included triple, quintuple or decuple horizontal jumps using the same leg or alternating legs and weightlifting at 70–90% intensity; to improve maximum sprint speed, vertical jumps over high hurdles and lifting lighter weights against time; and to improve sprint speed-endurance, horizontal with more ground contacts and vertical jumps over more, lower hurdles, and even lighter timed weightlifting.

TG used horizontal and vertical jumping and throwing exercises in circuit formats at least twice a week from the beginning of his preparation. They were performed mostly on the same day as short sprint training, in order to work the same theme and give more recovery time between the session in order to promote recovery.

3.3.6 Resisted and Assisted Sprinting

Resisted sprinting is considered the most specific strengthening exercise for sprinting and has been found in various forms. Hill running has been found to be more efficient than flat runs only to improve both acceleration and maximum sprint velocity (Golokhvastov 1959). Sprinting uphill or with a weight belt has been described as early as 1949 in Soviet training plans (Chomenkov 1955); tire sleds and pulleys a decade later, designed as special exercised for start and acceleration (Bartenyev 1965). TG used grass steep hills with 6 to 10 × 20 m, sled 6 to 8 × 30 m with 20 kg from December to March, one to three times a week.

Despite the intention to develop this specific and paramount actions for improving acceleration, it is quite difficult to check the real transfer during acceleration. For this reason and considering the essential training principle of specificity, Morin and colleagues thought that it was possible to stimulate and recreate the conditions during acceleration using a specific training such as "heavy sled training" which could allow to coaches to train in a very specific way the scenario during acceleration (Cross et al. 2015, 2017; Morin et al. 2016).

Resisted heavy sled training allows to reproduce the similarity of specific actions during acceleration related to angles for body orientation and, depending on the loading, challenges athletes in maintaining body alignment around the horizontal force vector being produced. These heavy loads will force the athlete to direct forces horizontally and have been shown to provide specific transfer to acceleration performance (Morin and Samozino 2016). Furthermore, from a practical standpoint, and considering that one of the main issues during acceleration sessions for team-sports and beginners in track and field is that subjects are not able to maintain body alignment and continue to apply horizontal forces longer, this training enable the movement to be slowed for technical purposes and can support transfer with athletes being able to explore projection angles and body alignment to combine with free sprinting unloaded acceleration runs.

A real example of the use of heavy sled training could be found on Morin et al. (2016) where authors performed a training during 8 consecutive weeks (16 sessions in total) with a progressive increase of heavy sled over sessions as follows: the heavy sled training program was assigned a mixed content of unresisted and resisted sprints with an increasing amount of resisted sprint over the training intervention: 5 heavy sled sprints out of 10 during sessions 1–4, 6 during sessions 5–8, 7 during sessions 9–12, and 8 during the last 4 sessions. The main outcome of this specific training intervention was that heavy sled training using much greater loads than traditionally recommended clearly increased maximal horizontal-force production and mechanical effectiveness (i.e., more horizontally applied force).

Authors would like to highlight that this new approach of heavy sled training for improving acceleration is one of the main practices with elite sprinters and team-sports in a daily basis supported with relevant competitive results with sprinters coached by authors.

Assisted sprinting has been used to go beyond the speed-strength continuum, by making the body light and thus allowing it to move faster. This sprint training method usually perform in sprinters has been the so called "overspeed" or "assisted sprinting" with the aim to reproduce similar conditions than maximal velocity but challenges athletes to maintain technique with lesser contact time trying to maintain or increase slightly stride frequency and maintain or also slightly increase of stride length, taking the athlete into a domain of movement management they would not otherwise have access to. In Soviet Union, the intent was originally to break the so-called "speed barrier" which is the point where sprinters can no longer improve their running velocity (Ozolin 1949): Towed by a moto, sprinting on a slope track or following the rhythm imposed by drumbeats artificially helped to increase running pace. However, East German coaches came from a different observation: sprinters timed in 30 m or flying 30 m would run about 5% slower at training compared to competition (Gundlach 1964). Using artificial traction with sprinters were allowing to run as fast as the goal pace planned for championships. Athletes reported that they were able to sprint with a better muscular relaxation, the effective pulling force was 2–2.5 kg, and they couldn't keep normal running technique with a greater traction. The author of the experiment stated that since artificial traction only assists effect on push-off phase of the running movement, the runner is forced to accelerate the forward swing of the unsupported leg by voluntary muscle work, thus a training effect was expected from this.

TG didn't employ such devices, and only took advantage of the huge winds that usually blow on his training track. Favorable wind has the effect of lowering the air resistance when the sprinter is sprinting, and a wind of 2 m/s (inside the limit of wind-assistance in competition according to World Athletic rules) can provide an advantage of 0.1 s or 1% at 100 m (Linthorne 1994). Chronologically, getting help from the wind to increase running speed was occurring in Spring and Summer, after the resisted sprinting workload held in Fall and Winter. This planification is similar to what was done to prepare 1964 Olympic Games by East German sprinters (1968): using sled pull sprinting from December to April, and artificial traction from March to September.

3.4 Training Weekly Schedule

The organization of workload is similar among all-time best sprinters in that they keep at least 48 h between workouts of the same type. For consecutive strength

Table 3 Overview of training weekly schedule mainly used for world class sprinters

Monday	Tuesday	Wednesday	Thursday	Friday	Saturday	Sunday
Tyson Gay (USA) 100 m 9.69 in 2009, 2nd all time (coach Lance Brauman)						
Strength Sprint	Strength Speed end	Active recovery	Strength Sprint	Strength Speed end	Active recovery	Rest
Asafa Powell (JAM) 100 m 9.72 in 2008, 4th all time (coach Stephen Francis)						
Sprint Strength	Speed end Sprint techn	Speed end Strength	Streng. end Sprint techn	Sprint Strength	Strength Endurance	Rest
F. Griffith-Joyner (USA), 100 m 10.49 in 1988, world record (Bob Kersee, Al Joyner)						
Speed end Strength	Speed end Strength	Rest	Speed end Strength	Speed end Strength	Speed End	Rest
Carmelita Jeter (USA) 100 m 10.64 in 2009, 2nd all time (John Smith)						
Strength Speed end	Strength Starts	Speed end	Strength Starts	Strength Sprint	Rest	Rest
Christine Arron (France) 100 m 10.73 in 1998, 6th all time (Jacques Piasenta)						
Strength	Rest	Sprint	Strength	Rest	Sprint	Endurance

training days, they usually split the work by focusing on lower body on day 1 and upper body on day 2. Most of the athletes perform two sessions per day, with notable exception of Christine Arron in 1998, whose workouts were of longer duration (up to 3 h) (Table 3).

4 Filling Gaps

In the area of sports training it is common to distance oneself from its essence, which is no other than to optimize physical performance with the right stimuli for the subject in question, in order to make the necessary adaptations for a given situation. Although the practical evidence concerning to sprint training has been supported by evidence-based, it is worth to note that there are some filling gaps should be taken into account in order to improve the stimulus proposed with various exercises and/or training tasks for optimizing performance. Going back to that discussed at the beginning of the chapter, understanding sprint thorough the velocity–time curve and applied force during sprint phases has been the rationale for organizing training methods during decades, there is a novel and scientifically validated approach called "Force–velocity profile in sprinting (Morin and Samozino 2016)" (F–v) which could provide very useful and practical information, allowing the selection of the most appropriate exercises for improving sprint performance to specifically stimulate the different zones of velocity–time curve considering kinetics and kinematics accordingly, enabling the design of better training programs for different modalities by considering which component (derived from the calculation of F–v profile such as F0, RF, Pmax, DRF, and v0) should be emphasized.

Therefore, one of the main challenges that coaches and practitioners should address is how to specifically target the different components of the F–v profile, and, depending on the orientation of the F–v profile towards force or velocity, the targeted programs for individuals could differ depending on whether a force-oriented or velocity-oriented profile dominates. Thus, F–v profile is the future for monitoring and to structure and organize sprint training, being the complement to all have exposed during our chapter. Furthermore, given the simplicity of obtaining the data required for this method, such as velocity–time measurements with an adequate sampling rate, for training and assessment purposes this profiling method would be easily used by strength and conditioning coaches and practitioners with timing gates, a radar gun, or even a recently validated iPhone app (MySprint) (Romero-Franco et al. 2016).

With this rationale, we should stay away from imitating procedures or "miracle" solutions since, regardless of the type of training used (sprint training in this case), we must make sure it is the most suitable one at every given moment. We must not consider various sprint training methods as better or worse, but rather as suitable or not for each specific context and needs.

The features explained above concerning F–v profiling could provide both useful information for sport practitioners and a simple, accessible, yet accurate method for more individualized monitoring and training of physical and technical capabilities. This method can be easily implemented on a regular basis and can therefore be used for long-term monitoring and training processes (Morin and Samozino 2016).

5 Take Home Messages and Practical Resources

1. Take into account the need to understand the sprint nature from the entire velocity–time curve, force production and muscular implications as the basis to select the most adequate sprint training methods for acceleration or maximal velocity purposes.
2. Heavy sled training as a new sprint training method for improving acceleration via an increase of maximal horizontal-force production and mechanical effectiveness (i.e., more horizontally applied force) is one of the main practices with elite sprinters and team-sports in a daily basis.
3. Use the various sprint training methods according to the enhancement sought (training in the speed-endurance continuum via sprint speed or sprint speed endurance training and training in the speed-strength continuum) making sure the kinetics and kinematics of the method and selected exercise are appropriate for each specific situation considering the sport demands and moment of season. It is likely highly recommended to use the long to short distance (LtS) approach

for speed endurance along the training season together with short to long distances (StL) for acceleration and maximum speed work.

References

Alabin V (1976) Speed and strength of the sprinter. Legk Atleika 248(1):13

Bartenyev L (1965) Start, acceleration and special exercises for sprinters. Legk Atleika 120(5):11–14

Birkemeyer H (1966) Training analysis of our most successful sprinters. Leipzig

Bompa T (1983) Theory and methodology of training. IA K& H, Dubuque, USA

Bondarchuk A (2010) Transfer of training in sports, vol II. Ultimate Athlete Concept ed, Chicago, USA, 48 p

Bushnell T, Hunter I (2007) Differences in technique between sprinters and distance runners at equal and maximal speeds. Sport Biomech 6(3):261–268

Chevychalov G, Gorozhanin V (1965) They write about the sprint again. Legk Atleika 125(10):7–10

Chomenkov L (1955) 100m and 200m races. Phys Cult Sport

Cormie P, McGuigan MR, Newton RU (2010) Adaptations in athletic performance after ballistic power versus strength training. Med Sci Sports Exerc 42(8):1582–1598

Cromwell D (1929) The background for the track athlete. Life B (ed) Boy's Life 19(11):20

Cromwell D (1939) The sprint races. International Sports, Indianapolis, USA

Cross MR, Brughelli M, Brown SR, Samozino P, Gill ND, Cronin JB et al (2015) Mechanical properties of sprinting in elite rugby union and rugby league. Int J Sports Physiol Perform 10 (6):695–702

Cross MR, Brughelli M, Samozino P, Brown SR, Morin J-B (2017) Optimal loading for maximising power during sled-resisted sprinting. Int J Sports Physiol Perform 12(8):1–25

Cuthbbert B (1966) Golden girl, the autobiography of betty cuthbert. Pelham Books LTD. London, Great Britain, pp 78–80

De Villarreal ES, Requena B, Cronin JB (2012) The effects of plyometric training on sprint performance: a meta-analysis. J Strength Cond Res

Debaere S, Delecluse C, Aerenhouts D, Hagman F, Jonkers I (2013) From block clearance to sprint running: characteristics underlying an effective transition. J Sports Sci 31(2):137–149

Delecluse C, Van Coppenolle H, Willems E, Van Leemputte M, Diels R, Goris M (1995) Influence of high-resistance and high-velocity training on sprint performance. Med Sci Sports Exerc 27 (8):1203–1209

Deutsch MU, Kearney GA, Rehrer NJ (2007) Time—motion analysis of professional rugby union players during match-play. J Sports Sci 25(4):461–472

Erbstösser H (1968) The design of the annual training 1963/64 in the men's sprint. Leipzig DHfK

Faude O, Koch T, Meyer T (2012) Straight sprinting is the most frequent action in goal situations in professional football. J Sports Sci 30(7):625–631

Fillin V (1955) Training methods for the best foreign sprinters. Legk Atleika 1(1):26–28

Francis C (1990) Speed trap, inside the biggest scandal in olympic history. Lester é Orpen Dennys, Toronto, Canada, 306 p

Francis C (1992) The charlie francis training system. TBLI Publications Inc. Ottawa, Canada

Gabbett TJ (2012) Sprinting patterns of National Rugby League competition. J Strength Cond Res 26(1):121–130

Garcelon W (1894) Some points on sprinting. Levingst Evening J Apr 17

Golokhvastov D (1959) Running uphill. Legk Atleika 49(6):16

Graubner R, Nixford E (2009) Biomechanical analysis of the sprint and hurdle events at the 2009 IAAF World Championships in Athletics. New Stud Athl 26(1–2):19–53

Gundlach H (1964) Research on the use of artificial traction in speed training for sprinters. Theory Pract Compet Sport 2(5):85–88
Gundlach H (1968) Application of high-speed intensity in sprinting. Theory Pract Compet Sport 6 (3):114
Gundlach H (1969) Test procedure for checking sprint speed. Theory Pract Compet Sport 18 (3):225
Handreck K (1965) My experience in advanced training for sprinters. Theory Pract Compet Sport 3(1):31–69
Harre D (1971) Principles of sports training. Sportverlag, Berlin, Germany
Harris G, Stone M, O'Bryant H, Proulx C, Johnson R (2000) Short term performance effects of high speed, high force or combined weight training. J Strength Cond Res 14(1):4–120
Hayes B (1992) Run, bullet, run, the rise, fall and recovery of Bob Hayes. Harper & Row, New York, USA
Hess W (1979) Developing the speed ability in the annual structure of the short sprinter, taking special consideration on the volume and intensity of the special training area. Martin Luther University, Halle, Germany
Horrigan J (1990) The fastest man in the world, Coach Charlie Francis reveals Ben Johnson's weight training connection. Iron Mag. 6:100–102
Hunter JP, Marshall RN, McNair PJ (2004) Interaction of step length and step rate during sprint running. Med Sci Sports Exerc 36(2):261–271
Hunter JP, Marshall RN, McNair PJ (2005) Relationships between ground reaction force impulse and kinematics of sprint-running acceleration. J Appl Biomech 21(1):31–43
Jaenicke L (1981) The typical microcyclic load evolution of a sprinter in a preparation and competition period. Leipzig DHfK, Germany
Jesse J (1971) Strength, power and muscular endurance for runners and hurdlers. Athletic Press, San Marino, USA, pp 41–42
Kersee B (1989) Philosophy of running training and methodology, women sprinters and heptathlon. Track F Quart Rev 1:3–5
Korobkov G, Fillin V (1956) The way to improve sprinters skills. Legk Atleika 8(1):8–9
Lang R (1975) Physiological performance examinations in the athletics sprint. Theory Pract Compet Sport 13(7):114–127
Letzelter M (1986) Krafttraining (strength training). Gmbh RTV. Hamburg, Germany
Letzelter M (2010) Sprinttraining, volume II. Schors-Verlag Niedernhausen/TS, Hamburg, Germany, pp 272–276
Letzelter M, Faubel G (1973) The influence of selected power properties on sprint performance. Leistungssport 6:424–430
Letzelter M, Schilling T (1974) Maximum strength and power with female sprinters of different performance levels. Leibesübungen 27(6):146–150
Linthorne N (1994) The effect of wind on 100-m sprint times. J Appl Biomech 10(2):110–131
Mann RA, Moran GT, Dougherty SE (1986) Comparative electromyography of the lower extremity in jogging, running, and sprinting. Am J Sports Med 14(6):501–510
Matveyev L (1965) The problem of sport training periodization. Physical Culture and Sport, Moscow, Russia
McBride JM, Triplett-McBride T, Davie A, Newton RU (2002) The effect of heavy- versus light-load jump squats on the development of strength, power, and speed. J Strength Cond Res 16(1):75–82
McDonald-Beiley E (1953) If it's speed you're after. Co SP&, Chicago, USA, 112 p
Mero A, Komi PV, Gregor RJ (1992) Biomechanics of Sprint Running. Sport Med 13(6):376–392
Morin JB, Samozino P (2016) Interpreting power-force-velocity profiles for individualized and specific training. Int J Sports Physiol Perform 11(2):267–272
Morin JB, Samozino P, Bonnefoy R, Edouard P, Belli A (2010) Direct measurement of power during one single sprint on treadmill. J Biomech 43(10):1970–1975
Morin J-B, Edouard P, Samozino P (2011) Technical ability of force application as a determinant factor of sprint performance. Med Sci Sports Exerc 43(9):1680–1688

Morin JB, Bourdin M, Edouard P, Peyrot N, Samozino P, Lacour JR (2012) Mechanical determinants of 100-m sprint running performance. Eur J Appl Physiol 112(11):3921–3930
Morin J-B, Gimenez P, Edouard P, Arnal P, Jiménez-Reyes P, Samozino P et al (2015a) Sprint acceleration mechanics: the major role of hamstrings in horizontal force production. Front Physiol 6:404
Morin JB, Slawinski J, Dorel S, de villareal ES, Couturier A, Samozino P et al (2015b) Acceleration capability in elite sprinters and ground impulse: push more, brake less? J Biomech 48(12):3149–3154
Morin J-B, Petrakos G, Jimenez-Reyes P, Brown SR, Samozino P, Cross MR (2016) Very-heavy sled training for improving horizontal force output in soccer players. Int J Sports Physiol Perform:1–13
Murphy M (1894) College athletics training. Spalding's Athletic Library, Chicago, USA, 39 p
Nagahara R, Matsubayashi T, Matsuo A, Zushi K (2014a) Kinematics of transition during human accelerated sprinting. Biol Open 3(8):689–699
Nagahara R, Naito H, Morin J-B, Zushi K (2014b) Association of acceleration with spatiotemporal variables in maximal sprinting. Int J Sports Med 35(9):755–761
Ozolin N (1945) The secret of strength and endurance. Young Guard Edition, Moscow, Russia
Ozolin N (1949) Track and field training: fundamentals of training methods. Physical Culture and Sport, Moscow, Russia
Petrovsky V (1973) Training and Monitoring. Legk Atleika 212(1):10
Petrovsky V (1974) How much? or how? Legk Atleika 232(9):14
Petrovsky V (1978) Short distance running. Physical Culture Sport, Moscow, Russia
Rabita G, Dorel S, Slawinski J, Sàez-de-Villarreal E, Couturier A, Samozino P et al (2015) Sprint mechanics in world-class athletes: a new insight into the limits of human locomotion. Scand J Med Sci Sport 25(5):583–594
Romero-Franco N, Jiménez-Reyes P, Castaño-Zambudio A, Capelo-Ramírez F, Rodríguez-Juan JJ, González-Hernández J et al (2016) Sprint performance and mechanical outputs computed with an iPhone app: Comparison with existing reference methods. Eur J Sport Sci 17(4):1–7
Ronnestad BR, Kvamme NH, Sunde A, Raastad T (2008) Short-term effects of strength and plyometric training on sprint and jump performance in professional soccer players. J Strength Cond Res 22(3):773–780
Rumpf MC, Lockie RG, Cronin JB, Jalilvand F (2016) Effect of different sprint training methods on sprint performance over various distances: a brief review. J Strength Cond Res 30(6):1767–1785
Samozino P, Rabita G, Dorel S, Slawinski J, Peyrot N, Saez de Villarreal E, et al (2016) A simple method for measuring power, force, velocity properties, and mechanical effectiveness in sprint running. Scand J Med Sci Sports 26(6):648–658
Schache AG, Brown NAT, Pandy MG (2015) Modulation of work and power by the human lower-limb joints with increasing steady-state locomotion speed. J Exp Biol 218(15):2472–2481
Schimpchen J, Skorski S, Nopp S, Meyer T (2016) Are 'classical' tests of repeated-sprint ability in football externally valid? A new approach to determine in-game sprinting behaviour in elite football players. J Sports Sci 34(6):519–526
Sheppard JM, Dingley AA, Janssen I, Spratford W, Chapman DW, Newton RU (2011) The effect of assisted jumping on vertical jump height in high-performance volleyball players. J Sci Med Sport 14(1):85–89
Simperingham KD, Cronin JB, Ross A (2016) Advances in sprint acceleration profiling for field-based team-sport athletes: utility, reliability validity and limitations. Sport Med 46 (11):1619–1645
Sleivert G, Taingahue M (2004) The relationship between maximal jump-squat power and sprint acceleration in athletes. Eur J Appl Physiol 91(1):46–52
Spencer M, Bishop D, Dawson B, Goodman C (2005) Physiological and metabolic responses of repeated-sprint activities:specific to field-based team sports. Sports Med 35(12):1025–1044

State O (1955) Strength training for athletes. Amateur Athletics Association, London, Great Britain, 90 p

Tepper E (1989) On the state of development and the training system in the GDR women's short sprint. In: European Athletics Coaching Association Congress. Bad Blankenburg

Ter Ovanesyan A (1946) The use of strength exercises during training in sprinting, jumping, throwing. Theory Pract Phys Cult 9(10):449–456

Turova G (1955) Experience in increasing the volume of a sprinter's running training. Legk Atleika 2(2):9–11

Vazel P (2018) History of the reaction time in sprint events. In: Biomechanical report for the IAAF World Indoor Championships 2018, 60 metres women. International Association of Athletics Federations, Birmingham, UK

Veale JP, Pearce AJ, Carlson JS (2007) Player movement patterns in an elite junior Australian rules football team: an exploratory study. J Sports Sci Med 6(2):254–260

Verkhoshanskiy Y (1970) Fundamentals of special strength training in sports. In: Physical culture and sport. Moscow, Russia

Verkhoshanskiy Y (1974) Jump in the training of sprinters. Legk Atleika 232(9):16–17

Vigne G, Gaudino C, Rogowski I, Alloatti G, Hautier C (2010) Activity profile in elite Italian soccer team. Int J Sports Med 31(5):304–310

Walsh J (1859) 'Stonehenge". British rural sports. Routledge, Warnes & Routledge F, London, Great Britain

Westhall C (1863) The modern method of training. Beeton, London, Great Britain, 36 p

Weyand PG, Sternlight DB, Bellizzi MJ, Wright S (2000) Faster top running speeds are achieved with greater ground forces not more rapid leg movements. J Appl Physiol 89(5):1991–1999

Wheeler C (1868) Sportascrapiana. Marshal & Co, London, Great Britain, 180 p

Wild J, Bezodis N, Blagrove R, Bezodis I (2011) A biomechanical comparison of accelerative and maximum velocity sprinting : specific strength training considerations. UK Strength Cond Assoc 21(21):23–36

Worman J (1894) Answers to correspondents. Outing Mag

Yu J, Sun Y, Yang C, Wang D, Yin K, Herzog W et al (2016) Biomechanical insights into differences between the mid-acceleration and maximum velocity phases of sprinting. J Strength Cond Res 30(7):1906–1916

Zabierzowski Z (1962) Short distance races. Sport & Tourism, Warsaw, Poland, 70 p

Zatsiorsky V (1966) Motor abilities of athletes. In: Fizkultura l sport. Moscow, Russia, 70 p

Pedro Jiménez-Reyes is a senior lecturer at the Center for Sport Studies of Rey Juan Carlos University (Madrid, Spain). He obtained a Track and Field Coach National Diploma in 2010 and obtained his Ph.D. in High Performance in Sport in 2010 at the University of Pablo de Olavide (Seville, Spain). His field of research is mainly monitoring training, running, and maximal power movements (sprint, jumps), with application in the field of sport training. He has published several peer-reviewed journal articles, trains elite athletes, and collaborates and plays a consultant role with elite sprinters, and team sports in Spain.

Pierre-Jean Vazel is an athletics coach in Metz, France. Since 2004, he coached Nigerian Olusoji Fasuba to the 100 m African Record (9.85 in 2006) and 60 m world indoor title (2008), and other international medalists from France, China, Senegal and Greece, including Christine Arron. He has been correspondent for the International Association of Athletics Federations (IAAF) website and Le Monde. As a coach or chronicler, he covered two Olympic Games, thirteen World Athletics Championships, and over 500 events. He has lectured on the history of sprint training and science in Canada, France, Great Britain, Norway, and Sweden, and participated to the IAAF biomechanics projects at the 2017 and 2018 world championships.

Resistance Training Using Flywheel Resistance Training Devices

Francisco Javier Nuñez

Abstract It is known that muscle fibres have a greater capacity to generate force during the eccentric phase of a movement compared with the concentric phase of the same movement. Thus, some authors argue that training protocols where exercise is overloaded during the eccentric phase of movement achieve greater strength gains than those in which the load is constant during the eccentric and concentric phases. Different training devices, using the inertia of rotating flywheels, have been designed to increase the eccentric overload during movement. The use of flywheel resistance training devices during short periods of training may improve muscle force more than traditional methodologies. In this chapter we will respond to the main questions about how to use the flywheel device and discuss how to optimize athletes' performance with it.

Keywords Rotary inertia · Flywheel paradigm · Eccentric-overload · Variability

1 Introduction

Resistance training produces muscular adaptations at multiple structural and functional levels (Aagaard et al. 2001; Higbie et al. 1996; Moritani and deVries 1979; Narici et al. 1989), and is an important part of most training regimens in competitive sports (Nuñez et al. 2017). Chronic resistance exercise has been shown to enhance neural, hypertrophic, and strength adaptations within the first 4–8 weeks of training (Schuenke et al. 2012; Staron et al. 1994). Independently of the exercise design, it appears that training protocols based on the concentric phase of a movement, or only the eccentric phase, increase both muscle mass and strength in a similar manner (Higbie et al. 1996; Colliander and Tesch 1990; Jones and Rutherford 1987; Seger et al. 1998), but when concentric and eccentric actions are

F. J. Nuñez (✉)
Universidad Pablo de Olavide, Physical Performance & Sports Research,
41013 Seville, Spain
e-mail: fjnunsan@upo.es

A. Muñoz-López et al. (eds.), *Resistance Training Methods*, Lecture Notes
in Bioengineering, https://doi.org/10.1007/978-3-030-81989-7_7

combined in the same movement the increases are magnified (Higbie et al. 1996; Hather et al. 1991; Hortobagyi et al. 2001). It is known that muscle fibres have a greater capacity to generate force during the eccentric phase of a movement compared with the concentric phase of the same movement (Colliander and Tesch 1990; Crenshaw et al. 1995; Enoka 1996; Gual et al. 2016; Katz 1939; Komi and Buskirk 1972; Westing and Seger 1989; Westing et al. 1991). Thus, some authors argue that training protocols where exercise is overloaded during the eccentric phase of movement achieve greater strength gains than those in which the load is constant during the eccentric and concentric phases (Hortobagyi et al. 2001; Brandenburg and Docherty 2002; Doan et al. 2002), giving rise to the concept of "eccentric overload".

Different training devices, using the inertia of rotating flywheel(s), have been designed to increase the eccentric overload during movement (Berg and Tesch 1994, 1998). This technology generates resistance by opposing the athlete's effort with the inertial force generated by a lightweight rotating flywheel, such that the same inertia must be overcome during each repetition by means of accommodated loading (Norrbrand et al. 2008; Nuñez and Saez de Villarreal 2017). Recently, there has been increasing interest in the use of flywheel resistance training devices to improve athletes' performance (Nuñez and Saez de Villarreal 2017; Hoyo et al. 2015a; Sabido et al. 2017). There are many questions associated with the use of such devices. Does the variability in force-power production between different repetitions of the same exercise in such devices produce benefits or undesired adaptations in an athlete's performance? Is performance improved by the use of eccentric overload during the exercise? Do different flywheel devices provide similar force profiles? Are these improvements higher than those obtained with traditional methodologies? In this chapter we will respond to these questions and discuss how to optimize the use of flywheel devices to improve athletes' performance.

1.1 Theoretical Background

Flywheel resistance training devices have been designed to use the moment of inertia of a rotating flywheel to provide a maximal resistance load during the concentric and eccentric phases of movement (Nuñez and Saez de Villarreal 2017). These devices provide a source of linear resistance from a tether wrapped around a shaft, where the concentric action unwinds the tether, spinning the flywheel in one direction, and the eccentric action occurs during rewinding, spinning the flywheel the opposite way (Nuñez et al. 2017). The kinetic energy from the concentric phase of the exercise is transferred to the eccentric phase, where an equal impulse is necessary to halt the rotation of the moment of inertia (Nuñez et al. 2017). To be efficient using a rotary inertia device, the athlete tends to apply force at maximal velocity during the concentric phase of the movement, and halt the rotation during the eccentric phase of the movement to produce force enhancement in the subsequent concentric phase (Nuñez and Saez de Villarreal 2017). However, the athlete can halt the rotation during the eccentric phase by: (a) increasing the damping time in the eccentric phase so that this load will be attenuated; (b) equalizing the duration of the concentric and eccentric phases, which would result in similar concentric and eccentric loads; and (c) reducing the duration of the eccentric phase with respect to the concentric phase, which can be achieved by accompanying movement without slowing the movement in the first part and reducing the energy of the shaft only at the end of the action (Nuñez et al. 2019). These different possibilities for halt the rotation during the eccentric phase is one of the main characteristics of this training system since the load that the athlete receives in each repetition depends directly on how he has executed his previous repetition, causing a high variability in force-power production between different repetitions of the same exercise (Moras et al. 2018; Sabido et al. 2018). In accordance with the execution model presented, in a pilot study that we developed using a flywheel device to resist squat-training in an athlete, we were able to verify that the eccentric phase offers higher intra-set variability than the concentric phase in both force and velocity (unpublished data).

Berg and Tesch (1994) indicated that by allowing a short period of not resisting immediately after completion of the concentric phase of the movement, a sudden steep rise in force will occur while attempting to stop the movement at the end of eccentric phase of the movement. The eccentric overload in flywheel resistance training devices was described by Tesch et al. (2004) in an open kinetic chain exercise (i.e., knee extension) when the average eccentric force was greater than the concentric force, in a 30-degree window from 5 degrees above the angle at which the transition from eccentric to concentric occurred. There are few studies that corroborate the existence of average eccentric overload in the same exercise and with the same devices (Norrbrand et al. 2010). However, Tous-Fajardo et al. (2006), using other open kinetic chain exercise (i.e., leg curl), found an eccentric overload existed for peak force values but not for the average. Surprisingly, it is very common to find conclusions in the scientific literature based on closed kinetic chain exercises (i.e., half-squat) with flywheel resistance training devices that claim the production of eccentric overload by controlling the execution technique (i.e., delaying the braking action in the eccentric phase or indicating to the participants that they should apply braking forces near the end of the ECC phase), without checking whether eccentric overload is actually being produced (Hoyo et al. 2015a, b; Sabido et al. 2017, 2018; Fernandez-Gonzalo et al. 2014a; Romero-Rodriguez et al. 2011). Recently Nuñez et al. (2020), by delaying the braking action in the eccentric phase using a flywheel resistance training device, showed that the peak force during the eccentric phase of a closed kinetic chain exercise (i.e., half-squat) was delayed with respect to the start of the next concentric phase of the movement by approximately 340–700 ms, and they never obtained eccentric overload for the peak and mean forces. In a closed kinetic chain exercise (i.e., half-squat), during the concentric phase of the movement, besides generating the moment of inertia, athletes must mobilize their body mass, while in the eccentric phase they must only stop the generated moment of inertia. This could explain how achieving eccentric overload in exercises not so dependent on gravity (i.e., knee extension or leg curl) would be favoured with respect to exercises where the displacement is done against the force of gravity. It is recommended not to assume the production of eccentric overload by controlling the execution technique: it is necessary to check whether eccentric overload is being produced, in order to use this variable to improve athletes' performance. A relatively recent systematic review with meta-analysis verified that: the ability to produce an eccentric overload with a flywheel system appears to require some experience in using such devices; significant increases in muscle mass and strength were noted with the use of flywheel systems over short periods of training; and the increase in strength was influenced by the existence of eccentric overload during the exercise, while muscle mass was not influenced by the existence of eccentric overload (Nuñez and Saez de Villarreal 2017). This information should be considered by strength and conditioning professionals with respect to the technique of using flywheel-based systems to optimize the increase in muscle volume and force during short periods of training (Nuñez and Saez de Villarreal 2017).

Different flywheel resistance training devices have been designed to improve athletes' performance. One of the main difference between them is whether the linear resistance is provided by a tether wrapped around a horizontal cylinder-shaped shaft or a vertical cone-shaped shaft (Nuñez et al. 2020). Both kinds of devices can generate considerable increases in kinetic and kinematic variables, as moments of inertia are increased (Sabido et al. 2017; Carroll et al. 2018; Martinez-Aranda and Fernandez-Gonzalo 2017). It is hypothesized that traction through a cone-shaped vertical axis, from its base (i.e., a larger diameter) to its vertex (i.e., a smaller diameter), favours the acceleration of movement in the concentric phase and the deceleration of movement in the eccentric phase with respect to traction through a horizontal axis in the form of a cylinder (i.e., the same, smaller diameter throughout its length) (Nuñez et al. 2020). A recent study confirmed this hypothesis and showed that a horizontal cylinder-shaped system generates a higher mean force and impulse during the concentric and eccentric phases of the movement, and a higher peak force during the eccentric phase of the movement, than a cone-shaped shaft with the same moment of inertia as a measure of training intensity (Nuñez et al. 2020). However, a cone-shaped shaft generates higher peak velocity during the concentric and eccentric phases of the movement than a horizontal cylinder-shaped shaft (Nuñez et al. 2020). In addition, during a flywheel exercise, a vertical cone-shaped shaft offers a higher intra-set variability in velocity and lower intra-set variability in force than a horizontal cylinder-shaped shaft (unpublished data).

Increases in muscle force were noted through the use of flywheel resistance training devices during short periods of training and these improvements were higher than those with traditional methodologies (Norrbrand et al. 2008, 2010; Onambele et al. 2008). Norrbrand et al. (2008), using flywheel training with a horizontal cylinder-shaped device, reported a higher increase, of 11.6% in the maximal voluntary contraction at 90° knee-flexion, compared to a weight stack system. Similarly, Onambele et al. (2008) reported a higher increase in maximal voluntary contraction (17%) using a flywheel knee extension device than was achieved using a traditional machine (8%). A relatively recent study using a flywheel resistance training device showed that for the same exercise executed at a similar concentric propulsive velocity, vertical cone-shaped devices allowed a greater mean eccentric force than free-weights in every repetition performed (Nuñez et al. 2017). The velocity of eccentric contraction has a high correlation to sport performance and the athlete's ability to sequence stiff muscle and compliant tendon for power production, depending on the eccentric utilization ratio (Katz 1939; Komi and Buskirk 1972). Therefore, it could be inferred that the high eccentric loading of flywheel devices in comparison with free weights would provide added training benefits. The other main findings of this study were that players developed a higher peak velocity and higher propulsive acceleration with a flywheel device than with free weights (Nuñez et al. 2017). During the free-weight exercise there was a substantial reduction in the barbell displacement, while the displacement in the flywheel system remained constant (Nuñez et al. 2017). In flywheel resistance training devices the range of motion, and particularly the end of the movement, is determined by the length of the rope. Nuñez et al. (2017) argue that a higher displacement would have increased the time for force application and explain the substantial increase of peak velocity and higher propulsive acceleration compared with free weights. Thus, athletes required to generate high speeds against moderate to high loads may benefit from the inclusion of flywheel devices in their training regimens in comparison with traditional methodology. Similarly, some research suggests that in addition to the neuromuscular aspects, flywheel devices may also improve target metabolic aspects such as anaerobic energy pathways (Nuñez et al. 2017; Caruso and Hernandez 2002; Caruso et al. 2006; Dudley et al. 1991). So, if anaerobic fitness can be developed while increasing the structural and mechanical performance of the muscle-tendon unit, there is considerable scope for implementing flywheel resistance training to improve fatigue resistance in a range of sport modalities. Flywheel devices could offer a safer and more efficient alternative for power and power-endurance training as long as the user is prepared for the reciprocal eccentric recoil inherent with rotary inertia training. Further long-term studies should be carried out to further assess the safety and efficiency that flywheel training can provide athletes.

2 From Practice

Most movements in the field require players to produce forces in variable and unpredictable contexts, with an emphasis on the eccentric component (Tous-Fajardo et al. 2016). Understanding the concept of variability as the normal variations that occur in motor performance across multiple repetitions of a task (Stergiou and Decker 2011), several studies support the use of flywheel resistance training devices for improving physical performance that transfers to sports, because such devices offer a high variability in the force-power production during an exercise (Moras et al. 2018; Sabido et al. 2018) and show significantly higher ECC force values compared to free weights (Nuñez et al. 2017). Some degree of motor variability can be beneficial as it allows a system that is more adaptable to internal and external perturbations, enabling training programs that prepare athletes better for variable and unpredictable contexts (Moras et al. 2018). However, it is known that very high levels of variability may cause undesirable adaptations (Moreno and Ordoño 2009). We already know that using a flywheel device to resist squat-training produced a high intra-set variability in the application of force and velocity expressed (Unpublished data). So, if we design training protocols with different exercises (which produce a high inter-exercise variability) using flywheel devices (which produce a high inter-repetition variability), will this cause undesired adaptations? Gonzalo-Skok et al. (2017) compared two protocol with the same training volume based on six different exercises or only one using the same flywheel device. The results showed that a training protocol that offered a high inter-repetition variability (i.e., used a flywheel device) and a high inter-exercise variability (i.e., six different exercises) was more effective at improving performance than the training protocol that offered a high inter-repetition variability too but a low inter-exercise variability (i.e., only one exercises), however none caused no undesired adaptations.

Increases in strength are influenced by the existence of eccentric overload during the exercise, while the muscle mass is not (Nuñez et al. 2017). This theoretical contribution requires a practical qualification. It is true that training protocols that allow for a short pause with no resistance immediately after completion of the concentric phase of the movement and an attempt to stop the movement at the end of the eccentric phase result in greater strength improvement than protocols that don't. However, it is also true that this execution model (i.e., reducing the time of the eccentric phase with respect to that of the concentric phase) distorts the execution technique, so it would be more appropriate to use it in specific scenarios where the objective of increasing muscular strength takes priority over the technical execution, which favours optimization of performance. In all the protocols that use flywheel resistance training devices (regardless of the existence of eccentric overload) there is an increase in force greater than can be achieved by traditional means. This may be caused by the increased eccentric load produced by these devices in comparison with the more traditional methodology (Nuñez et al. 2017). This increase in the eccentric load does not require that there be an eccentric overload and therefore does not require that the exercise execution technique be distorted. If we want our athletes to improve their ability to generate force and increase their muscle mass without detracting from their usual execution technique, we must ask them not to accentuate the production of eccentric overload. On the other hand, if we need an athlete to increase his muscular strength in a short space of time and we do not care if he alters his execution technique, we will ask him to produce eccentric overload, which must be verified during training, and it will be much more feasible to achieve it in open kinetic chain exercises than closed.

It is known that sports performance has a high correlation with the velocity of eccentric contractions (Katz 1939; Komi and Buskirk 1972). If, during your flywheel exercise, you want to favour the speed of the movement over the force used to perform it, you should use vertical cone-shaped devices; however, if your goal is to achieve the greatest forces, you should use horizontal cylinder-shaped devices (Nuñez et al. 2020). Independently of using vertical cone-shaped or horizontal cylinder-shaped device, we know that, to be effective with this device, the subject must apply force throughout the course of the entire concentric phase of the movement, thus producing force for a longer time period than in other systems (Nuñez et al. 2017). This longer period of force application may provide training benefits, improving the ability to accelerate from the stop position over short distances (e.g., sprint 0–10 m) (Nuñez et al. 2019; Gonzalo-Skok et al. 2017). Several studies have confirmed the efficacy of this training for improving short-distance acceleration in movements with different force vector applications, such as the countermovement jump (Gual et al. 2016; Hoyo et al. 2015b, c; Tous-Fajardo et al. 2016) and change of direction (Hoyo et al. 2015a, b, c; Tous-Fajardo et al. 2016).

3 Filling Gaps

Several studies, using flywheel resistance training devices, have noted improved maximum voluntary contraction in the ECC phase of movement without producing modifications in the electromyographic activity of the muscle (Tesch et al. 2004; Norrbrand et al. 2010). What structure generates this increase in eccentric force? From our experience, the effects of these training processes on the structural system can be attributed almost exclusively to an increase in muscle mass. In animal subjects, more than 30 years ago, it was demonstrated that during the stretch-shortening cycle, tension is not only transmitted longitudinally to the direction of muscle fibre contraction, but also transmitted transversely through the cytoskeleton that supports the exercised muscles (Street 1983). More recent findings in humans have demonstrated that training protocols that continuously exercise the muscle in the rapid stretch-shortening cycle and favour the eccentric phase of the movement generate the production of large quantities of type I collagen in response to load (Langberg et al. 2007). Therefore, it is possible that the increase in force in the eccentric phase of the movement, without changes in the EMG activity of the muscle, does not depend solely on the muscle structure, and that it may be highly influenced by the capacity of the tendinous-fascial tissue. In line with this hypothesis, Seynnes et al. (2007) concluded that each training protocol can generate different fascia-tendon-muscle architectures depending on the stimulation provided to the different structures. The same authors, using a 5-week flywheel training protocol in which the subject was asked to perform a knee flexion-extension movement in a system that favoured the production of eccentric overload, modified the pennation angle of the vastus lateralis quadriceps by 7.7% (Seynnes et al. 2007). Sanz-Lopez et al. (2016), with a 5-week flywheel training protocol, based on a half squat exercise in a system that favoured the eccentric phase of the movement, showed a modification of the medial calf pennation angle of 9.9% and an increase in cross-section of the Achilles tendon of 42%. Several studies have confirmed the efficacy of flywheel resistance training for improving or benefiting muscle volume (Norrbrand et al. 2008; Tesch et al. 2004; Seynnes et al. 2007; Alkner and Tesch 2004; Fernandez-Gonzalo et al. 2014b; Nuñez et al. 2018). However, our experience shows that this type of training system also has a very direct effect on the fascial cytoskeleton, giving greater consistency or stiffness to the exercised muscles and being one of the determining factors in increasing muscles' eccentric performance.

In contrast, flywheel resistance training devices offer athletes a greater increase force production, over short periods of training, than do traditional methodologies (Nuñez et al. 2017). However, it is possible that this gain could be influenced by the longer duration of force application in flywheel devices than in traditional systems. This longer duration of force application would not provide training adaptations when the time of the application of the force is minimal (e.g., sprint launches) (Moras et al. 2018). Therefore, we recommend the use of an exercise, immediately after each set of flywheel exercises, where the application of the force tends to be minimal (e.g., a plyometric exercise) to avoid these undesired adaptations.

Take-Home Messages and Practical Resources

1. Flywheel resistance training devices provide a source of linear resistance from a tether wrapped around a shaft, where concentric action unwinds the tether, spinning the flywheel in one direction, and eccentric action occurs during rewinding, Spinning the flywheel in the opposite direction.
2. Linear resistance is provided from a tether wrapped around a horizontal cylinder-shaped or a vertical cone-shaped shaft.
3. To be efficient using flywheel resistance training devices, force must be applied at maximal velocity during the concentric phase of movement and used to halt the rotation during the eccentric phase of the movement, in each repetition.
4. Flywheel resistance training devices produce a high variability in the application of force and velocity expressed in each repetition.
5. To produce eccentric overload, athletes have to allow for a short period of non-resistance immediately after completion of the concentric phase of the movement and before attempting to stop the movement at the end of the eccentric phase of the movement.
6. The ability to produce an eccentric overload in a flywheel system appears to require some experience in using such devices, and it is much more feasible to achieve it in open kinetic chain exercises that are not so dependent on gravity (i.e., knee extension or leg curl) than in closed kinetic chain exercises in which the displacement is done against the force of gravity (i.e., squat).
7. The use of flywheel resistance training devices during short periods of training may improve muscle force more than traditional methodologies.
8. The increase in muscle force is influenced by the existence of eccentric overload during the exercise.
9. Horizontal cylinder-shaped flywheel shafts generate higher mean force and impulse during the concentric and the eccentric phases of the movement, and a

higher peak force during the eccentric phase of the movement, than cone-shaped shafts.

10. Cone-shaped shafts generate higher peak velocity during the concentric and the eccentric phases of the movement than horizontal cylinder-shaped shafts.
11. The increase in muscle mass is not influenced by the existence of eccentric overload during the exercise. We must consider the possible effects of flywheel resistance training on the fascial cytoskeleton structure, giving greater consistency or stiffness to the exercised muscles.
12. The use of flywheel resistance training devices may require the athlete to engage in a longer period of force application that does not provide training adaptations for exercises where the time of application of the force is minimal. Therefore, we recommend using an exercise, immediately after each set of flywheel exercises, where the application of force is minimal.

References

Aagaard P, Andersen JL, Dyhre-Poulsen P, Leffers AM, Wagner A, Magnusson SP et al (2001) A mechanism for increased contractile strength of human pennate muscle in response to strength training: changes in muscle architecture. J Physiol 534(Pt. 2):613–623

Alkner BA, Tesch PA (2004) Efficacy of a gravity-independent resistance exercise device as a countermeasure to muscle atrophy during 29-day bed rest. Acta Physiol Scand 181(3):345–357

Berg HE, Tesch A (1994) A gravity-independent ergometer to be used for resistance training in space. Aviat Space Environ Med 65(8):752–756

Berg HE, Tesch PA (1998) Force and power characteristics of a resistive exercise device for use in space. Acta Astronaut 42(1–8):219–230

Brandenburg JP, Docherty D (2002) The effects of accentuated eccentric loading on strength, muscle hypertrophy, and neural adaptations in trained individuals. J Strength Cond Res 16 (1):25–32

Carroll KM, Wagle JP, Sato K, Taber CB, Yoshida N, Bingham GE et al (2018) Characterising overload in inertial flywheel devices for use in exercise training. Sports Biomech 1–12

Caruso JF, Hernandez DA (2002) Net caloric cost of a 3-set flywheel ergometer resistance exercise paradigm. J Strength Cond Res 16(4):567–572

Caruso JF, Herron JC, Capps LB, Coday MA, Ramsey CA, Drummond JL (2006) Blood lactate responses to leg presses performed against inertial resistance. Aviat Space Environ Med 77 (7):707–712

Colliander EB, Tesch PA (1990) Effects of eccentric and concentric muscle actions in resistance training. Acta Physiol Scand 140(1):31–39

Crenshaw AG, Karlsson S, Styf J, Backlund T, Friden J (1995) Knee extension torque and intramuscular pressure of the vastus lateralis muscle during eccentric and concentric activities. Eur J Appl Physiol Occup Physiol 70(1):13–19

Doan BK, Newton RU, Marsit JL, Triplett-McBride NT, Koziris LP, Fry AC et al (2002) Effects of increased eccentric loading on bench press 1RM. J Strength Cond Res 16(1):9–13

Dudley GA, Tesch PA, Miller BJ, Buchanan P (1991) Importance of eccentric actions in performance adaptations to resistance training. Aviat Space Environ Med 62(6):543–550

Enoka RM (1996) Eccentric contractions require unique activation strategies by the nervous system. J Appl Physiol (1985) 81(6):2339–2346

Fernandez-Gonzalo R, Lundberg TR, Alvarez-Alvarez L, de Paz JA (2014a) Muscle damage responses and adaptations to eccentric-overload resistance exercise in men and women. Eur J Appl Physiol 114(5):1075–1084
Fernandez-Gonzalo R, Irimia JM, Cusso R, Gustafsson T, Linne A, Tesch PA (2014b) Flywheel resistance exercise to maintain muscle oxidative potential during unloading. Aviat Space Environ Med 85(7):694–699
Gonzalo-Skok O, Tous-Fajardo J, Valero-Campo C, Berzosa C, Bataller AV, Arjol-Serrano JL et al (2017) Eccentric-overload training in team-sport functional performance: constant bilateral vertical versus variable unilateral multidirectional movements. Int J Sports Physiol Perform 12(7):951–958
Gual G, Fort-Vanmeerhaeghe A, Romero-Rodriguez D, Tesch PA (2016) Effects of in-season inertial resistance training with eccentric overload in a sports population at risk for patellar tendinopathy. J Strength Cond Res 30(7):1834–1842
Hather BM, Tesch PA, Buchanan P, Dudley GA (1991) Influence of eccentric actions on skeletal muscle adaptations to resistance training. Acta Physiol Scand 143(2):177–185
de Hoyo M, de la Torre A, Pradas F, Sañudo B, Carrasco L, Mateo-Cortes J et al (2015a) Effects of eccentric overload bout on change of direction and performance in soccer players. Int J Sports Med 36(4):308–314
de Hoyo M, Pozzo M, Sañudo B, Carrasco L, Gonzalo-Skok O, Dominguez-Cobo S et al (2015b) Effects of a 10-week in-season eccentric-overload training program on muscle-injury prevention and performance in junior elite soccer players. Int J Sports Physiol Perform 10(1): 46–52
de Hoyo M, Sañudo B, Carrasco L, Dominguez-Cobo S, Mateo-Cortes J, Cadenas-Sanchez MM et al (2015c) Effects of traditional versus horizontal inertial flywheel power training on common sport-related tasks. J Hum Kinet 47:155–167
Higbie EJ, Cureton KJ, Warren GL 3rd, Prior BM (1996) Effects of concentric and eccentric training on muscle strength, cross-sectional area, and neural activation. J Appl Physiol (1985) 81(5):2173–2181
Hortobagyi T, Devita P, Money J, Barrier J (2001) Effects of standard and eccentric overload strength training in young women. Med Sci Sports Exerc 33(7):1206–1212
Jones DA, Rutherford OM (1987) Human muscle strength training: the effects of three different regimens and the nature of the resultant changes. J Physiol 391:1–11
Katz B (1939) The relation between force and speed in muscular contraction. J Physiol 96(1): 45–64
Komi PV, Buskirk ER (1972) Effect of eccentric and concentric muscle conditioning on tension and electrical activity of human muscle. Ergonomics 15(4):417–434
Langberg H, Ellingsgaard H, Madsen T, Jansson J, Magnusson SP, Aagaard P et al (2007) Eccentric rehabilitation exercise increases peritendinous type I collagen synthesis in humans with Achilles tendinosis. Scand J Med Sci Sports 17(1):61–66
Martinez-Aranda LM, Fernandez-Gonzalo R (2017) Effects of inertial setting on power, force, work, and eccentric overload during flywheel resistance exercise in women and men. J Strength Cond Res 31(6):1653–1661
Moras G, Fernandez-Valdes B, Vazquez-Guerrero J, Tous-Fajardo J, Exel J, Sampaio J (2018) Entropy measures detect increased movement variability in resistance training when elite rugby players use the ball. J Sci Med Sport 21(12):1286–1292
Moreno FJ, Ordoño E (2009) Aprendizaje motor y síndrome general de adaptación. Motricidad European J Hum Mov 22:1–19
Moritani T, deVries HA (1979) Neural factors versus hypertrophy in the time course of muscle strength gain. Am J Phys Med 58(3):115–130
Narici MV, Roi GS, Landoni L, Minetti AE, Cerretelli P (1989) Changes in force, cross-sectional area and neural activation during strength training and detraining of the human quadriceps. Eur J Appl Physiol Occup Physiol 59(4):310–319
Norrbrand L, Fluckey JD, Pozzo M, Tesch PA (2008) Resistance training using eccentric overload induces early adaptations in skeletal muscle size. Eur J Appl Physiol 102(3):271–281

Norrbrand L, Pozzo M, Tesch PA (2010) Flywheel resistance training calls for greater eccentric muscle activation than weight training. Eur J Appl Physiol 110(5):997–1005

Nuñez FJS, Saez de Villarreal E (2017) Does flywheel paradigm training improve muscle volume and force? A meta-analysis. J Strength Cond Res 31(11):3177–3186

Nuñez FJ, Suarez-Arrones LJ, Cater P, Mendez-Villanueva A (2017) The high-pull exercise: a comparison between a versapulley flywheel device and the free weight. Int J Sports Physiol Perform 12(4):527–532

Nuñez FJ, Santalla A, Carrasquila I, Antonio Asian J, Ignacio Reina J, Jesus Suarez-Arrones L (2018) The effects of unilateral and bilateral eccentric overload training on hypertrophy, muscle power and COD performance, and its determinants, in team sport players. Plos One 13(3)

Nuñez FJ, Hoyo M, López AM, Sañudo B, Otero-Esquina C, Sanchez H et al (2019) Eccentric-concentric ratio: a key factor for defining strength training in soccer. Int J Sports Med 40(12):796–802

Nuñez FJ, Galiano C, Muñoz-López A, Floria P (2020) Is possible an eccentric overload in a rotary inertia device? Comparison of force profile in a cylinder-shaped and a cone-shaped axis devices. J Sports Sci 1–5

Onambele GL, Maganaris CN, Mian OS, Tam E, Rejc E, McEwan IM et al (2008) Neuromuscular and balance responses to flywheel inertial versus weight training in older persons. J Biomech 41(15):3133–3138

Romero-Rodriguez D, Gual G, Tesch PA (2011) Efficacy of an inertial resistance training paradigm in the treatment of patellar tendinopathy in athletes: a case-series study. Phys Ther Sport 12(1):43–48

Sabido R, Hernandez-Davo JL, Botella J, Navarro A, Tous-Fajardo J (2017) Effects of adding a weekly eccentric-overload training session on strength and athletic performance in team-handball players. Eur J Sport Sci 17(5):530–538

Sabido R, Hernandez-Davo JL, Pereyra-Gerber GT (2018) Influence of different inertial loads on basic training variables during the flywheel squat exercise. Int J Sports Physiol Perform 13(4): 482–489

Sanz-Lopez F, Berzosa Sanchez C, Hita-Contreras F, Cruz-Diaz D, Martinez-Amat A (2016) Ultrasound changes in achilles tendon and gastrocnemius medialis muscle on squat eccentric overload and running performance. J Strength Cond Res 30(7):2010–2018

Schuenke MD, Herman JR, Gliders RM, Hagerman FC, Hikida RS, Rana SR et al (2012) Early-phase muscular adaptations in response to slow-speed versus traditional resistance-training regimens. Eur J Appl Physiol 112(10):3585–3595

Seger JY, Arvidsson B, Thorstensson A (1998) Specific effects of eccentric and concentric training on muscle strength and morphology in humans. Eur J Appl Physiol Occup Physiol 79(1):49–57

Seynnes OR, de Boer M, Narici MV (2007) Early skeletal muscle hypertrophy and architectural changes in response to high-intensity resistance training. J Appl Physiol (1985) 102(1):368–373

Staron RS, Karapondo DL, Kraemer WJ, Fry AC, Gordon SE, Falkel JE et al (1994) Skeletal muscle adaptations during early phase of heavy-resistance training in men and women. J Appl Physiol (1985) 76(3):1247–1255

Stergiou N, Decker LM (2011) Human movement variability, nonlinear dynamics, and pathology: is there a connection? Hum Mov Sci 30(5):869–888

Street SF (1983) Lateral transmission of tension in frog myofibers: a myofibrillar network and transverse cytoskeletal connections are possible transmitters. J Cell Physiol 114(3):346–364

Tesch PA, Ekberg A, Lindquist DM, Trieschmann JT (2004) Muscle hypertrophy following 5-week resistance training using a non-gravity-dependent exercise system. Acta Physiol Scand 180(1):89–98

Tous-Fajardo J, Maldonado RA, Quintana JM, Pozzo M, Tesch PA (2006) The flywheel leg-curl machine: offering eccentric overload for hamstring development. Int J Sports Physiol Perform 1(3):293–298

Tous-Fajardo J, Gonzalo-Skok O, Arjol-Serrano JL, Tesch P (2016) Enhancing change-of-direction speed in soccer players by functional inertial eccentric overload and vibration training. Int J Sports Physiol Perform 11(1):66–73

Westing SH, Seger JY (1989) Eccentric and concentric torque-velocity characteristics, torque output comparisons, and gravity effect torque corrections for the quadriceps and hamstring muscles in females. Int J Sports Med 10(3):175–180

Westing SH, Cresswell AG, Thorstensson A (1991) Muscle activation during maximal voluntary eccentric and concentric knee extension. Eur J Appl Physiol Occup Physiol 62(2):104–108

Francisco Javier Nuñez is a professor of football in the Sports Faculty at Universidad Pablo de Olavide, Seville (Spain). He is the editor at PlosOne and, at the research level, focuses on the demands of competition in football and other team sports. He has been CEO of training companies such as VF Sport and Nuon Sport. As practitioner, he is an specialist in strength training that has worked in strength and conditioning departments of professional football teams such: Valencia CF, Villarreal CF, Real Betis, Real Zaragoza, and Racing de Santander.

Variable Resistance Training Methods

Fernando Hernández-Abad

Abstract Variable resistance training devices have become a widespread tool in strength programs. One of the advantages of using this type of training device is the increase or decrease of the external resistance throughout the range of movement. In addition to the good results obtained by the use of elastic bands and chains, which have become widely popular due to their positive adaptations in the expression of strength, versatility, and ease of use, the use of conical pulleys has been added in recent years. In this chapter, we will describe how to approach training with variable resistance training devices, although they are a very ecological resource within the training process and have a wider range of possibilities than the most commonly used. We will take a journey from a traditional vision to offer a more contemporary vision, based on our personal experience, both in high performance and in the rehabilitation and prevention of sports injuries. In addition, there will be a brief introduction and contextualization of the concepts of dynamic rotational stability and vector diversification, proposals to optimize the use of training resources with variable resistance.

Keywords Variable resistance training · Elastic band-resisted training · Dynamic rotational stability training · Vector diversification training · Conic pulley · Force-vector training

1 Introduction

"The development of muscle strength is based on a combination of morphological and neural factors, including the cross-sectional area of the muscle and its architecture, muscle-tendon stiffness, recruitment of motor units, frequency of activation,

F. Hernández-Abad (✉)
Faculty of Health Sciences, Universidad Europea de Canarias, La Orotava, Spain
e-mail: fernandojesus.hernandez@universidadeuropea.es

F. Hernández-Abad
FSI, Sport Research Lab, Granada, Spain

A. Muñoz-López et al. (eds.), *Resistance Training Methods*, Lecture Notes in Bioengineering, https://doi.org/10.1007/978-3-030-81989-7_8

synchronisation of motor units and neuromuscular inhibition" (Suchomel et al. 2018). When designing intervention strategies, there are numerous variables that we use to configure the training load. We have variables such as intensity and volume, which depend on each other and in turn on factors such as speed of contraction, psychological aspects, recovery between sets, order of training contents or frequency of training sessions (Benedict 1999). However, improving an athlete's motor expression, especially in a team sport, with non-reproducible motor situations dependent on opponents, team-mates and various contextual factors, requires other types of variables that help us to equalize the stimuli according to the motor situations that the athlete will face. In this situation, a contextualization of the motor output in absolute terms becomes too reductionist to be efficient. For this reason, and if we understand that "in strength lies the genesis of motricity and therefore in its optimisation lies that of the movements" (Tous 2017), relative values acquire another dimension within the design of intervention strategies and the management of loads in strength training. This forces us to adopt non-conventional variables such as fluctuation of the stimulus; variability of the external load or the biomechanical model of execution; variation in the interserial or repetition range of movement; hierarchical order of activation; and a good number of factors inherent to movement that allow us to adapt the stimuli to athletes' needs. Therefore, both qualitative and quantitative manipulation of any of these factors will affect the others directly or indirectly. The main aim is that such manipulation produces positive adaptations in the body, and by extension in motor performance.

The training devices that can be used to manipulate these non-conventional variables during strength training are those whose resistance is not constant throughout the range of movement. Variable resistance training devices have a resistance that has to be overcome throughout the range of movement (Fleck and Kraemer 2014; McMaster et al. 2009) or per unit of time throughout a repetition. This corresponds to the "criterion of the specificity of the force to the type of movement" (García Manso 1999), making it difficult to think of any sport situation as existing in a stable environment, where the applications of force are constant.

In this chapter we will consider how to approach training with variable resistance devices, which have a wider spectrum of possibilities than the more commonly used devices. We will take a look from a traditional point of view and then offer a more contemporary vision, based on our personal experience, both in high performance and in the rehabilitation and prevention of sports injuries.

2 From Theory

The increase or decrease in resistance throughout the range of movement is one of the advantages of variable resistance training devices. They allow the force–range of movement (ROM) curve to be modified for a proposed exercise. Broadly speaking we can describe 3 main types of force–ROM curve: ascending, descending and bell-shaped (Fleck and Kraemer 2014; Kulig et al. 1984). With the use of

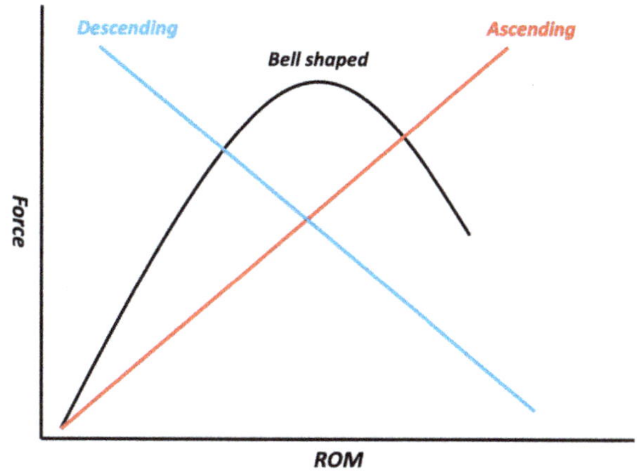

Fig. 1 Curves that commonly represent the expression of force along the range of motion

elastic bands and chains, either exclusively or combined with traditional training methods with free weights and variable resistance tools, we are dealing with an ascending curve (Wallace et al. 2018) (Fig. 1).

The most popular materials for obtaining this type of stimulus are elastic bands, chains or machines with asymmetrical cams (McMaster et al. 2009; Wallace et al. 2018). Although the literature on the effects of training with cam machines is limited, there are numerous studies that demonstrate the positive effects of the use of elastic bands (García-López et al. 2016; Soria-Gila et al. 2015; Kashiani et al. 2020) and chains (Soria-Gila et al. 2015; Kashiani et al. 2020), compared to training using constant resistance (Joy et al. 2016; Rivière et al. 2017; Rhea et al. 2009), and although the use of chains and elastic bands seems to be equally effective (Wallace et al. 2018; Jones 2014), the most popular and widely used are elastic bands. They are easily portable, can be easily attached to various training devices with a high degree of adaptability to the athlete's context, and are affordable. With regard to the parameters of force expression where the best adaptations are obtained by the combined use of free weights and elastic bands, there are studies that refer to significant improvements in the average and peak speed of load displacement and in the average and peak power with respect to constant resistance with free weights (Heelas et al. 2019), as well as in the expression of the force-time slope (Stevenson et al. 2010) and speed in eccentric conditions, which would contribute to improving the efficiency of the stretching-shortening cycle (Wallace et al. 2018). With regard to the training of young athletes, training with elastic bands has been shown to cause similar adaptations in variables such as power, improvements in 1RM and speed of throwing with different positions of the upper limb (Aloui et al. 2019). When athletes are moving loads at high speed, during the concentric phase they spend a good deal of time slowing down the resistance. Variable resistance makes it easier to reduce this deceleration phase (Wallace et al. 2018) in favour of the propulsive phase, encouraging the athlete to apply force for longer during the range of movement. This becomes especially interesting when there is evidence of possible adaptations in reflex-type responses, even improving electromechanical delay (Smith et al. 2019); this is of particular interest, and we consider it to be a field of study that could bring significant advantages in explosive sports, in a type of adaptation that is difficult to improve. However, field experience suggests that the selection of both tools and content must always be subject to their effectiveness in achieving the desired objectives. It has been found that the introduction of this type of variable resistance in certain technical situations can be counterproductive, depending on the objectives (Nijem et al. 2016).

Variable resistance, especially the use of elastic bands, can be a very suitable resource in the early stages of an athlete's return to play after injury. This is especially the case when it comes to avoiding overloading in certain ranges of movement where there is no mechanical advantage and braking the inertia of the movement with an extra load can impose an added risk. Thus, in exercises such as leg presses, squats, or bench presses, the first stages of the concentric phase would not be subjected to as much load as the last ones. This is also true for the eccentric phase, where the first stages would be subjected to more load and the later ones to less, facilitating the eccentric-concentric transition in this type of exercise. The principle of a progressive increase in load is then achieved, not only in the training process or in the return to play process, but also according to the force moment of each phase of the range of movement, which is very important if we want to minimise risks when executing certain exercises. It also seems to be a suitable training tool for improving neuromuscular activation in these processes and when working with particularly sensitive populations such as children or the elderly (Melchiorri and Rainoldi 2011). With respect to this last group, a study has recently been published showing very positive adaptations in a wide variety of functional tests, after three months of training following knee replacement surgery (Liao et al. 2020).

Training using elastic bands or conical pulleys allows movements to be executed bilaterally and unilaterally (Núñez et al. 2018), which is very interesting in terms of rehabilitation processes (Lorenz 2014), optimisation of performance (Núñez et al. 2018; Gonzalo-Skok et al. 2017), and in intervention strategies to compensate for asymmetries between limbs (Núñez et al. 2018).

3 From Practice

During task design it is important to identify risk factors that can generate cognitive biases in the decision-making process. This has led us to think that the structural configuration of training devices conditions our ability to design tasks with a specific objective, regardless of the area of intervention. Training devices with variable resistance, especially elastic and conical-pulley inertial devices, make this process much easier, since they allow us to design stimuli by adapting the force vector and the plane of movement according to the motor expression requirements of the sport (Gonzalo-Skok et al. 2017). From our point of view, these types of

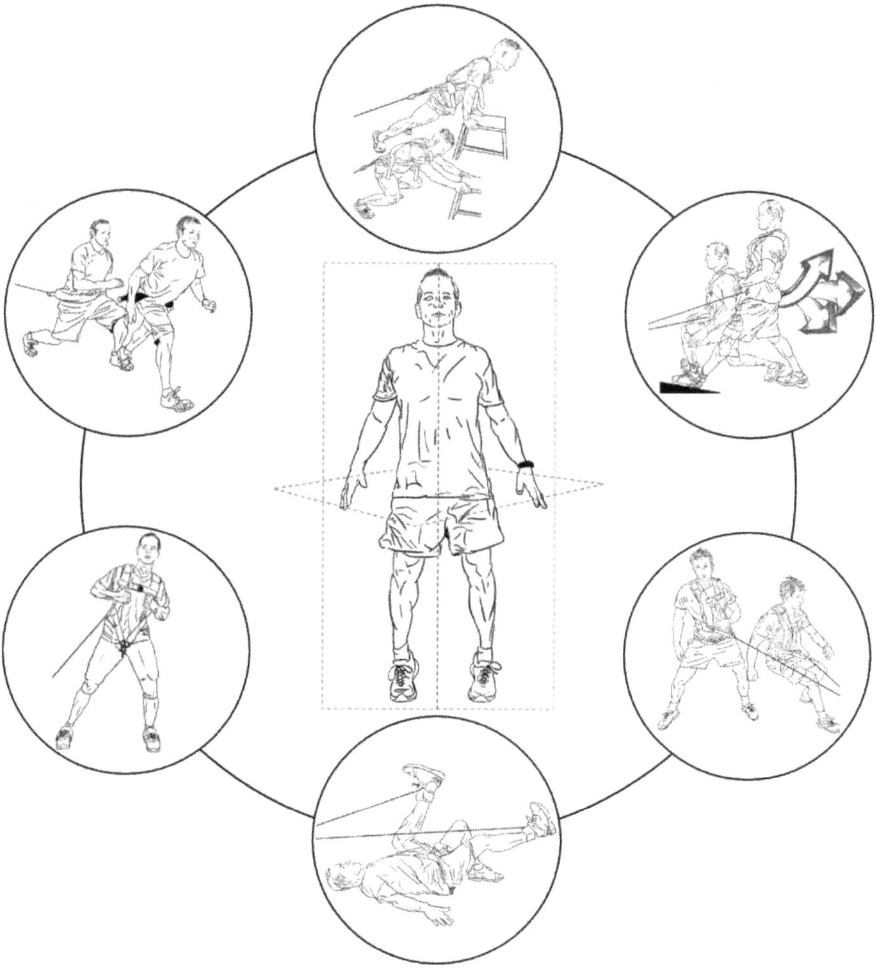

Fig. 2 Some planes and axes of load that can be worked on using variable resistance training

devices enable the design of tasks that are very well adapted to the type of expression of force desired for our athletes, not only with a high degree of dynamic correspondence, but also focusing on any anatomical deficiencies so that they are correctly integrated into the movement (Núñez et al. 2018) (Fig. 2).

Another very important concept that contributes to the optimization of motor output, and which can be efficiently addressed using variable resistance, is dynamic rotational stability. Dynamic rotational stability is the correct response to actions with a rotational component, especially in a monopodal stand, in one or several anatomical areas, in a synergistic way and in the appropriate hierarchical order (Hernandez-Abad 2021). Two joint complexes have generated special interest because of their biomechanical characteristics and the importance of their contribution to mechanical advantage during movement. These are the joint complexes of the hip and the shoulder, also including the myofascial complex that gives stability to the spine, the passive and active structures that shape the plantar arch and their relationship with those previously mentioned. Variable resistance is very beneficial for training in these areas, especially in terms of optimising the training of the sensorimotor system and thus improving neuromuscular responses. This is because of the possibility of working with different planes and non-vertical vectors of force that, during strength training, tend to limit this type of sensorimotor experience because of the biomechanical model used. When using these types of planes and force vectors, it is easier to adopt open biomechanical models (Fig. 3b) that do not need to be executed correctly to avoid triggering the risk of injury, especially when using relatively high resistance accompanied by a high degree of fatigue. This is a major advantage when it comes to adjusting strength training to the structural, neural and metabolic profiles required by athletes.

Variable resistance, in terms of both sports performance and injury prevention, and return to play, offers various advantages that are not always considered:

(1) The possibility of influencing the inclination of the body during the execution of an exercise not only allows net improvement of the propulsive phase of the movement and average speed; it also produces structural, neural and metabolic responses in anatomical areas that optimise performance. In the proposed exercise, engaging the intrinsic and extrinsic muscles of the foot requires greater support with the first metatarsal, providing stability of the plantar arch during execution (Fig. 4a).

(2) Variable resistance can also be used to generate variable perturbation along the ROM, which makes the reflex neuromuscular response of the stabilizing muscles very variable throughout the range (Fig. 4b).

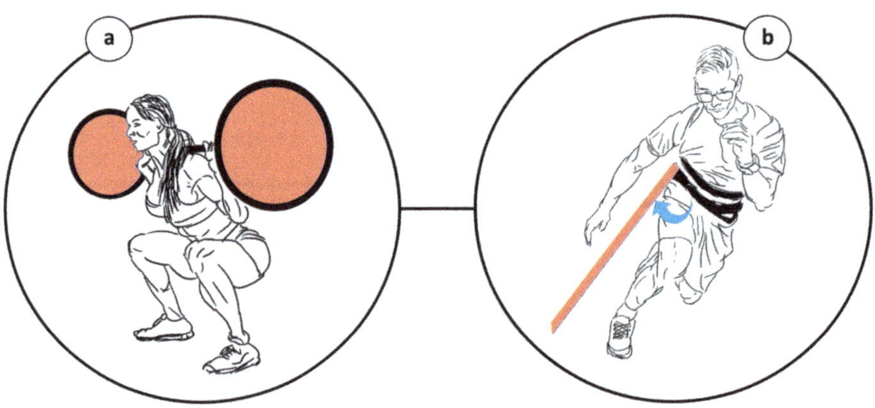

Fig. 3 **a** Closed biomechanical model. When moving high loads, departing from the proposed biomechanical model can lead to a high risk of injury. **b** Open biomechanical model. The proposed biomechanical model is susceptible to exposure to variability without significantly increasing the risk of injury

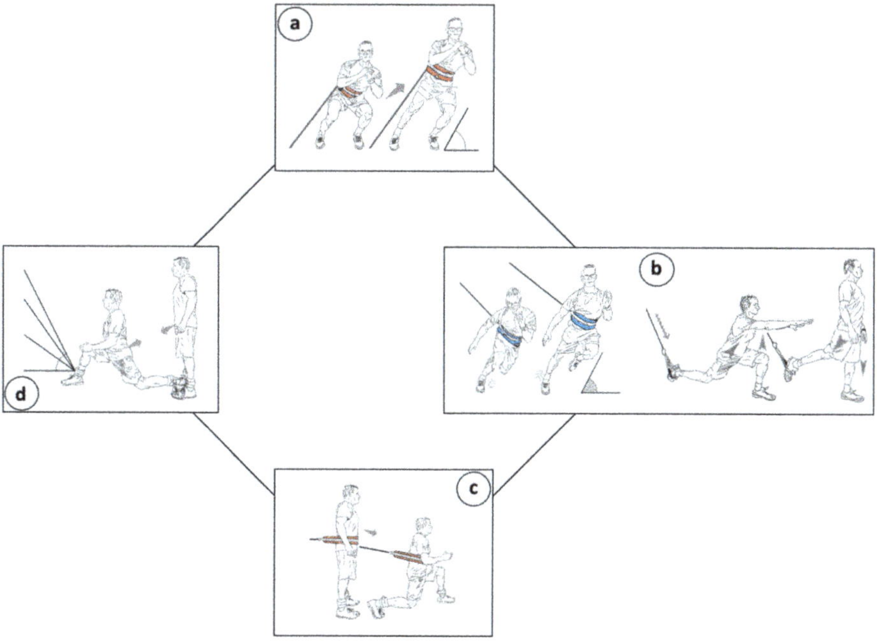

Fig. 4 Some uses of variable resistance in task design, in this example using elastic bands

(3) The management of variable resistance during the execution of an exercise such as that proposed not only generates adaptations during the propulsive phase of the movement, but also dissipates joint stress during landing of the leading leg, which allows reconsideration of the point in the readaptation phase at which this type of movement can be introduced (Fig. 4c).
(4) Using this type of task, not only can we train certain muscles in an eccentric regime with dynamic correspondence, it is also possible to pre-activate appropriate muscles before contact with or support from the ground. In the example proposed, the ischiosural muscles, acting synergistically with the anterior cruciate ligament, provide better knee stability at the start of weight-bearing (Fig. 4d).

Another concept contributed by our working group is that of vector diversification. This involves diversifying a primary force vector into two or more anatomical areas using a small pulley (Fig. 5a). Vector diversification can be considered as a variable when managing the training load with this type of resistance. In our opinion, it represents a step forward in neuromuscular training, since it can facilitate better coactivation of the muscles situated between the anatomical areas to which the primary vector has been diversified. This also promotes the participation of more anatomical areas during the process of pre-activation.

When one of the diversification points is fixed and the other is mobile, a simple hoist is generated, dividing the primary resistance, so that the expression of the

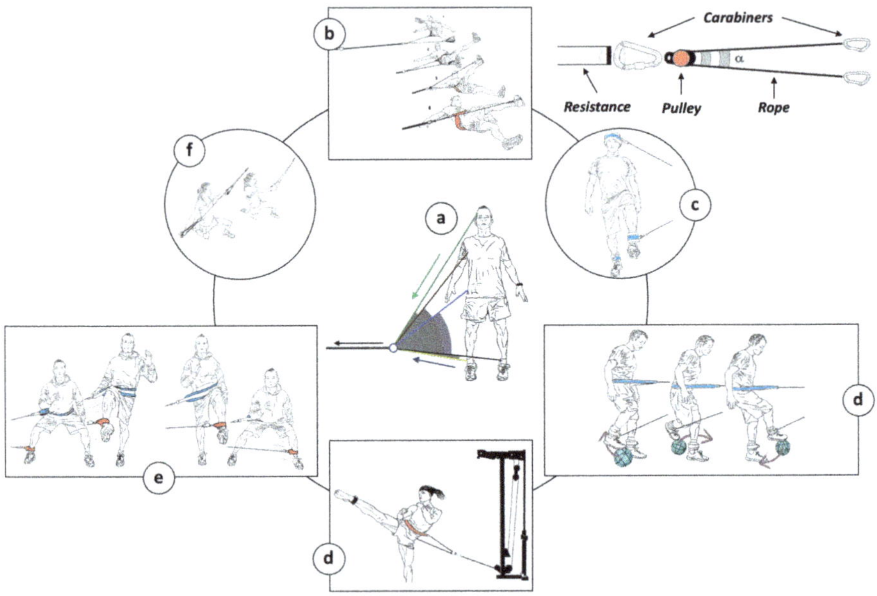

Fig. 5 Examples of vector diversification

force over time is smoothed out, allowing correct technical execution to be maintained at relatively high speeds (Fig. 5d). The distance between the diversified areas determines an angle (α in Fig. 5) that will make the primary resistance divide more or less; the greater the separation between diversified areas, the less the division. Another advantage is that as there is a mobile and a fixed point of attachment, the deformation of the elastic is lower in the same range of movement, and when using weight-guided machines the range of movement will not be limited by the length of the cable of that machine. In this case (Fig. 5d) although we were originally working with a constant resistance machine, the fact that the alpha angle changes along the range of movement converts it into a variable resistance (with limited magnitude), changing the type of stimulus for which the machine is designed. If the angle decreases along the ROM the resistance will be downward (corresponding to the reality of a technical motion such as a kick). If, on the other hand, the angle increases along the ROM the resistance will be upward.

Designing tasks that challenge athletes' neuromuscular responsiveness is relatively easy by diversifying vectors with variable resistance devices. For example, consider a player who executes an open or crossed exit, where one leg applies force in a closed kinetic chain and the other leg does so in an open kinetic chain. By diversifying the vectors, we can add load without affecting the motor action of the leg in the open kinetic chain, while at the same time increasing the rotational moment of force on the leg working in a closed kinetic chain (Fig. 5e).

4 Filling Gaps

During strength training, especially in team sports, the focus has been largely on absolute parameters; net values of a biomechanical pattern with low mechanical and dynamic correspondence to the technical sport movement (Fig. 3a). Variable resistance training equipment is a very useful resource for adapting these net values to the specific characteristics of the sport and its strength requirements. To this end and based on our field experience, it is important to take into account a number of considerations when designing tasks.

4.1 *Determining What Type of Variable Resistance Training Device to Use*

When deciding which variable resistance device to use, it is important not to solely consider the adaptations in terms of peak speed, power, or length of the propulsive phase. It is also important to take into account other factors, such as the capacity to adapt the training device to the specific planes and axes of load of the sport; the dynamic correspondence; the capacity to generate perturbation during the execution of the exercise; and the potential to provide fluctuation in the external resistance, all without increasing the risk of injury during the execution, independently of the expression of force being worked on (Fig. 6).

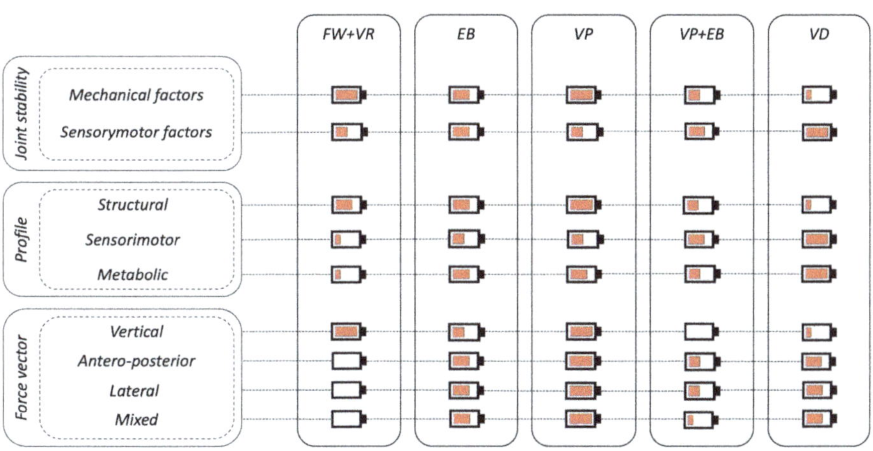

Fig. 6 Variable resistance and adaptability to the target in task design. FW + RV (free weight + variable resistance); EB (elastic bands); VP (versa pulley); VP + EB (versa pulley + variable resistance); VD (vector diversification)

4.2 Considering the Planes and Force Vectors When Designing the Task

With regard to decision making about the planes and vectors of force application during the design of tasks using variable resistance devices, experience tells us that although it is true that when we are searching for certain types of structural adaptations there are non-specific force vectors that can be beneficial, such decision making should be largely influenced by the degree of dynamic correspondence with the sport activity in question. This is important with regard to the integration of anatomical areas in time and form.

4.3 Managing Non-conventional Variables During the Execution of the Task

In the day-to-day life of a physical trainer it is very difficult to control every training variable, and experience tells us that it is also not necessary to have complete control over each variable for a training to be well structured. This chapter has already mentioned some factors that are particularly difficult to measure, but this is not exclusive to strength training sessions; it is also true of technical-tactical training and competitive activity, especially in team sports. From our point of view, with regard to the design of tasks, we should consider how we can intervene in these variables (Fig. 7). These variables may correspond to the competitive activity itself (Fig. 7):

(1) Range of movement within the series itself.
(2) Fluctuation of the external resistance.
(3) Perturbation during repetition (on the training device or on the athlete him/herself).

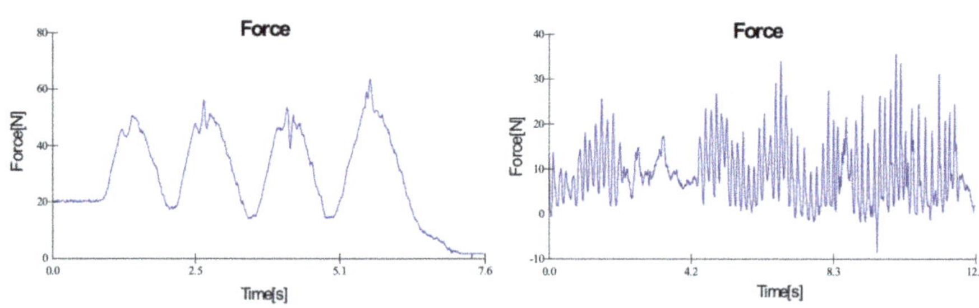

Fig. 7 Example of the dynamometric signal for the same exercise, bilateral standing rowing with elastic + suspension training device, applying a disturbance to the elastic itself during exercise execution (right), and without applying it (left)

(4) Introduction of balls or sports equipment during the execution of the exercise (Moras et al. 2018; Fernández-Valdés et al. 2020).
(5) Dynamic correspondence within the series itself.

4.4 Vector Diversification

When training with variable resistance, conducting generic families of movement in a closed kinetic chain, both elastics and conical pulleys are a very good option. However, when we are working with angular movements of a free extremity, the slope of the force–ROM curve tends to be exponential due to the speed of execution that this type of movement often requires. This, besides carrying a certain risk, doesn't correspond with the characteristics of the resistance to be overcome during the development of competitive motor action, especially for elastics, where at the end of the movement there will be more resistance to overcome. This is where we should consider the hoists used in the diversification of vectors, which can be a very useful tool for smoothing out the force–ROM gradient in this type of exercise, achieving the desired expression of force while keeping athletes within a safe spectrum, especially for those joint complexes that need coactivation of multiple muscles for optimisation of the output.

5 Take-Home Messages and Practical Resources

1. Variable resistance is a very ecological stimulus for athletes, which can not only help to achieve positive adaptations in the propulsive phase of movement, but will also help to dissipate stress in certain anatomical areas, generate selective preactivations, and produce specific structural, neural and metabolic adaptations to the conditional profile of the sport discipline.
2. When working with variable resistances, it is important to consider axes and planes of load in order to be able to orientate the stimulus appropriately, and to take advantage of open biomechanical patterns, which will adapt the mechanical and dynamic correspondence of the proposed task to the sporting action in which performance is to be improved.
3. This correct choice of plane and vector of force and the manipulation of non-conventional variables such as the fluctuation of external resistance; perturbations, variability of the range of movement; and bilateral and unilateral bearings, will promote a better link between the locomotor system, the nervous system and the metabolic system, giving greater specificity to intervention strategies.

4. Dynamic rotational stability is a condition inherent to human motor skills, which is why we believe it should be considered in the strength training process, with vector diversification being a very useful variable during its optimisation, helping us to coactivate more anatomical areas per unit of time.

References

Aloui G, Hermassi S, Hammami M, Gaamouri N, Bouhafs EG, Comfort P et al (2019) Effects of an 8-week in-season upper limb elastic band training programme on the peak power, strength, and throwing velocity of junior handball players. Sportverletzung-Sportschaden 33(3):133–141
Benedict T (1999) Manipulating resistance training program variables to optimize maximum strength in men: a review. J Strength Cond Res 13(3):289–304
Fernández-Valdés B, Sampaio J, Exel J, González J, Tous-Fajardo J, Jones B et al (2020) The influence of functional flywheel resistance training on movement variability and movement velocity in elite rugby players. Front Psychol 11:1205
Fleck SJ, Kraemer W (2014) Designing resistance training programs, 4°. Champaign, Il: Human Kinetics
García Manso JM (1999) La fuerza. Fundamentación, valoración y entrenamiento. Gymnos, Madrid
García-López D, Hernández-Sánchez S, Martín E, Marín PJ, Zarzosa F, Herrero AJ (2016) Free-weight augmentation with elastic bands improves bench press kinematics in professional rugby players. J Strength Cond Res 30(9):2493–2499
Gonzalo-Skok O, Tous-Fajardo J, Valero-Campo C, Berzosa C, Bataller AV, Arjol-Serrano JL et al (2017) Eccentric-overload training in team-sport functional performance: constant bilateral vertical versus variable unilateral multidirectional movements. Int J Sports Physiol Perform 12(7):951–958
Heelas T, Theis N, Hughes JD (2019) Muscle activation patterns during variable resistance deadlift training with and without elastic bands. J Strength Cond Res 17:1–6
Hernandez-Abad F (2021) Estabilidad dinámica rotacional. Más allá del entrenamiento en suspensión. Mastercede, Barcelona
Jones M (2014) Effect of compensatory acceleration training in combination with accommodating resistance on upper body strength in collegiate athletes. Open Access J Sport Med 5:183–189
Joy JM, Lowery RP, Oliveira De Souza E, Wilson JM (2016) Elastic bands as a component of periodized resistance training. J Strength Cond Res 30(8):2100–2106
Kashiani AB, Soh KG, Soh KL, Ong SL, Kittichottipanich B (2020) Comparison between two methods of variable resistance training on body compoosition, muscular strength and functional capacity among untrained males. Malays J Mov Heal Exerc 9(1):45–66
Kulig K, Andrews J, Hay J (1984) Human strength curves. Exerc Sport Sci Rev 12(1):417–466
Liao C-D, Tsauo J-Y, Chiu Y-S, Ku J-W, Huang S-W, Liou T-H (2020) Effects of elastic resistance exercise after total knee replacement on muscle mass and physical function in elderly women with osteoarthritis. Am J Phys Med Rehabil 99(5):381–389
Lorenz D (2014) Clinical suggestion variable resistance training using elastic bands to enhance lower extremity corresponding author. Int J Sports Phys Ther 9(3):410–414
McMaster DT, Cronin J, McGuigan M (2009) Forms of variable resistance training. Strength Cond J 31(1):50–64
Melchiorri G, Rainoldi A (2011) Muscle fatigue induced by two different resistances: elastic tubing versus weight machines. J Electromyogr Kinesiol [internet] 21(6):954–959

Moras G, Fernández-Valdés B, Vázquez-Guerrero J, Tous-Fajardo J, Exel J, Sampaio J (2018) Entropy measures detect increased movement variability in resistance training when elite rugby players use the ball. J Sci Med Sport [internet] 21(12):1286–1292
Nijem RM, Coburn JW, Brown LE, Lynn SK, Ciccone AB (2016) Electromyographic and force plate analysis of the deadlift performed with and without chains. J Strength Cond Res 30 (5):1177–1182
Núñez FJ, Santalla A, Carrasquila I, Asian JA, Reina JI, Suarez-Arrones LJ (2018) The effects of unilateral and bilateral eccentric overload training on hypertrophy, muscle power and COD performance, and its determinants, in team sport players. PLoS ONE 13(3):1–13
Rhea MR, Kenn JG, Dermody BM (2009) Alterations in speed of squat movement and the use of accommodated resistance among college athletes training for power. J Strength Cond Res 23 (9):2645–2650
Rivière M, Louit L, Strokosch A, Seitz L (2017) Variable resistance training promotes greater strength and power adaptations than traditional resistance training in elite youth rugby league players. J Strength Cond Res 31(4):947–955
Smith CM, Housh TJ, Hill EC, Keller JL, Anders JPV, Johnson GO et al (2019) Variable resistance training versus traditional weight training on the reflex pathway following four weeks of leg press training. Somatosens Mot Res [internet] 36(3):223–229
Soria-Gila MA, Chirosa IJ, Bautista I, Baena S, Chirosa JL (2015) Effects of variable resistance training on maximal strength: a meta-analysis. J Strength Cond Res 29(11):3260–3270
Stevenson MW, Warpeha JM, Dietz CC, Giveans RM, Erdman AG (2010) Acute effects of elastic bands during the free-weight barbell back squat exercise on velocity, power, and force production. J Strength Cond Res 24(11):2944–2954
Suchomel TJ, Nimphius S, Bellon CR, Stone MH (2018) The importance of muscular strength: training considerations. Sport Med 48(4):765–785
Tous J (2017) Todo es fuerza. In: Seirul·lo Vargas F (ed) El entrenamiento en los deportes de equipo, 1°. Mastercede, Barcelona
Wallace BJ, Bergstrom HC, Butterfield TA (2018) Muscular bases and mechanisms of variable resistance training efficacy. Int J Sport Sci Coach 13(6):1177–1188

Fernando Hernández-Abad is a lecturer in Physical Activity and Sport Sciences at the European University of the Canary Islands (Spain), where he also is a headmaster of the Master's Degree in Sports Training and Return to Play. He is the director of the return to play area in Sanaya sport center. He is a member of the Athlete Care Unit at the San Juan De Dios Hospital. Within his training and sports rehabilitation record, he has contributed with "Dynamic Rotational Stability," especially in task design processes with a new variable "vector diversification." He has trained a large number of physical trainers, strength specialists, and physiotherapists in the main European football leagues. He is an external consultant for clubs and federations, an expert in strength training, integration of CORE work and prevention and rehabilitation of injuries, especially in football.

Monitoring Training and Testing

Velocity-Based Training for Monitoring Training Load and Assessing Training Effects

Fernando Pareja-Blanco and Irineu Loturco

Abstract Currently, velocity-based training (VBT) is one of the hot topics in sport science and among strength and conditioning coaches. However, its wide use has spread some misunderstandings of the fundamental concepts of this methodology. It should be highlighted that this is not a new training method, but rather, a new approach that enables more accurate, frequent, and objective control of resistance training intensity and volume. The VBT approach is no other thing than recording lifting velocity every repetition during resistance training. The quantification of actual repetition velocities achieved during resistance training sessions provides a more consistent and precise understanding of training effects, opening up the possibility to establish causal relationships between stimuli and response, which is one of the main and most important targets of research and practice in sport science. As such, VBT can be defined as a resistance training method that uses movement velocity to improve training process and enhance training effects, via a deeper understanding of the input signal (actual training load) and the output signal (changes in performance). Through this chapter we will see how VBT contributes to improve the resistance training methodology, as well as discuss its potential benefits, limitations, and practical implications.

Keywords Strength training · Movement velocity · Velocity loss · Training prescription · Dose-response

F. Pareja-Blanco (✉)
Physical Performance and Sports Research Center, Pablo de Olavide University, Seville, Spain
e-mail: fparbla@upo.es

I. Loturco
NAR—Nucleus of High Performance in Sport, São Paulo, Brazil

A. Muñoz-López et al. (eds.), *Resistance Training Methods*, Lecture Notes in Bioengineering, https://doi.org/10.1007/978-3-030-81989-7_9

1 Introduction

Velocity-based training (VBT) is one of the hot topics in scientific literature and among strength and conditioning coaches. However, its wide use has spread some misunderstandings of the fundamental concepts of this methodology. It should be highlighted that it is not a new training method, but a novel approach that allows accurate and objective control of resistance training (RT) intensity and volume as well as of training effects. Put simply, VBT consists of measuring and recording every repetition during RT. The quantification of repetition velocities during RT provides a more precise understanding of the resistance exercise stimuli and their possible effects (i.e., training adaptations). This fact opens up the possibility to establish causal relationships between stimulus and response, which is certainly one of the main targets of research in sport science. Specifically, VBT can be defined as an RT approach that uses movement velocity to optimize training process by means of a deeper information about the “input signal” (actual training load) and the “output signal” (actual changes in performance). Through this chapter, it addresses how VBT contributes to improve the RT methodology and provide some practical applications and examples of its use.

2 From Theory

2.1 Lifting at Maximal Velocity is the Essential Premise of Velocity-Based Training

As described in the Introduction section, VBT can be used whenever the repetition velocity is recorded. The essential premise of VBT is that the concentric phase of each repetition should always be performed at maximal voluntary velocity. This should be interpreted as an important aspect of VBT rather than a shortcoming, since the velocity at which the loads are lifted determines the training intensity and, consequently, the training effects. In this context, RT intensity is more than solely the magnitude of the load (%1RM) being lifted, as the velocity at which loads are actually moved affects the acute and chronic training responses (Pareja-Blanco et al. 2014; Gonzalez-Badillo et al. 2014). With regard to the influences of movement velocity on training intensity, the greater the velocity against a given load, the higher the applied force (Schilling et al. 2008). As a result, the greater the velocity against a given load, the lower the difference between the force value achieved with this “relative load” and the force applied against the maximum load that can be lifted in a specific exercise (i.e., lower strength deficit) (Loturco et al. 2021a). Concerning training effect, it has been shown that performing squat repetitions at maximal concentric velocity compared to “intentionally slower velocity” (i.e., half-velocity) led to greater gains in squat performance (one-repetition maximum “1RM” strength as well as the velocity developed against any load, from light to heavy) and vertical jump height (Fig. 1a) (Pareja-Blanco et al. 2014). Similar

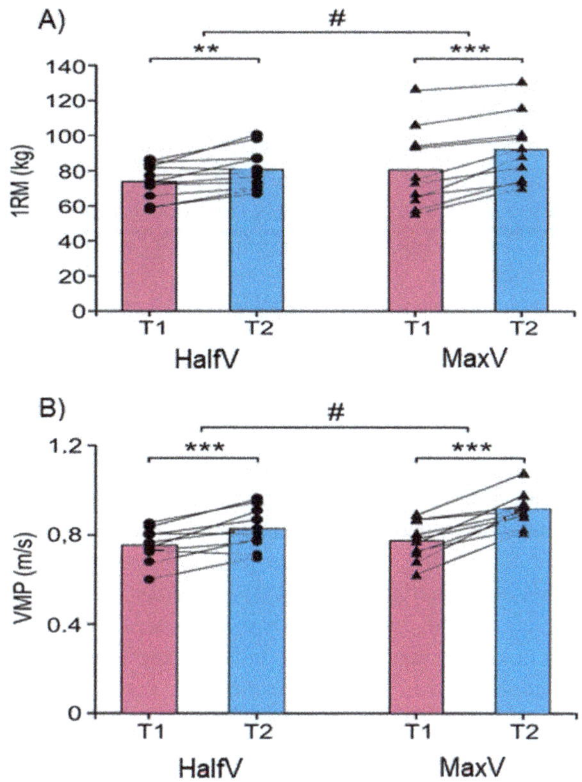

Fig. 1 Effect of the maximal voluntary velocity (MaxV) compared to half-maximal velocity (HalfV) bench-press training on: **a** one-repetition maximum (1RM) and **b** average velocity attained against absolute loads common to pre-(T1) and post-(T2) test in the bench-press progressive loading test. Signficant group × time interactions: # $P < 0.05$. Intra-group significant differences from pre- to post-training: **$P < 0.01$, ***$P < 0.001$. Adapted from Gonzalez-Badillo et al. (2014)

findings were observed for the bench-press exercise (Fig. 1b) (Gonzalez-Badillo et al. 2014). Moreover, when the load is lift at maximal intended velocity, this results in higher recruitment of motor units and level of neural activation (Desmedt and Godaux 1977; Behm and Sale 1993), despite spending shorter time under tension (Schilling et al. 2008). Therefore, for maximizing training adaptations when using VBT, it is necessary to lift the loads at the maximal intended velocity.

2.2 *Using Movement Velocity to Determine Loading Intensity*

A common concern of strength and conditioning coaches is how to objectively and precisely quantify and monitor the actual training load undertaken by trainees. Among the main RT variables (Suchomel et al. 2018; Kraemer and Ratamess 2004), exercise intensity is certainly one of the most critical, since this determines the direction and extent of the resulting adaptations (Bird et al. 2005; Fry 2004). The most common reference to prescribe RT intensity is the 1RM value (i.e., the maximum load that can be lifted successfully once in a given exercise) (Kraemer and Ratamess 2004; Buckner et al. 2017). Accordingly, RT intensity is expressed as percentages of 1RM (%1RM) (Scott et al. 2016). Likewise, exercise intensity may be prescribed by using an indirect measurement: the maximal number of repetitions that can be performed when using submaximal loads (MNR, e. g., 5RM, 10RM, 15RM). This method is based on the relationship already observed between the % 1RM and the MNR, which established a "RM continuum" (Brzycki 1993; Reynolds et al. 2006), allowing coaches to estimate the 1RM through the use of a derived equation. Despite their extensive utilization, these methods present important shortcomings, such as: (1) very time-consuming and laborious procedures (Loturco et al. 2016; Rontu et al. 2010); (2) high risk of injury when performed incorrectly (Chapman et al. 1998; Braith et al. 1993); (3) they can induce high levels of muscle damage, potentially hampering physical and technical performance on the following days (Niewiadomski et al. 2008). Furthermore, it is known that the MNR that can be performed with a given relative load differs between athletes (Richens and Cleather 2014; Sánchez-Moreno et al. 2021a) and, thus, a given MNR load may represent different %1RM for different subjects, leading to erroneous estimation of training intensity. In addition, the 1RM value can either fluctuate on a daily basis or change throughout the training program, which may produce that the prescribed load may not match the intended %1RM. Consequently, there is a need to implement faster, safer, and more practical methods to estimate both maximum (i.e. 1RM) and relative (i.e. %1RM) RT loads. The possibility of using a more precise and consistent "non-exhaustive approach" to determine the 1RM would allow coaches to frequently correct and adjust training loads even during habitual training sessions. In this sense, a pioneering study reported a very strong relationship ($R^2 = 0.98$) between %1RM and lifting velocity in the BP exercise (Fig. 2) (Gonzalez-Badillo and Sánchez-Medina 2010). Subsequently, several investigations confirmed that lifting velocity is a valid measure to objectively and frequently quantify training intensity in different exercises. From these data, multiple predictive equations have been developed, allowing practitioners to estimate loading magnitude (%1RM) under different training settings. This close relationship enables practitioners to rapidly and accurately determine the individual's current 1RM values and what percentages of 1RM are being used as soon as the first repetition of the set is performed with maximal voluntary velocity, without exposing trainees to the exhaustive and time-consuming traditional 1RM or MNR measurements (Gonzalez-Badillo and Sánchez-Medina 2010; Gonzalez-Badillo et al. 2011).

Given these relevant findings, it is not surprising the increased interest in applying the load-velocity relationship as a tool to define and monitor RT intensity. However, prescribing RT intensity using barbell velocity as a reference has a drawback: movement velocity is exercise-dependent and, consequently, a given velocity may represent different intensities for distinct exercises. Thus, in order to properly use the movement velocity as a tool to prescribe and estimate absolute and relative loads for a given exercise, the load–velocity relationship should be properly established for this exercise. As a consequence, the load-velocity relationship has been analyzed in a myriad of exercises, such as: full-squat (Sánchez-Medina et al. 2017) and different squat-variants (Loturco et al. 2016; Conceicao et al. 2016; Martinez-Cava et al. 2019), leg-press (Conceicao et al. 2016), deadlift (Benavides-Ubric et al. 2020), hip-thrust (Hoyo et al. 2019), bench-press (Gonzalez-Badillo and

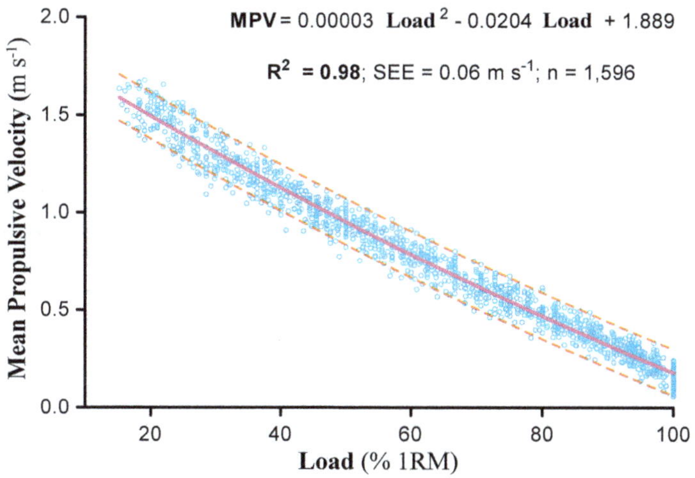

Fig. 2 Relationship between relative load (%1RM) and mean propulsive velocity in the bench-press exercise. Solid line shows the fitted curve to the data, and the dotted lines indicate the 95% confidence limits of the estimation. Adapted from Sánchez-Medina (2010)

Sánchez-Medina 2010), bench-pull (Sánchez-Medina et al. 2014), military-press (Balsalobre-Fernández et al. 2018), pull-up (Sánchez-Moreno et al. 2017), prone row (Loturco et al. 2021b), among others. Coaches and practitioners should regularly use these findings on a daily basis, adjusting the absolute load (kg) to match the repetition velocity associated with the %1RM that is intended for the RT session (Table 1).

Table 1 Mean propulsive velocity ($m \cdot s^{-1}$) attained against each percent 1RM in different resistance exercises

Load (%1RM)	Bench-press	Bench-pull	Pull-up	Full-squat	Deadlift
40	1.11	1.35		1.13	1.09
45	1.02	1.28		1.07	1.02
50	0.94	1.20		1.00	0.96
55	0.85	1.13		0.93	0.90
60	0.77	1.06		0.87	0.83
65	0.69	0.99	0.83	0.80	0.77
70	0.61	0.92	0.74	0.73	0.71
75	0.53	0.85	0.66	0.67	0.64
80	0.46	0.79	0.57	0.60	0.58
85	0.38	0.72	0.48	0.53	0.52
90	0.31	0.65	0.39	0.46	0.45
95	0.24	0.59	0.30	0.40	0.39
100	0.17	0.52	0.20	0.33	0.33

Values based on Sánchez-Medina et al. (2014); Sánchez-Moreno et al. (2017); Pareja-Blanco et al. (2020d); Benavides-Ubric et al. (2020)

2.3 *Using Movement Velocity as a Measure of Level of Effort Within the Set*

Another important training variable that should be considered when designing RT programs is the training volume (Spiering et al. 2008; Schoenfeld et al. 2017). Regarding the related-training volume parameters, the level of effort—defined as the actual number of repetitions performed within the set in relation to the maximum number of repetitions (MNR) that can be completed (Sánchez-Medina and Gonzalez-Badillo 2011)—has been claimed as a critical factor in determining adaptations to strength training (Pareja-Blanco et al. 2020a, 2017a). Accordingly, the degree of fatigue and, consequently, the training adaptations, will be substantially different when performing, for example, four out of eight repetitions with a given load [4(8)] compared to completing all possible repetitions [8(8)]. In this regard, the level of effort has traditionally been determined by fixing beforehand a specific number of repetitions to complete in each exercise set when using a submaximal load (i.e., %1RM) (Sánchez-Medina and Gonzalez-Badillo 2011; Pareja-Blanco et al. 2020b). However, the MNR that can be completed with a given %1RM shows large inter-individual variability (coefficient of variation [CV]: 8.6–33.1%) (Sánchez-Moreno et al. 2021a; Gonzalez-Badillo et al. 2017). This fact may lead to different levels of effort among athletes performing the same number of repetitions per set even when using a similar %1RM, because the number of repetitions that remain undone (i.e., repetitions left in reserve) may considerably differ between them (Moran-Navarro et al. 2019; Rodriguez-Rosell et al. 2020a). These considerations suggest that it is necessary to find better ways to objectively monitor the actual level of effort experienced by athletes during RT sets. Accordingly, rather than completing a fixed and predetermined number of repetitions in each set, each training set should be stopped as soon as a certain level of performance impairment is detected, which will depend on the specific target being pursued (Pareja-Blanco et al. 2020a, 2017a, 2020c). When performing each repetition at maximal voluntary effort during a training set, the force applied and, hence, movement velocity and power output will progressively decline, as a consequence of the development of fatigue (Fig. 3) (Ortega-Becerra et al. 2021; Izquierdo et al. 2006). In this regard, the velocity loss (VL) incurred within the set, which is calculated as the relative difference between the fastest (usually the first) and the last repetition performed, has been proposed as an objective, practical, and non-invasive indicator of neuromuscular fatigue during RT (Sánchez-Medina and Gonzalez-Badillo 2011). Indeed, high relationships (R^2 = 0.83–0.94) have been observed between the repetition VL and different mechanical and metabolic measures of fatigue (Sánchez-Medina and Gonzalez-Badillo 2011). Likewise, it has been observed strong relationships (R^2 = 0.92–0.97) between the VL incurred within the set and the percentage of repetitions completed with regard to MNR (%Rep, i.e. proximity to muscle failure) for relative loads ranging between 50–85% 1RM in the bench-press and full-squat exercises (Sánchez-Moreno et al. 2021a; Gonzalez-Badillo et al. 2017; Rodriguez-Rosell et al. 2020a). Moreover, the %Rep completed to a given magnitude of VL presented a low inter-individual variability (CV: 2.5–12.1%) and a high reliability (intra-individual CV: 2.1–6.6%) (Gonzalez-Badillo et al. 2017). Therefore, the % Rep for a given magnitude of VL is very similar for all individuals, regardless the MNR that could be completed. These findings allow practitioners to estimate with a high precision the %Rep that has already been performed and/or as a result how many repetitions are left in reserve, as soon as a given VL magnitude is attained in the exercise set. However, the pattern of repetition velocity decline seems to be exercise-specific (Rodriguez-Rosell et al. 2020a). According to previous research, when athletes have completed approximately 50% of the MNR, the VL magnitude is about 20–25% in the bench-press, full-squat, and pull-up exercises (Rodriguez-Rosell et al. 2020a). Figure 5 depicts the %Rep corresponding for different magnitudes of VL in bench-press, full-squat, and pull-up exercises. Thus, instead of previously defining a fixed number of repetitions to perform against a given load, practitioners may use different VL thresholds to obtain more accurate information about the actual degree of fatigue incurred during the set as well as to prescribe more effective and tailored training sessions (Table 2).

2.4 *Effort Index as a New Method to Quantify Training Load During Resistance Training*

As it has been described in the two previous epigraphs, the close relationship between bar velocity and relative intensity in different exercises, and between the magnitude of VL within the set and the percentage of performed repetitions (%Rep) against a given load confirm that VBT is a valid and accurate approach for

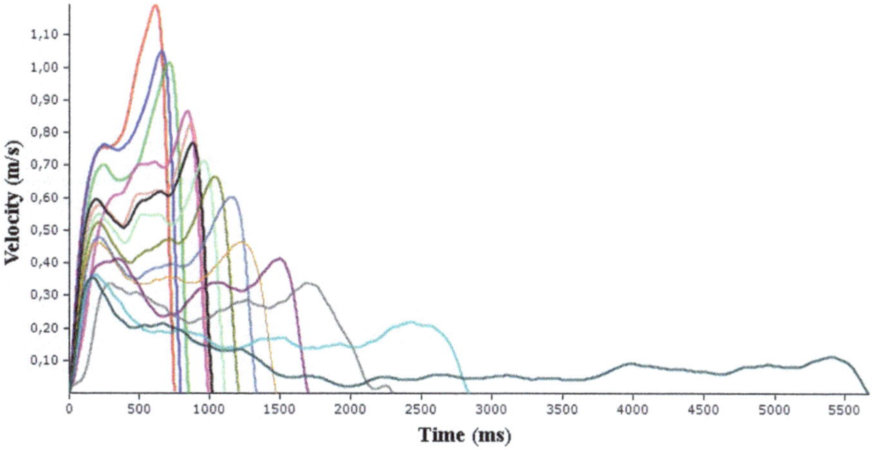

Fig. 3 Evolution of lifting velocity throughout a set in the bench-press exercise conducted up to the task failure

Table 2 Percentage of completed repetitions out of the maximum possible number when a given magnitude of mean propulsive velocity loss is reached in each set to failure in the bench press, full-squat and pull-up exercises

Velocity loss (%)	Percentage of completed repetitions out of the maximum possible number when a given velocity loss is reached (%)		
	Bench-press	Full-squat	Pull-up
10	20–25	25–30	25–30
15	30–35	35–40	30–35
20	35–40	45–50	40–45
25	45–50	55–60	50–55
30	50–55	60–65	55–60
35	55–60	65–70	65–70
40	65–70	75–80	70–75
45	70–75	80–85	75–80
50	75–80	85–90	80–85
55	80–85	90–95	85–90
60	85–90	90–100	90–95
65	95–100	95–100	95–100

Values based on González-Badillo et al. (2017); Sánchez-Moreno et al. (2017); Rodríguez-Rosell et al. (2020a)

monitoring RT intensity and volume. Therefore, instead of prescribing a certain amount of weight to be lifted and a given number of repetitions per set, training prescription may be based on the first repetition's velocity and the magnitude of VL over the set. This also opens up the possibility of quantifying the fatigue induced during RT by the interaction of these two parameters. In this regard, the product of the first repetition's mean velocity and the magnitude of VL induced within the set, termed as the "effort index", was shown to be strongly associated with certain fatigue indicators, such as the VL against a given absolute load and post-exercise blood lactate concentrations (Fig. 4) (Rodriguez-Rosell et al. 2018). Moreover, the effort index showed moderate to strong relationships with jump height loss and with relative changes in some surface electromyography (EMG) variables, such as the instantaneous median and mean frequency of the EMG power spectrum (Rodriguez-Rosell et al. 2020b). Previous studies have already reported that the higher the VL with a specific load, the higher the muscle fatigue induced by the training set (Sánchez-Medina and Gonzalez-Badillo 2011; Weakley et al. 2019a). Nonetheless, the effort index allows the comparison of fatigue incurred during the training session even when different relative loads are employed. When comparing similar VL thresholds between loads of different magnitudes, it is possible to observe that lower relative loads will produce higher degrees of fatigue (Rodriguez-Rosell et al. 2018). This may be explained by the fact that the number of repetitions performed to achieve a given percentage of VL in the set is lower as the %1RM increases. Likewise, the lower number of repetitions required to attain a given VL with heavier loads may be explained by the fact that the higher the load, the higher the degree of effort of each repetition. In practical terms, this indicates

Fig. 4 Relationships between effort index and loss of velocity pre-post exercise against the V1·m·s^{-1} load, jump height loss and post-exercise lactate concentration for the squat exercise. Each data point corresponds to one different resistance exercise protocol. Different colors are used to differentiate between the different relative intensities analyzed: 50% 1RM (yellow), 60% 1RM (green), 70% 1RM (blue), and 80% 1RM (red). Adapted from Rodríguez-Rosell (2017)

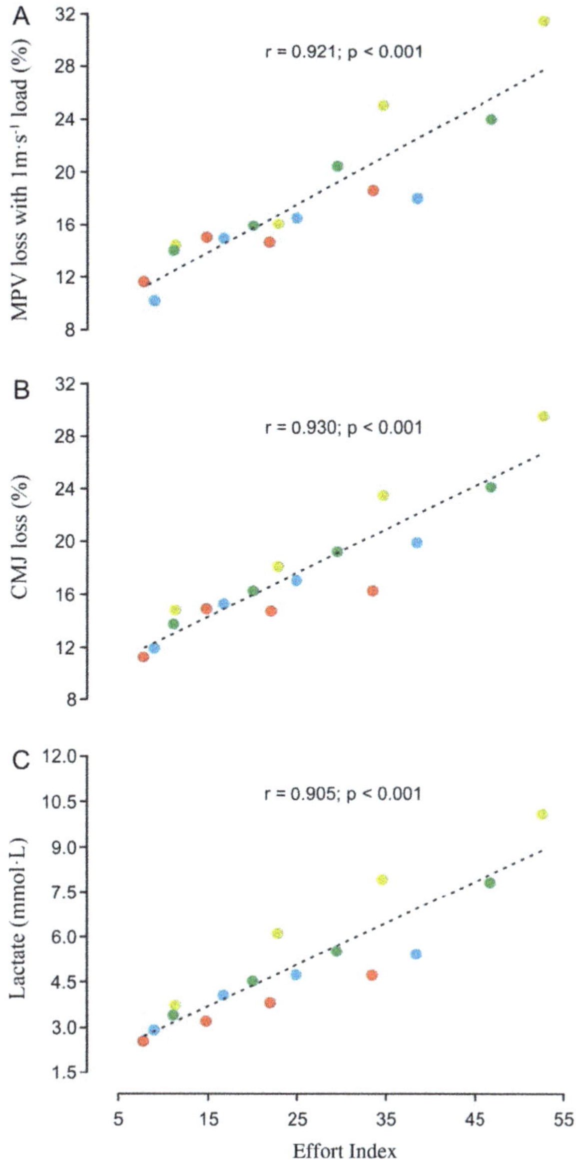

that heavier loads will result in greater magnitudes of VL for a similar number of repetitions (Sánchez-Medina and Gonzalez-Badillo 2011; Rodriguez-Rosell et al. 2020a). The strong correlations observed between effort index and different fatigue indicators suggest that, for a given effort index, the degree of fatigue experienced is equivalent, regardless of the first repetition's velocity and the magnitude of VL incurred in the set. However, it should be highlighted that, although similar degrees of fatigue are experienced by trainees, the mechanisms underlying fatigue likely differ between training stimulus with similar effort index but with different %1RM and VL magnitudes.

3 From Practice

3.1 How the Use of Movement Velocity as a Measure of Training Intensity Can Be Easily Implemented on a Daily Basis

As explained in the epigraph 2.2, the very close relationship between relative load and bar velocity results in mean velocity values strongly associated with distinct % 1RM. Therefore, intensity can be prescribed on a daily basis, using a very precise and time-saving approach, by adjusting the absolute load (kg) to match the repetition velocity associated with the respective percentage of 1RM intended for a given training session. However, it should be acknowledged that individual factors such as age, sex, or training status may potentially influence the prediction equation.

3.1.1 Individual Load-Velocity Relationship

The 1RM and %1RM estimations calculated from the velocity attained with distinct submaximal loads are usually based on general load-velocity equations. These equations assume that the velocity associated with each relative load is the same for all individuals; however, some inter-individual aspects, such as execution technique, length of members, fiber type composition, among others, may slightly compromise these estimations. Accordingly, the individual load-velocity relationship has been proposed to overcome these limitations. In this regard, it has been shown that individual load-velocity relationships provide more accurate predictions of %1RM than general equations (Benavides-Ubric et al. 2020; Pestana-Melero et al. 2018; Garcia-Ramos et al. 2018). Indeed, González-Badillo and Sánchez-Medina (2010) in their seminal work already reported almost perfect individual load-velocity relationships for a sample comprising 120 healthy men aged 24.3 ± 5.2 years, with a training experience of 1.5–4 years ($R^2 = 0.996 \pm 0.003$; range: 0.983–0.999; CV = 0.3%). The standard test carried out to determine the individual load-velocity relationship entails of recording lifting velocity with several submaximal loads and, afterward, modeling the load-velocity relationship through a linear or polynomial (depends on the exercise) adjustment to estimate the 1RM as the load corresponding to the velocity at which the 1RM is attained.

In an attempt to find the potential mechanisms underlying the small differences in the load-velocity relationship among individuals, the influence of different

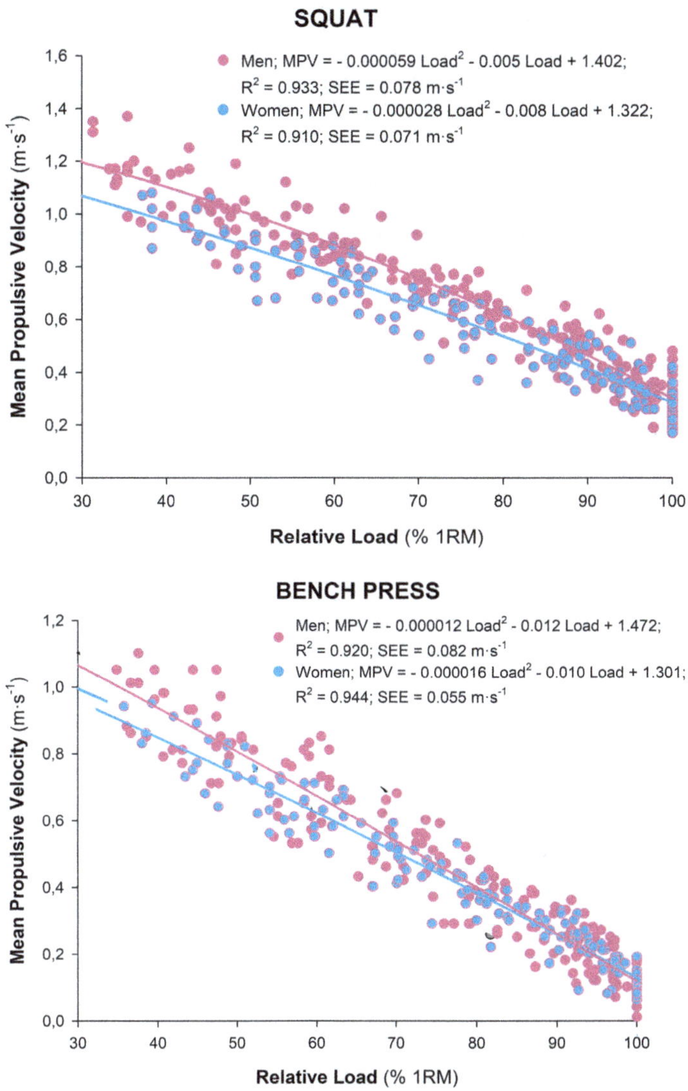

Fig. 5 Relationships between relative load and bar velocity for the full-squat and bench-press exercises. Data obtained from raw load-velocity values derived from the progressive loading tests performed on the sample of 25 men and 13 women. Black circles represent men and grey circles represent women. MPV: mean propulsive velocity. SEE: standard error of estimate. Adapted from Pareja-Blanco et al. (2020d)

strength levels and changes in performance have been investigated. In this regard, the load-velocity relationship seems similar for age-matched individuals with different strength levels (Loturco et al. 2016,2017; Gonzalez-Badillo and Sánchez-Medina 2010; Torrejón et al. 2019) and for the same individuals with increased strength levels after a RT program (Gonzalez-Badillo and Sánchez-Medina 2010; Sánchez-Moreno et al. 2017). Notwithstanding, it has been observed that the velocity associated with each %1RM, particularly with light and moderate loads, is higher in men compared to women in both upper- and lower-body exercises (Fig. 5) (Balsalobre-Fernández et al. 2018; Torrejón et al. 2019; Pareja-Blanco et al. 2020d; García-Ramos et al. 2019). Nevertheless, the velocity of the 1RM did not differ between sexes (Balsalobre-Fernández et al. 2018; Pareja-Blanco et al. 2020d; García-Ramos et al. 2019). Therefore, the general equations previously published for some exercises (Loturco et al. 2016,2017; Gonzalez-Badillo and Sánchez-Medina 2010; Sánchez-Medina et al. 2017) may not be suitable for women. Accordingly, if the training load is prescribed with the same velocity value for both men and women, the training stimulus for both sexes may be different. As a consequence, although the lifting velocity is a valid tool to estimate the relative load in both sexes, a more precise prediction can be obtained when a sex-specific equation is used. The fact that women attain similar %1RM at slower velocities than men may indicate that women have higher strength deficit than men, which should be understood as the percentage of maximal strength potential which is not used (or applied) during a given motor task (Loturco et al. 2021a). However, the underlying mechanisms of differences in the load–velocity relationship between sexes require further investigation. Similarly, young men attain higher velocities for every %1RM than middle-aged men (Fernandes et al. 2018). These findings highlight the need to individualize the load-velocity relationship to the trained population.

3.1.2 Using Bar Velocity on a Daily Basis

The VBT approach allows coaches and athletes to monitor performance by measuring the velocity of the barbell (or load) which is being moved during a lift. In this regard, an increase in the bar velocity may indicate an increase in maximum strength, while an immediate decrease in bar velocity may suggest an impaired ability to apply force (possible due to fatigue). As a result, the velocity against a submaximal absolute load (kg) can be regularly monitored during the warm-up of key RT exercises, thus providing objective and accurate information about the changes in the athlete's fitness/fatigue status. Moreover, by monitoring repetition velocity, it is possible to determine in real-time whether the proposed load (kg) actually represents the effort (%1RM) that was intended in the training session. This allows making adjustments to the training load accounting for day-to-day fluctuations, resulting in more tailored and effective training programs (Dorrell et al. 2020).

The use of feedback during RT is a powerful tool for promoting both acute and chronic improvements in certain physical capacities (Nagata et al. 2020; Weakley et al. 2019b). VBT provides feedback in real-time, which can enhance motivation and, hence, the performance attained in each repetition. In this regard, Nagata et al. (2020) reported that providing verbal feedback of bar velocity after each loaded jump squat repetition resulted in acute and long-term improvements in physical performance. Therefore, this approach may serve as a motivational tool not only to increase training accuracy and quality, but also, to optimize gains in physical performance.

3.2 *Effects of Different Velocity Loss Thresholds During Resistance Training*

As previously mentioned, the level of effort, which is defined as the relationship between the actual number of repetitions performed in the set with regard to the MNR that can be completed (Sánchez-Medina and Gonzalez-Badillo 2011), should be prescribed by the VL attained within the RT set, in order to prevent the drawbacks of determining beforehand a number of repetitions to be performed with a given load (Gonzalez-Badillo et al. 2017). Long-term adaptations to RT seem to be dependent on the VL incurred within the training set during a training program. In this regard, high VL thresholds ($\geq$40%), which accumulate high levels of RT volume by performing more fatiguing and slower repetitions, maximize muscle hypertrophy (Pareja-Blanco et al. 2017a, 2020a, c), but these VL thresholds can also induce negative neuromuscular adaptations (Pareja-Blanco et al. 2020a). These high hypertrophic responses may be due to the high exercise-induced metabolic and mechanical stress, elevated growth-promoting hormones responses, and high levels of muscle damage (Schoenfeld 2010; Goldberg et al. 1975), which are typically observed in RT protocols with higher VL thresholds (Sánchez-Medina and Gonzalez-Badillo 2011; Pareja-Blanco et al. 2020b; Moran-Navarro et al. 2017). In contrast, moderate VL thresholds (i.e. 10–20%) evoke positive neuromuscular adaptations, as well as increases in maximal neuromuscular excitation and enhancements in muscle stiffness indicators (Pareja-Blanco et al. 2020c; Rodriguez-Rosell et al. 2020c). Likewise moderate VL thresholds (10–25%) seem to maximize strength gains (Pareja-Blanco et al. 2017a, 2020a, c; Rodriguez-Rosell et al. 2020c). Indeed, higher VL thresholds do not induce further 1RM strength gains than lower VL thresholds and could even result in lower strength gains, especially in high-velocity actions against low-loads or unloaded conditions or when short periods of time are available to apply force (i.e. early rate of force development) (Pareja-Blanco et al. 2020b, c). This phenomenon may be related to a reduction in the IIX fiber type pool, which present faster cross-bridge cycling rates (compared to type I fibers) (Bottinelli et al. 1996) subsequent to an RT program using a 40% VL, while a 20% VL preserved this pool (Pareja-Blanco et al. 2017a). Accordingly, the dose–response relationship between the level of effort within the set, quantified by the VL, and performance adaptations has been suggested to exhibit an inverted U-shaped curve (Fig. 6) (Pareja-Blanco et al. 2020a, c; Rodriguez-Rosell et al. 2020c). This relationship indicates that a progressive increase in the level of effort within the set will be accompanied by an increase in strength gains, up to a certain limit, beyond which an increase will not produce additional benefits in terms of muscle strength, which may be even detrimental for strength gains. In this regard, moderate VL thresholds (10–25%) should be used to maximize athletic performance, since low VL thresholds (<10%) seem to induce very low levels of fatigue to maximize adaptations, while high VL thresholds ($\geq$40%) do not produce further strength gains (Pareja-Blanco et al. 2020a,c). Hence, the training efficiency is higher for moderate VL thresholds compared to higher VL magnitudes, since moderate VL thresholds evoke similar or even greater benefits. Furthermore, higher VL thresholds (40%) result in greater fatigue and slower rate of recovery than lower VL thresholds (20%) (Pareja-Blanco et al. 2019). Accordingly, moderate VL thresholds induce similar or even greater strength gains than higher VL magnitudes with the advantage of requiring shorter recovery times. In summary, a minimal VL threshold is necessary and effective to optimize strength gains. However, performing additional repetitions under higher magnitudes of VL does not seem to elicit additional strength gains and may even induce "suboptimal adaptations".

3.3 How the Use of Movement Velocity as a Measure of Level of Effort Can Be Practically Implemented on a Daily Basis

It has been shown that both velocity and power could be better preserved during the set by using the VL threshold as a training approach (Weakley et al. 2019a). In addition, the number of repetitions required for attaining a given magnitude of fatigue within the sets may be reduced throughout the training session, which can be individualized when using the VL approach (Weakley et al. 2019a). Accordingly, prescribing level of effort incurred within the set using VL thresholds may enhance exercise quality by mitigating neuromuscular fatigue as well as enables strength and conditioning professionals taking into account factors associated with individual differences in strength performance, daily readiness, or within-session fatigue development. Monitoring of repetition velocity is currently possible by means of the ever increasing number of commercially available portable measuring systems (linear position and velocity transducers, accelerometers, mobile apps, and inertial measurement units). This approach is of great practical relevance for practitioners aiming to find optimal and more time-efficient RT stimuli. In a previous study, a professional soccer team was divided into two groups that trained the squat exercise with similar relative loads but under distinct VL thresholds: one group trained under a VL of 15% (VL15) and the other under a VL equal to 30% (VL30) (Pareja-Blanco et al. 2017b). The VL15 group obtained similar or even greater gains than VL30 in all tested physical qualities (i.e., jump, sprint, and muscle strength) despite the fact that they performed a considerably lower number of repetitions (60% of the repetitions completed by the VL30 group) (Pareja-Blanco et al. 2017b). Likewise, another study compared the effects of two VL thresholds (10% vs. 20%) during a matched-intensity and matched-volume jump squat training

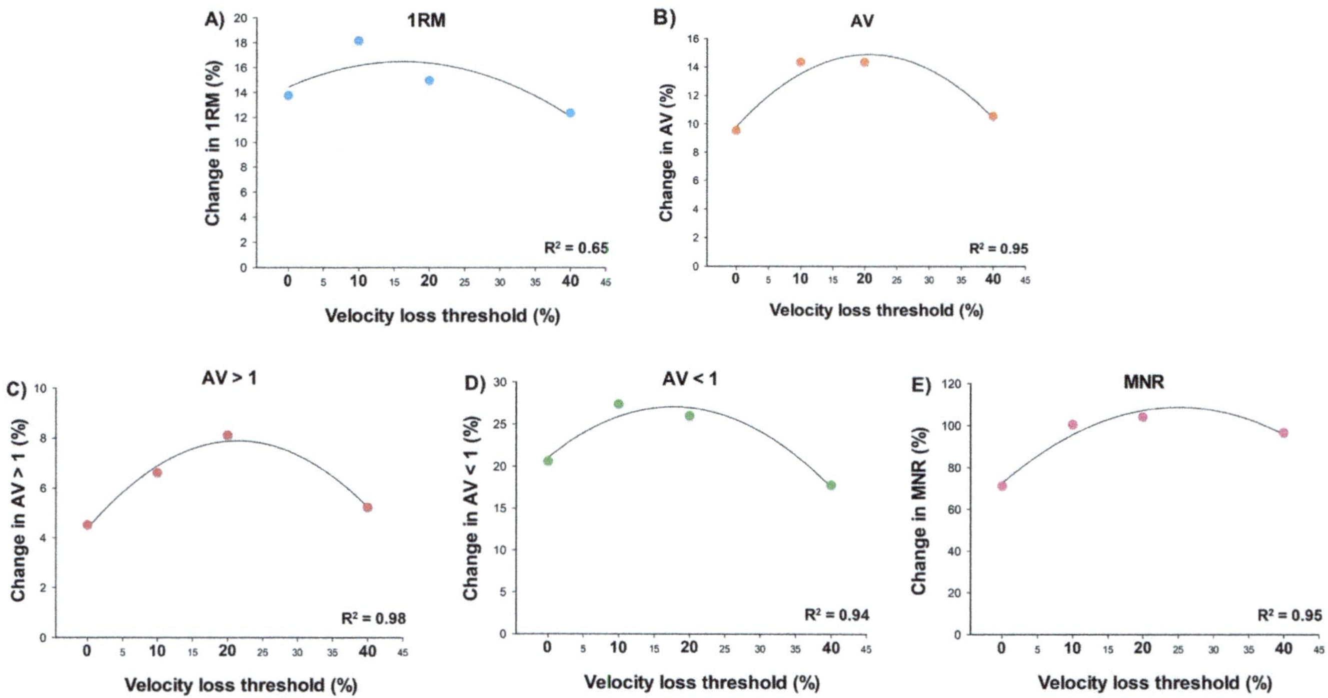

Fig. 6 Relationship between velocity loss in the set in the squat and change obtained in 1RM (**a**), average velocity attained against absolute loads common to pre and post in the progressive loading test (AV; **b**), average velocity attained against absolute loads common to pre and post that were moved faster than 1 $m{\cdot}s^{-1}$ (AV > 1; **c**), average velocity attained against absolute loads common to Pre and Post that were moved slower than 1 $m{\cdot}s^{-1}$ (AV < 1; **d**), and maximal number of repetitions to failure (MNR; E) after the resistance training programs from 70 to 85% 1RM in the full squat exercise. Adapted from Pareja-Blanco et al. (2020a)

program on athletic performance (Perez-Castilla et al. 2018). Both VL thresholds (10% and 20%) resulted in similar performance gains (i.e., jump height, maximal power output and maximal velocity); however, no changes were observed for maximal strength and sprint capacities (Perez-Castilla et al. 2018). Importantly, moderate VL thresholds (10–20%) in the squat exercise induce greater enhancements in high-speed actions, such as jumping and sprinting, than higher VL thresholds (30–40%) (Pareja-Blanco et al. 2017a,b; Rodriguez-Rosell et al. 2020c), which is of great importance for athletes aiming to maximize their performance without generating an excessive degree of fatigue during RT which could interfere, for example, on some technical and tactical skills. Moreover, since the change in lifting velocity against a given load is directly dependent on the force applied onto this load, an increase in the lifting velocity may be considered as an indicator of strength improvement (Loturco et al. 2021c). From a practical standpoint, this means that this approach may allow for the measurement of the estimated 1RM in every training session. Thus, based on the current evidence in this regard, it is possible to state that moderate VL magnitudes induce faster and greater strength gains than higher VL thresholds (Fig. 7), which is very relevant for sports requiring the maintenance of high levels of strength throughout the season.

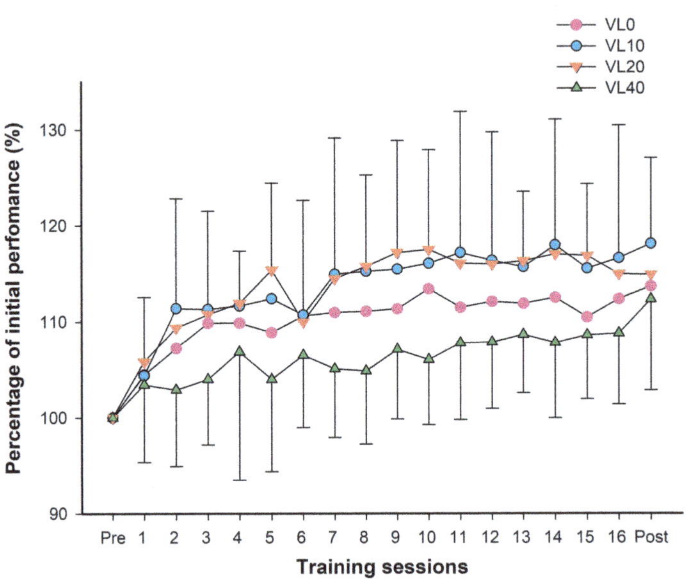

Fig. 7 Evolution of the 1RM strength in the squat exercise in each training session expressed as a percentage of the initial pre-training level for each experimental group. 0, 10, 20, or 40 indicate the session from which the respective group attained significant improvements ($P < 0.05$) in 1RM strength compared with their pretraining values. Data are mean ± SD, N = 55. VL0: group that trained with a mean VL of 0% in each set (n = 14); VL10: group that trained with a mean VL of 10% in each set (n = 14); VL20: group that trained with a mean VL of 20% in each set (n = 13); VL40: group that trained with a mean VL of 40% in each set (n = 14). Adapted from Pareja-Blanco et al. (2020a)

4 Filling Gaps

4.1 Programming Using Velocity-Based Training

Programming is considered the manipulation of RT variables (i.e. exercise choice and order, relative intensity, volume, ratio between work and rest time, among others) with the intention of maximizing athletic performance (Cunanan et al. 2018). There are different programming models concerning the evolution of relative loads and volume throughout the training program as: linear programming (LP), in which training intensity gradually increases while volume decreases (Haff and Triplett 2015); undulating programming (UP), in which both volume and intensity increase or decrease repeatedly throughout the training program (Rhea et al. 2002); reverse programming (RP), in which training intensity gradually decreases and volume increases (Rhea et al. 2003); and constant programming (CP), in which both intensity and volume remain constant throughout the training program (Fig. 8a) (Schiotz et al. 1998). Although a myriad of works have already tried to analyze the effects of different programming models on strength gains (Williams et al. 2017; Harries et al. 2015; Loturco et al. 2013), previous research comparing different programming models have determined RT intensity based on the percentage-based training method, which, as previously described, cannot ensure that the actual load corresponds to the scheduled one, due to day-to-day fluctuations in performance. Since the VBT approach can account for these fluctuations by monitoring velocity during the warm-up and throughout training sessions, it is possible to determine and adjust the RT intensity and level of effort with high accuracy, in real time and on a daily basis. In this regard, a pioneering study compared the effects of four programming models (LP vs. UP vs. RP vs. CP) using VBT approach to prescribe both training intensity (50–85% 1RM) and level of effort (20% VL) (Riscart-Lopez et al. 2021). After 8 training weeks, the four programming models resulted in similar improvements in squat performance (i.e., 1RM strength and load-velocity spectrum), and jumping and sprinting capacities (Riscart-Lopez et al. 2021). However, the time-course of improvements in squat strength varied between groups, being less pronounced in the UP (Fig. 8b) (Riscart-Lopez et al. 2021). Another recent investigation also compared the effects of two different VBT programming models on physical performance. Both VBT programs ("LP" and "UP") used similar relative intensities (ranging from 50 to 80% 1RM) and level of effort (15% VL) (Rodriguez-Rosell et al. 2021). The LP model resulted in greater strength gains than UP (Rodriguez-Rosell et al. 2021). Moreover, the enhancements in jumping ability and 1RM strength were earlier and uninterrupted for LP in comparison with DP (Rodriguez-Rosell et al. 2021). This finding is of great practical relevance when selecting RT models according to time-related criteria (i.e., prioritizing or not prioritizing faster and greater adaptations in strength-related capacities). As such, the use of the VBT approach to determine RT intensity and level of effort allows for an accurate, flexible, and tailored prescription of different programming models, which certainly increases training efficiency and performance gains.

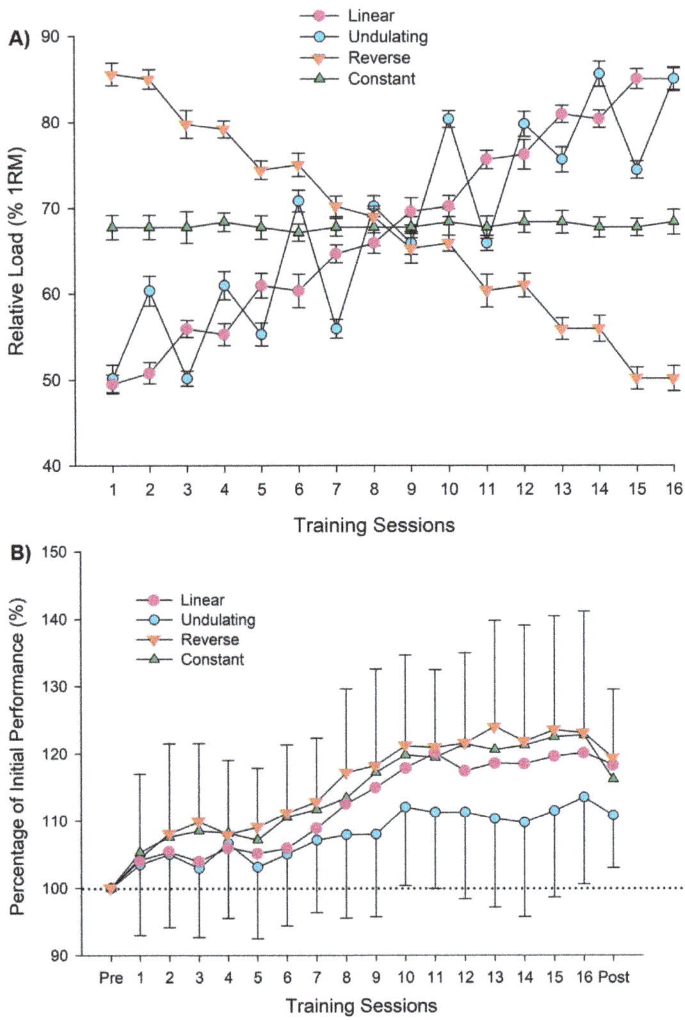

Fig. 8 **a** Example of using different programming models via the velocity-based training approach. **b** Evolution of 1RM across training sessions with respect to baseline performance for each experimental group. Data are presented as mean ± SD. LP = linear programming (n = 11); UP = undulating programming (n = 10); RP = reverse programming (n = 11); CP = constant programming (n = 11). Adapted from Riscart-Lopez et al. (2020)

4.2 How to Design Concurrent Training Implementing the Velocity-Based Training Approach

Concurrent training consists of combining endurance and strength training (Hickson 1980). This training modality is regularly used in sports disciplines requiring high levels of strength and endurance (Garcia-Pallares and Izquierdo

2011). Due to time constraints or/and sport demands, elite athletes are frequently required to conduct both strength and endurance trainings within short periods of time. The exercise sequence (i.e., strength prior to endurance training or vice versa) and the extent of fatigue induced by each exercise may be critical for maximizing endurance and strength development (Doma et al. 2019), since the residual fatigue generated by the previous exercise may compromise the quality in the following exercise (Eddens et al. 2018). In this regard, a recent study compared the acute effects of training sequence and level of effort during RT when using four different training protocols, as follows: endurance training (ET) (running 10 min at 90% of maximal aerobic velocity) followed by RT (3 squat sets with 60% 1RM) with 20% and 40% VL, respectively [1) ET + RT20 and 2) ET + RT40]; and RT with 20% and 40% VL, respectively, followed by ET [3) RT20 + ET and 4) RT40 + ET] (Najera-Ferrer et al. 2021). In summary, higher VL magnitude (40% VL) during RT resulted in higher metabolic and mechanical stress (i.e. greater blood lactate and higher impairments in jump and strength capacities), and greater performance decrements in endurance-related variables (i.e., impaired running time and increased ventilatory equivalents) (Table 3) (Najera-Ferrer et al. 2021). However, endurance performance was not affected when RT with a moderate VL was performed before ET (i.e. RT20 + ET) (Najera-Ferrer et al. 2021). These findings suggest that high-fatigue resistance exercise prior to endurance exercise should be avoided to prevent decreased performance in subsequent endurance performance. In addition, the quality of the "fastest repetition" during resistance exercise can be reduced when endurance exercise is performed prior to resistance exercise (Najera-Ferrer et al. 2021). As such, both endurance and strength training may significantly impair subsequent exercise performance. However, setting a moderate VL threshold (20% VL) for RT may avoid performing an excessive number of repetitions, which could compromise in a greater extent the quality of the subsequent endurance exercise (Najera-Ferrer et al. 2021). These findings are relevant for designing concurrent training schemes capable of reducing the fatigue accumulated between sequential sessions of resistance and endurance training. With regard to long-term adaptations, Sánchez-Moreno et al. (2021b) analyzed and compared the effects of three distinct 8-week training programs: an "isolated" endurance training program (ET) and two concurrent training schemes, which differed in the magnitudes of velocity loss within the RT sets: 15% (VL15) versus 45% (VL45) (Sánchez-Moreno et al. 2021b). Both concurrent training interventions produced greater gains in all strength-related variables and in jump height compared to the ET group. On the other hand, performing higher total volume during RT (VL45) did not elicit additional strength gains compared to the intervention that incurred a lower level of effort within the RT set (VL15). Furthermore, as expected, 8 weeks of isolated ET resulted in decreased strength performance (Sánchez-Moreno et al. 2021b). Lastly, although all groups improved certain endurance-related parameters (i.e., maximal aerobic speed; vVO_2max), notably, the VL15 group obtained greater increases than the ET group in these respective variables (Sánchez-Moreno et al. 2021b). Therefore, establishing moderate VL thresholds (i.e. VL15) during RT when combined with endurance training could be a good strategy for concomitantly optimizing strength and endurance adaptations, as this probably results in reduced residual fatigue and increased training efficiency (when compared to higher magnitudes of VL; e.g., VL45) (Sánchez-Moreno et al. 2021b).

Table 3 Physiological and mechanical characteristics of four concurrent training protocols differing in the exercise sequence and the velocity loss incurred within the resistance exercise set

Variable	E + R20	E + R40	R20 + E	R40 + E	P-value protocol effect
Endurance training					
90% VVO_2max ($km \cdot h^{-1}$)	14.1 ± 1.3	14.1 ± 1.3	14.1 ± 1.3	14.1 ± 1.3	1.00
Running time (s)	600 ± 0[3]	600 ± 0[3]	561 ± 70[3]	450 ± 131	0.002
HR peak (BPM)	180.7 ± 9.3	181.4 ± 9.9	184.1 ± 11.2	183.2 ± 8.4	0.06
Rel-HR peak (%)	96.5 ± 2.9	96.8 ± 2.1	98.2 ± 2.5	97.8 ± 2.1	0.07
VO_2 peak ($ml \cdot kg \cdot min^{-1}$)	48.5 ± 4.6	47.4 ± 5.6	48.1 ± 5.0	47.7 ± 7.7	0.90
Rel-VO_2 peak (%)	90.5 ± 8.7	88.3 ± 9.6	89.5 ± 8.1	88.8 ± 14.1	0.86
RER peak	1.11 ± 0.05	1.11 ± 0.06	1.11 ± 0.06	1.08 ± 0.08	0.28
VE peak ($l \cdot min^{-1}$)	131.8 ± 15.8	129.3 ± 12.5	133.9 ± 14.6	134.6 ± 20.9	0.48
VE/VO_2 peak	35.1 ± 3.7	34.9 ± 3.2[3]	36.0 ± 3.2	37.3 ± 3.9	0.006
VE/VCO_2 peak	32.8 ± 3.1[3]	33.7 ± 3.0	34.2 ± 3.0	36.0 ± 3.5	0.007
Resistance training					
Load (kg)	63.1 ± 15.4	61.8 ± 14.7	61.5 ± 16.0	62.1 ± 13.8	0.74
Rep/Set (N)	9.1 ± 4.6[1,3]	14.4 ± 6.0	10.0 ± 4.9[1,3]	18.3 ± 9.0	0.001
Fastest-V ($m \cdot s^{-1}$)	0.96 ± 0.07	0.94 ± 0.07[2,3]	1.00 ± 0.02	0.99 ± 0.02	0.04
Slowest-V ($m \cdot s^{-1}$)	0.76 ± 0.09[1,3]	0.59 ± 0.09	0.82 ± 0.03[1,3]	0.56 ± 0.05	<0.001
Mean-V ($m \cdot s^{-1}$)	0.85 ± 0.08[1,3]	0.75 ± 0.07	0.88 ± 0.02[1,3]	0.78 ± 0.04	<0.001
VL (%)	20.5 ± 2.9[1,3]	36.1 ± 5.7	18.7 ± 2.4[1,3]	36.9 ± 2.3	<0.001

Adapted from Najera-Ferrer et al. (2021). Data are expressed as mean ± standard deviation. E + R20: endurance (10 min at 90% vVO_2max) + resistance (60% 1RM with 20% VL); E + R40: endurance (10 min at 90% vVO_2max) + resistance (60% 1RM with 40% VL); R20 + E: resistance (60% 1RM with 20% VL) + endurance (10 min at 90% vVO_2max); R40 + E: resistance (60% 1RM with 40% VL) + endurance (10 min at 90% vVO_2max); 90% vVO_2max: 90% of speed associated with the maximum oxygen uptake; running time: time that was maintained the endurance exercise protocol; HR peak: peak heart rate attained during the endurance exercise protocol; Rel-HR peak: peak heart rate, relative to maximum heart rate, attained during the endurance exercise protocol; VO_2 peak: peak oxygen uptake attained during the endurance exercise protocol; Rel-VO_2 peak: peak oxygen uptake, relative to maximal oxygen uptake, attained during the endurance exercise protocol; RER peak: peak respiratory exchange ratio attained during the endurance exercise protocol; VE peak: peak of ventilation volume attained during the endurance exercise protocol; VE/VO_2 peak: peak of ventilatory equivalent for oxygen attained during the endurance exercise protocol; VE/VCO_2 peak: peak of ventilatory equivalent for carbon dioxide attained during the endurance exercise protocol; Load: absolute load employed in the resistance exercise protocol; Rep/set: repetitions performed in each set; Fastest-V: highest velocity measured in the three sets; Slowest-V: lowest velocity measured in the three sets; Mean-V: mean velocity of all repetitions during the three sets; VL: mean percent loss in velocity over the three sets; Velocities correspond to the mean concentric propulsive velocity of each repetition. Statistically significant differences with E + R40 protocol: [1]$P < 0.05$. Statistically significant differences with R20 + E protocol: [2]$P < 0.05$. Statistically significant differences with R40 + E protocol: [3]$P < 0.05$

Muscular endurance is also a target of many training programs. In this regard, it has traditionally suggested that higher number of repetitions per set during RT should be completed to maximize improvements in this regard (Kraemer and Ratamess 2004; Campos et al. 2002). However, it has been observed that the gains in muscular endurance in the squat exercise—as assessed by the maximal number of repetitions with 70% 1RM until the lifting velocity falls below 0.50 $m \cdot s^{-1}$—do not seem to depend exclusively on the total training volume accumulated. Indeed, after 8 weeks of a squat training program with loads ranging from 70 to 85% 1RM, both moderate and high magnitudes of VL led to similar increases in muscular endurance capacity (VL0: 70.9%; VL10: 100.2%; VL20: 103.8%; VL40: 96.3%). However, importantly, the VL40 group accumulated a substantially higher number of repetitions throughout the 8-week intervention (Pareja-Blanco et al. 2020a). Indeed, it appears that enhancements in muscular endurance depends, at least partially, on 1RM increments, since a significant relationship (r = 0.63) was previously reported between 1RM and muscular endurance changes in the squat exercise (Rodriguez-Rosell et al. 2020c). Moreover, some exercises, such as the pull-up, are often assessed on the basis of the MNR completed up to failure, using only the body mass as workload. Also for this exercise without additional overload, a training program with 25% of VL (VL25) resulted in greater gains in both pull-up strength and endurance compared to a training program with 50% of VL (VL50), despite the VL50 group performed more repetitions than VL25 (556 vs. 363 repetitions) across the entire training program (it is important to highlight that these extra repetitions were performed at low and very low velocities [0.6–0.3 $m \cdot s^{-1}$]) (Sánchez-Moreno et al. 2020). The greater strength gains attained by VL25 group evoked that the relative intensity representing the body mass in pull-up for this group was reduced by 7% in relation to their maximum strength levels (from 70 to 63% of 1RM) (Sánchez-Moreno et al. 2020). In this sense, the greater increase observed in MNR by the VL25 group might be explained by the increase in muscle strength and, as a result, the decrease in the relative intensity (%1RM) represented by their BM, since the lower the relative intensity, the higher the MNR that can be performed. Therefore, even for "body-load exercises", using a specific VL threshold during training programs can be an effective strategy to avoid performing an excessive number of deliberately slow and fatiguing repetitions which may even hamper training effectiveness.

5 Take-Home Messages

VBT is a novel practical, and effective training approach that provides coaches and researchers with accurate and immediate information regarding RT intensity, volume and RT effects. By means of VBT, it is possible to dictate and (re) adjust RT loads in real time, accounting for individual fluctuations in both strength and fatigue levels, which regularly occur on a set by set basis. The application of VL thresholds along with the concept of the level of effort attained during the set increases substantially the efficiency of VBT, allowing practitioners to better and more precisely determine and achieve their training objectives. It is essential to understand that similar relative loads will result in different adaptations and responses to RT when different magnitudes of VL are being considered. Strength and conditioning coaches and sport scientists should use the information summarized here to create more effective VBT programs, taking into consideration that:

1. The very strong relationship that exists between relative load (%1RM) and movement velocity in multiple exercises allows the accurate determination of the 1RM value without the necessity of performing 1RM or XRM tests. This close and very stable relationship supports the use of VBT.
2. Higher VL thresholds within the set do not necessarily lead to superior adaptations in strength-related capacities and may even impair athletic performance by generating excessive degrees of fatigue.
3. The effort index has been shown to be strongly associated with distinct fatigue markers and, hence, can be used as a viable and additional tool for controlling and monitoring RT fatigue. However, it should be noted that the mechanisms underlying fatigue may differ between RT sessions with similar effort index but under different intensities and magnitudes of VL.
4. Recent evidence suggests that different VBT programming models lead to similar improvements in physical performance. This highlights the importance of properly defining and monitoring the magnitudes of relative loads and VL which, in essence, will determine the extent and directions of RT adaptations.

References

Balsalobre-Fernández C, García-Ramos A, Jiménez-Reyes P (2018) Load–velocity profiling in the military press exercise: effects of gender and training. Int J Sports Sci Coaching 13(5):743–750

Behm DG, Sale DG (1993) Intended rather than actual movement velocity determines velocity-specific training response. J Appl (1985) Physiol 74(1):359–368

Benavides-Ubric A, Diez-Fernandez DM, Rodriguez-Perez MA, Ortega-Becerra M, Pareja-Blanco F (2020) Analysis of the load-velocity relationship in deadlift exercise. J Sports Sci Med 19 (3):452–459

Bird SP, Tarpenning KM, Marino FE (2005) Designing resistance training programmes to enhance muscular fitness: a review of the acute programme variables. Sports Med 35(10):841–851

Bottinelli R, Canepari M, Pellegrino MA, Reggiani C (1996) Force-velocity properties of human skeletal muscle fibres: myosin heavy chain isoform and temperature dependence. J Physiol 495 (Pt 2):573–586

Braith RW, Graves JE, Leggett SH, Pollock ML (1993) Effect of training on the relationship between maximal and submaximal strength. Med Sci Sports Exerc 25(1):132–138

Brzycki M (1993) Strength testing-predicting a one-rep max from reps-to-fatigue. J Phys Edu Recreation Dance 64(1):88–90

Buckner SL, Jessee MB, Mattocks KT, Mouser JG, Counts BR, Dankel SJ et al (2017) Determining strength: a case for multiple methods of measurement. Sports Med 47(2):193–195

Campos GE, Luecke TJ, Wendeln HK, Toma K, Hagerman FC, Murray TF et al (2002) Muscular adaptations in response to three different resistance-training regimens: specificity of repetition maximum training zones. Eur J Appl Physiol 88(1–2):50–60

Chapman PP, Whitehead JR, Binkert RH (1998) The 225–1b reps-to-fatigue test as a submaximal estimate of 1-RM bench press performance in college football players. J Strength Conditioning Res 12(4):258–261

Conceicao F, Fernandes J, Lewis M, Gonzalez-Badillo JJ, Jimenez-Reyes P (2016) Movement velocity as a measure of exercise intensity in three lower limb exercises. J Sports Sci 34(12): 1099–1106

Cunanan AJ, DeWeese BH, Wagle JP, Carroll KM, Sausaman R, Hornsby WG 3rd et al (2018) The general adaptation syndrome: a foundation for the concept of periodization. Sports Med 48(4):787–797

de Hoyo M, Nunez FJ, Borja S, Gonzalo-Skok O, Muñoz-López A, Romero-Boza S et al (2019) Predicting loading intensity measuring velocity in barbell hip thrust exercise. J Strength Cond Res

Desmedt JE, Godaux E (1977) Ballistic contractions in man: characteristic recruitment pattern of single motor units of the tibialis anterior muscle. J Physiol 264(3):673–693

Doma K, Deakin GB, Schumann M, Bentley DJ (2019) Training considerations for optimising endurance development: an alternate concurrent training perspective. Sports Med 49(5):669–682

Dorrell HF, Smith MF, Gee TI (2020) Comparison of velocity-based and traditional percentage-based loading methods on maximal strength and power adaptations. J Strength Cond Res 34(1):46–53

Eddens L, van Someren K, Howatson G (2018) The role of intra-session exercise sequence in the interference effect: a systematic review with meta-analysis. Sports Med 48(1):177–188

Fernandes JFT, Lamb KL, Twist C (2018) A comparison of load-velocity and load-power relationships between well-trained young and middle-aged males during three popular resistance exercises. J Strength Cond Res 32(5):1440–1447

Fry AC (2004) The role of resistance exercise intensity on muscle fibre adaptations. Sports Med 34(10):663–679

Garcia-Pallares J, Izquierdo M (2011) Strategies to optimize concurrent training of strength and aerobic fitness for rowing and canoeing. Sports Med 41(4):329–343

Garcia-Ramos A, Pestana-Melero FL, Perez-Castilla A, Rojas FJ, Haff GG (2018) Differences in the load-velocity profile between 4 bench-press variants. Int J Sports Physiol Perform 13(3): 326–331

García-Ramos A, Suzovic D, Pérez-Castilla A (2019) The load-velocity profiles of three upper-body pushing exercises in men and women. Sports Biomechanics 1–13

Goldberg AL, Etlinger JD, Goldspink DF, Jablecki C (1975) Mechanism of work-induced hypertrophy of skeletal muscle. Med Sci Sports 7(3):185–198

Gonzalez-Badillo JJ, Sánchez-Medina L (2010) Movement velocity as a measure of loading intensity in resistance training. Int J Sports Med 31(5):347–352

Gonzalez-Badillo JJ, Marques MC, Sánchez-Medina L (2011) The importance of movement velocity as a measure to control resistance training intensity. J Hum Kinet 29A:15–19

Gonzalez-Badillo JJ, Rodriguez-Rosell D, Sánchez-Medina L, Gorostiaga EM, Pareja-Blanco F (2014) Maximal intended velocity training induces greater gains in bench press performance than deliberately slower half-velocity training. Eur J Sport Sci 1–10

Gonzalez-Badillo JJ, Yanez-Garcia JM, Mora-Custodio R, Rodriguez-Rosell D (2017) Velocity loss as a variable for monitoring resistance exercise. Int J Sports Med 38(3):217–225

Haff GG, Triplett NT (2015) Essentials of strength training and conditioning, 4th edn. Human Kinetics

Harries SK, Lubans DR, Callister R (2015) Systematic review and meta-analysis of linear and undulating periodized resistance training programs on muscular strength. J Strength Cond Res 29(4):1113–1125

Hickson RC (1980) Interference of strength development by simultaneously training for strength and endurance. Eur J Appl Physiol Occup Physiol 45(2–3):255–263

Izquierdo M, Gonzalez-Badillo JJ, Hakkinen K, Ibanez J, Kraemer WJ, Altadill A et al (2006) Effect of loading on unintentional lifting velocity declines during single sets of repetitions to failure during upper and lower extremity muscle actions. Int J Sports Med 27(9):718–724

Kraemer WJ, Ratamess NA (2004) Fundamentals of resistance training: progression and exercise prescription. Med Sci Sports Exerc 36(4):674–688

Loturco I, Ugrinowitsch C, Roschel H, Lopes Mellinger A, Gomes F, Tricoli V et al (2013) Distinct temporal organizations of the strength- and power-training loads produce similar performance improvements. J Strength Cond Res 27(1):188–194

Loturco I, Pereira LA, Cal Abad CC, Gil S, Kitamura K, Kobal R et al (2016) Using bar velocity to predict the maximum dynamic strength in the half-squat exercise. Int J Sports Physiol Perform 11(5):697–700

Loturco I, Kobal R, Moraes JE, Kitamura K, Cal Abad CC, Pereira LA et al (2017) Predicting the maximum dynamic strength in bench press: the high precision of the bar velocity approach. J Strength Cond Res 31(4):1127–1131

Loturco I, Pereira LA, Freitas TT, Bishop C, Pareja-Blanco F, McGuigan MR (2021a) Maximum strength, relative strength, and strength deficit: relationships with performance and differences between elite sprinters and professional rugby union players. Int J Sports Physiol Perform 1–6

Loturco I, Suchomel T, Kobal R, Arruda AFS, Guerriero A, Pereira LA et al (2021b) Force-velocity relationship in three different variations of prone row exercises. J Strength Cond Res 35(2):300–309

Loturco I, McGuigan M, Freitas TT, Valenzuela P, Pereira LA, Pareja-Blanco F (2021c) Performance and reference data in the jump squat at different relative loads in elite sprinters, rugby players, and soccer players. Biol Sport 38(2):219–227

Martinez-Cava A, Moran-Navarro R, Sánchez-Medina L, Gonzalez-Badillo JJ, Pallares JG (2019) Velocity- and power-load relationships in the half, parallel and full back squat. J Sports Sci 37(10):1088–1096

Moran-Navarro R, Perez CE, Mora-Rodriguez R, de la Cruz-Sanchez E, Gonzalez-Badillo JJ, Sánchez-Medina L et al (2017) Time course of recovery following resistance training leading or not to failure. Eur J Appl Physiol 117(12):2387–2399

Moran-Navarro R, Martinez-Cava A, Sánchez-Medina L, Mora-Rodriguez R, Gonzalez-Badillo JJ, Pallares JG (2019) Movement velocity as a measure of level of effort during resistance exercise. J Strength Cond Res 33(6):1496–1504

Nagata A, Doma K, Yamashita D, Hasegawa H, Mori S (2020) The effect of augmented feedback type and frequency on velocity-based training-induced adaptation and retention. J Strength Cond Res 34(11):3110–3117

Najera-Ferrer P, Perez-Caballero C, Gonzalez-Badillo JJ, Pareja-Blanco F (2021) Effects of exercise sequence and velocity loss threshold during resistance training on following endurance and strength performance during concurrent training. Int J Sports Physiol Perform 1–7

Niewiadomski W, Laskowska D, Gąsiorowska A, Cybulski G, Strasz A, Langfort J (2008) Determination and prediction of one repetition maximum (1RM): safety considerations J Hum Kinet 19(1):109

Ortega-Becerra M, Sánchez-Moreno M, Pareja-Blanco F (2021) Effects of cluster set configuration on mechanical performance and neuromuscular activity. J Strength Cond Res 35(2):310–317

Pareja-Blanco F, Rodriguez-Rosell D, Sánchez-Medina L, Gorostiaga EM, Gonzalez-Badillo JJ (2014) Effect of movement velocity during resistance training on neuromuscular performance. Int J Sports Med 35(11):916–924

Pareja-Blanco F, Rodriguez-Rosell D, Sánchez-Medina L, Sanchis-Moysi J, Dorado C, Mora-Custodio R et al (2017a) Effects of velocity loss during resistance training on athletic performance, strength gains and muscle adaptations. Scand J Med Sci Sports 27(7):724–735

Pareja-Blanco F, Sánchez-Medina L, Suarez-Arrones L, Gonzalez-Badillo JJ (2017b) Effects of velocity loss during resistance training on performance in professional soccer players. Int J Sports Physiol Perform 12(4):512–519

Pareja-Blanco F, Villalba-Fernandez A, Cornejo-Daza PJ, Sanchez-Valdepenas J, Gonzalez-Badillo JJ (2019) Time course of recovery following resistance exercise with different loading magnitudes and velocity loss in the set. Sports 7(3)

Pareja-Blanco F, Alcazar J, Sánchez-Valdepeñas J, Cornejo-Daza PJ, Piqueras-Sanchiz F, Mora-Vela R et al (2020a) Velocity loss as a critical variable determining the adaptations to strength training. Med Sci Sports Exerc 52(8):1752–1762

Pareja-Blanco F, Rodriguez-Rosell D, Aagaard P, Sánchez-Medina L, Ribas-Serna J, Mora-Custodio R et al (2020b) Time course of recovery from resistance exercise with different set configurations. J Strength Cond Res 34(10):2867–2876

Pareja-Blanco F, Alcazar J, Cornejo-Daza PJ, Sanchez-Valdepenas J, Rodriguez-Lopez C, Hidalgo-de Mora J et al (2020c) Effects of velocity loss in the bench press exercise on strength gains, neuromuscular adaptations and muscle hypertrophy. Scand J Med Sci Sports

Pareja-Blanco F, Walker S, Häkkinen K (2020d) Validity of using velocity to estimate intensity in resistance exercises in men and women. Int J Sports Med 41:1047–1055

Perez-Castilla A, Garcia-Ramos A, Padial P, Morales-Artacho AJ, Feriche B (2018) Effect of different velocity loss thresholds during a power-oriented resistance training program on the mechanical capacities of lower-body muscles. J Sports Sci 36(12):1331–1339

Pestana-Melero FL, Haff GG, Rojas FJ, Perez-Castilla A, Garcia-Ramos A (2018) Reliability of the load-velocity relationship obtained through linear and polynomial regression models to predict the 1-repetition maximum load. J Appl Biomech 34(3):184–190

Reynolds JM, Gordon TJ, Robergs RA (2006) Prediction of one repetition maximum strength from multiple repetition maximum testing and anthropometry. J Strength Cond Res 20(3):584–592

Rhea MR, Ball SD, Phillips WT, Burkett LN (2002) A comparison of linear and daily undulating periodized programs with equated volume and intensity for strength. J Strength Cond Res 16(2):250–255

Rhea MR, Phillips WT, Burkett LN, Stone WJ, Ball SD, Alvar BA et al (2003) A comparison of linear and daily undulating periodized programs with equated volume and intensity for local muscular endurance. J Strength Cond Res 17(1):82–87

Richens B, Cleather DJ (2014) The relationship between the number of repetitions performed at given intensities is different in endurance and strength trained athletes. Biol Sport 31(2):157–161

Riscart-Lopez J, Rendeiro-Pinho G, Mil-Homens P, Soares-daCosta R, Loturco I, Pareja-Blanco F, Leon-Prados JA (2020) Effects of four different velocity-based training programming models on strength gains and physical performance. J Strength Cond Res (Epub Ahead of Print)

Riscart-Lopez J, Rendeiro-Pinho G, Mil-Homens P, Soares-daCosta R, Loturco I, Pareja-Blanco F et al (2021) Effects of four different velocity-based training programming models on strength gains and physical performance. J Strength Cond Res 35(3):596–603

Rodríguez-Rosell D (2017) Movement velocity as a critical variable for prescribing and monitoring resistance exercise, and as a determinant factor of the resistance training induced-adaptations. Doctoral dissertation, Supervised by Gonzalez-Badillo, JJ, Universidad Pablo de Olavide

Rodriguez-Rosell D, Yanez-Garcia JM, Torres-Torrelo J, Mora-Custodio R, Marques MC, Gonzalez-Badillo JJ (2018) Effort index as a novel variable for monitoring the level of effort during resistance exercises. J Strength Cond Res 32(8):2139–2153

Rodriguez-Rosell D, Yanez-Garcia JM, Sánchez-Medina L, Mora-Custodio R, Gonzalez-Badillo JJ (2020a) Relationship between velocity loss and repetitions in reserve in the bench press and back squat exercises. J Strength Cond Res 34(9):2537–2547

Rodriguez-Rosell D, Yanez-Garcia JM, Mora-Custodio R, Torres-Torrelo J, Ribas-Serna J, Gonzalez-Badillo JJ (2020b) Role of the effort index in predicting neuromuscular fatigue during resistance exercises. J Strength Cond Res

Rodriguez-Rosell D, Yanez-Garcia JM, Mora-Custodio R, Pareja-Blanco F, Ravelo-Garcia AG, Ribas-Serna J et al (2020c) Velocity-based resistance training: impact of velocity loss in the set on neuromuscular performance and hormonal response. Appl Physiol Nutr Metab (Epub Ahead of Print)

Rodriguez-Rosell D, Martinez-Cava A, Yanez-Garcia JM, Hernandez-Belmonte A, Mora-Custodio R, Moran-Navarro R et al (2021) Linear programming produces greater, earlier and uninterrupted neuromuscular and functional adaptations than daily-undulating programming after velocity-based resistance training. Physiol Behav 113337

Rontu JP, Hannula MI, Leskinen S, Linnamo V, Salmi JA (2010) One-repetition maximum bench press performance estimated with a new accelerometer method. J Strength Cond Res 24(8): 2018–2025

Sánchez-Medina L (2010) Movement velocity as a determinant of the level of effort in resistance training. Doctoral dissertation, Supervised by Gonzalez-Badillo, JJ, Universidad Pablo de Olavide

Sánchez-Medina L, Gonzalez-Badillo JJ (2011) Velocity loss as an indicator of neuromuscular fatigue during resistance training. Med Sci Sports Exerc 43(9):1725–1734

Sánchez-Medina L, Gonzalez-Badillo JJ, Perez CE, Pallares JG (2014) Velocity- and power-load relationships of the bench pull vs. bench press exercises. Int J Sports Med 35(3):209–216

Sánchez-Medina L, Pallarés JG, Pérez CE, Morán-Navarro R, González-Badillo JJ (2017) Estimation of relative load from bar velocity in the full back squat exercise. Sports Med Int Open 01(02):E80–E88

Sánchez-Moreno M, Rodriguez-Rosell D, Pareja-Blanco F, Mora-Custodio R, Gonzalez-Badillo JJ (2017) Movement velocity as indicator of relative intensity and level of effort attained during the set in pull-up exercise. Int J Sports Physiol Perform 12(10):1378–1384

Sánchez-Moreno M, Cornejo-Daza PJ, Gonzalez-Badillo JJ, Pareja-Blanco F (2020) Effects of velocity loss during body mass prone-grip pull-up training on strength and endurance performance. J Strength Cond Res 34(4):911–917

Sánchez-Moreno M, Rendeiro-Pinho G, Mil-Homens PV, Pareja-Blanco F (2021a) Monitoring training volume through maximal number of repetitions or velocity-based approach. Int J Sports Physiol Perform 1–8

Sánchez-Moreno M, Rodriguez-Rosell D, Diaz-Cueli D, Pareja-Blanco F, Gonzalez-Badillo JJ (2021b) Effects of velocity loss threshold within resistance training during concurrent training on endurance and strength performance. Int J Sports Physiol Perform (Epub Ahead of Print)

Schilling BK, Falvo MJ, Chiu LZ (2008) Force-velocity, impulse-momentum relationships: implications for efficacy of purposefully slow resistance training. J Sports Sci Med 7(2): 299–304

Schiotz MK, Potteiger JA, Huntsinger PG, Donald C, Denmark LC (1998) The short-term effects of periodized and constant-intensity training on body composition, strength, and performance. J Strength Conditioning Res 12(3):173–178

Schoenfeld BJ (2010) The mechanisms of muscle hypertrophy and their application to resistance training. J Strength Cond Res 24(10):2857–2872

Schoenfeld BJ, Ogborn D, Krieger JW (2017) Dose-response relationship between weekly resistance training volume and increases in muscle mass: a systematic review and meta-analysis. J Sports Sci 35(11):1073–1082

Scott BR, Duthie GM, Thornton HR, Dascombe BJ (2016) Training monitoring for resistance exercise: theory and applications. Sports Med 46(5):687–698

Spiering BA, Kraemer WJ, Anderson JM, Armstrong LE, Nindl BC, Volek JS et al (2008) Resistance exercise biology: manipulation of resistance exercise programme variables determines the responses of cellular and molecular signalling pathways. Sports Med 38(7): 527–540

Suchomel TJ, Nimphius S, Bellon CR, Stone MH (2018) The importance of muscular strength: training considerations. Sports Med 48(4):765–785

Torrejón A, Balsalobre-Fernández C, Haff GG, García-Ramos A (2019) The load-velocity profile differs more between men and women than between individuals with different strength levels. Sports Biomechanics 18(3):245–255

Weakley J, Ramirez-Lopez C, McLaren S, Dalton-Barron N, Weaving D, Jones B et al (2019a) The effects of 10%, 20%, and 30% velocity loss thresholds on kinetic, kinematic, and repetition characteristics during the Barbell Back Squat. Int J Sports Physiol Perform 1–9

Weakley JJS, Wilson KM, Till K, Read DB, Darrall-Jones J, Roe GAB et al (2019b) Visual feedback attenuates mean concentric barbell velocity loss and improves motivation, competitiveness, and perceived workload in male adolescent athletes. J Strength Cond Res 33(9):2420–2425

Williams TD, Tolusso DV, Fedewa MV, Esco MR (2017) Comparison of periodized and non-periodized resistance training on maximal strength: a meta-analysis. Sports Med 47(10): 2083–2100

Fernando Pareja-Blanco is graduated in Physical Activity and Sports Science and holder of a Doctorate from the Pablo de Olavide University (UPO) of Seville, where, under the supervision of Professor Juan José González-Badillo, he wrote his Doctoral Thesis "Movement velocity as a determining factor in the adaptations brought through resistance training." He has published over 70 scientific articles in JCR journals in the Sports Science section one book and two book chapters, developing his scientific career at the Physical Performance and Sports Research Center at the UPO. He has spent periods working with researchers of international prestige, as Prof. José Antonio López Calbet (University of Las Palmas de Gran Canaria, Spain), Prof. Per Aagaard (Southern Denmark University, Denmark), Prof. Keijo Hakkinen (University of Jyvaskyla, Finland), and Ph.D. Irineu Loturco (Nucleus of High Performance; Brazil). He is currently an associate professor at the Faculty of Sport Sciences at UPO. From a practical point of view, he has worked as strength and conditioning consulting in different athletic modalities, such as soccer, basket, tennis, track and field, volleyball, swimming, rugby, taekwondo, weightlifting, and fencing, among others.

Irineu Loturco has a Ph.D. in sport science at the Pablo de Olavide University (UPO) of Seville and a postdoctoral degree in Mechatronics Engineering and Mechanical Systems at the Polytechnic School of the University of São Paulo. He is the director of the Nucleus of High Performance in Sport (NAR; São Paulo, Brazil), a high-performance training center that serves hundreds of top-level athletes from many different sports. He has worked as a strength and conditioning coach in major Brazilian soccer clubs and with different Olympic and Paralympic sports, especially national teams. Currently, he is a professor of Methodology of Sports Training at the Federal University of São Paulo (UNIFESP) and a visiting professor at the University of South Wales, in Pontypridd, UK. During his academic career, he has published more than 160 peer-reviewed articles. The vast majority of his studies involve professional athletes, many of them competing at the Olympic level.

Measuring and Testing with Flywheel Resistance Training Devices

Alejandro Muñoz-López and Fabio Yuzo Nakamura

Abstract The applied physics that explains the nature of flywheel resistance training devices highlights the importance of knowing the mechanical demands during the concentric or the eccentric phase of the movement. As shown in other isoinertial equipment, flywheel devices show a mechanical overload as the moment of inertia is increased. However, it is not easy to express the load intensity as a percentage of "something" (such as the one-maximum repetition in free weights). For this purpose, the peak acceleration can distinguish among different inertial loads and fatigue levels along with repetitions and sets. In relation, the load of a flywheel resistance training device has to be accounted for considering the shaft's width and the moment of inertia used. A conical pulley will vary the axis and, consequently, the load, over the exercise range of motion compared to a horizontal cylinder device. On those, higher peak torque can be found. Finally, caution should be taken when monitoring the eccentric overload, that has been a topic of interest. Not all the individuals experiment an eccentric overload, and it is important to highlight that the only way to achieve eccentric overload with flywheel devices is by modifying the tempo or the range of motion during the eccentric phase.

Keywords Rotary inertia · Eccentric overload · Acceleration · Power · Monitoring · Real-time feedback

---

A. Muñoz-López (✉)
Departamento de Motricidad Humana y Rendimiento Deportivo, Universidad de Sevilla, Seville, Spain
e-mail: amunoz26@us.es

F. Y. Nakamura
Research Center in Sports Sciences, Health Sciences and Human Development, CIDESD, University Institute of Maia, ISMAI, Maia, Portugal

A. Muñoz-López et al. (eds.), *Resistance Training Methods*, Lecture Notes in Bioengineering, https://doi.org/10.1007/978-3-030-81989-7_10

1 Introduction

Traditionally, free weights or other related isoinertial equipment have been used in resistance training programs. As previously discussed, other interesting training paradigms, such as flywheel training, have also been included in those programs for more than twenty years (Berg and Tesch 1994). The rotary nature of the movement in flywheel devices dramatically changes the way the kinetics and kinematics need to be assessed during exercise. One of the main differences is that flywheel devices are known as gravity-independent equipment, at least in reference to the inertial component of the movement (Sjöberg et al. 2020).

A known characteristic when measuring isoinertial gravity-dependent movements is the existence of a propulsive phase (Sanchez-Medina et al. 2010). In addition, there is also a decelerative phase, which is necessary in non-ballistic exercises in which there is a need to stop the movement due to the length of the limbs. In contrast, during ballistic exercises or jumps, the whole concentric phase is propulsive, and the peak speed is typically achieved almost at the end of the movement. When an exercise is performed in a flywheel device, it has been shown similar kinetic characteristics compared to ballistic exercises, especially if the individual performs the concentric phase at maximum voluntary contraction during the whole range of motion. Hence, in flywheel devices, the entire concentric phase can be considered propulsive, and this determines how the resulting mechanical data should be interpreted.

In addition to the concentric phase, the eccentric phase has traditionally been the focus of training when flywheel devices are used. It is possible to perform mechanical measures in the eccentric phase to assess the presence or absence of an eccentric overload. This increases the challenges involved in monitoring flywheel devices during training sessions.

Several kinds of monitoring technologies are used for such purposes. To date, the most frequently used technologies are force plates, strain gauges, electromyography (EMG) and encoders. Both linear and rotary encoders have been described in the literature, but due to the nature of the movements involved, the rotary option tends to be the most suitable. Basically, those technologies have been used to determine actual training intensity, the level of fatigue produced by the mechanical effort, individual capacities for eccentric training, movement variability, and/or to assess different training adaptations. There is an increasing interest in using those technologies to monitor mechanical output in real time. Both the intensity and volume of exercise can be quantified in real time, enabling individualization of training sessions to induce higher positive effects of resistance training programs in athletes and non-athletes.

In this chapter, we will cover how to effectively monitor important training and testing features in resistance training programs, focusing particularly on the eccentric characteristic of the flywheel training paradigm and the possibilities related to real-time effective monitoring.

2 From Theory

As with other training equipment, flywheel devices can be used to perform exercises in the vertical and the horizontal plane. Both options increase the possibilities of how, theoretically, the exercise outcomes can be monitored. However, the main difference when monitoring a flywheel device, compared to others, is related to the training intensity, typically expressed by the external load (e.g., the weight of the barbells). As shown in the first chapter of this book, the radius of the shaft, and its shape, will determine the resulting intensity during the exercise. For this reason, coaches and practitioners must be aware of the nature of rotary movement and the components of the flywheel that can be modified to determine the final mechanical output. It has been shown that changing the position at which the rope is recoiled (Vázquez-Guerrero et al. 2016) in a cone pulley will increase or decrease the radius of the cone, thus changing the exercise load. For example, moving the position to the top of the cone will decrease the maximum radius of the cone.

2.1 Flywheel Training Paradigm Components to be Considered

As previously noted, there are some important components of flywheel devices that can determine the final mechanical output of an exercise, regardless of the moment of inertia used. First, the radius width can be modified in cone pulley flywheel devices. Usually the lower the arm lever, the wider the radius at the bottom of the cone. It has been shown that by decreasing the radius in a cone pulley flywheel, mean and peak force will increase in both the concentric and eccentric phases (Vázquez-Guerrero et al. 2016). Second, the shaft type also influences exercise intensity, but not only due to the width. In both flywheel shaft shapes (cylinder-shaped and cone pulley), in addition to a different radius width (typically, a cylinder-shaped flywheel has a narrower radius), there is a variation in the radius width during the execution of the movement. For instance, while the beginning of the concentric phase corresponds to a wider radius, at the end of this phase, when the eccentric phase begins, a narrower radius applies. This ongoing modification of the radius width is more dramatic in cone pulley than in cylinder-shaped devices. Recently, we have shown that despite using the same moment of inertia, the shaft shape can determine the final mechanical output (Núñez et al. 2020). According to the laws of physics and the concept of *torque*, we found that a cylinder-shaped device elicited higher mean and peak forces, but a lower *speed*, compared to a cone pulley device. Finally, although flywheel devices are known as gravity-independent systems, gravity does have an influence when the exercise is executed in the vertical plane, compared to the horizontal one. For example, during the squat exercise, more than half of the load is provided by the bodyweight, while the entire load is provided by the flywheel itself in the leg press exercise (Sjöberg et al. 2020). In summary, the flywheel shaft radius, its shape and the plane of movement must all be considered to explain all the components of exercise intensity, in addition to the moment of inertia.

2.2 Flywheel Overloading Profile

One of the main and traditional objectives when monitoring a resistance training exercise is to characterize the mechanical demands over a range of external loads, which can later be used by individuals to control overload and expected adaptations. For this purpose, a progressive loading test can be performed. As explained previously, the absence of the concept of maximum repetition (RM) in flywheel devices makes the characterization of training intensity difficult. However, monitoring some mechanical variables (i.e., speed or acceleration) in flywheel devices can help to determine the training intensity for a given moment of inertia and device components (shaft shape, shaft radius, number of pulleys and plane of movement).

A typical purpose when a progressive loading test is performed is to find the load at which peak power output is achieved, also known as the optimal power training zone (Loturco et al. 2013). The power-velocity or power-load curve states that *power* increases with load (or increases together with decreases in *velocity*) until a load is reached that exceeds the highest possible power. Thus, light (i.e. <30% RM) and high (i.e. >70% RM) loads can elicit similar power values, although the peak power output is typically achieved with moderate loads (Cormie et al. 2011). This classification of low, medium or high loads has been standardized for gravity-dependent equipment, but for flywheel devices, there is no consensus and performance depends on each manufacturer's device characteristics and exercise component settings, as we showed above. Thus, in contrast to other equipment, peak power was found to be higher on a cylinder-shaped flywheel device when a load of 0.025 kg m^2 was used, decreasing to 0.1 kg m^2 in the squat exercise (Sabido et al. 2018; Worcester et al. 2020), and from 0.0125 to 0.1 kg m^2 in the leg extension exercise (Martinez-Aranda and Fernandez-Gonzalo 2017). All the previous data supported the notion that *speed* decreases and *force* increases with higher moments of inertia. This confirms that the training intensity is higher when higher moments of inertia are used in flywheel devices, given that the same shaft shape is used (Núñez et al. 2020).

The eccentric phase has also been the object of study during progressive loading tests. In contrast to the prior concentric findings, the best eccentric/concentric ratios (E:C-r) were observed with higher moments of inertia [i.e. 0.075 kg m^2 (Sabido et al. 2018)], which highlights the possible higher eccentric overload with greater loads (Martinez-Aranda and Fernandez-Gonzalo 2017). However, lower muscle activation was observed during the concentric phase with lower loads (i.e. 0.010 kg m^2), while lower muscle activation during the eccentric phase was observed with higher loads (i.e. 0.050 kg m^2) in the squat exercise using a cylinder-shaped flywheel (Carroll et al. 2019). Finally, the transition between phases, which is an opportunity to assess the stretch-shortening cycle, was higher with medium loads (Martinez-Aranda and Fernandez-Gonzalo 2017). Both eccentric overload and the stretch-shortening cycle showed marked differences between genders in a leg extension exercise using a cylinder-shaped flywheel (Martinez-Aranda and Fernandez-Gonzalo 2017).

2.3 *Typical Mechanical Variables Used*

One of the most feasible options for monitoring a resistance training exercise is to control the speed of the implement or object that is moved by the individual (González-Badillo and Sánchez-Medina 2010). In flywheel devices, in practice, the *speed* of the flywheel can also be measured. However, in the literature, other variables related to *force* (or *torque*) and *power* have more often been used. Some studies have analyzed the waveforms of those variables to understand the mechanical profile of both the concentric and eccentric phases. Studying the waveforms of the mechanical variables (Fig. 1) is useful to understand at which point in the range of motion the highest load is achieved. Those characteristics in flywheel devices differ from free weights because at the beginning of the concentric phase, the radius is wider, and usually speed is lower (Fig. 1a) and the acceleration is higher (Fig. 1b). In contrast, at the end of the concentric phase, a narrower radius allows the individual to produce a greater amount of force, despite a higher velocity, which is in contrast to an isoinertial gravity-dependent movement profile.

Figure 1c shows how power, calculated from raw data, varies throughout the concentric phase of a leg extension exercise using a cylinder-shaped shaft with a load of 0.025 and 0.050 kg m^2 (private unpublished data). Interestingly, in this device and exercise, peak angular acceleration occurs prior to peak power, followed by peak speed at the end of the concentric phase, which emphasizes that the whole movement is propulsive. From these three typical mechanical outcomes in training, we found that *acceleration* (a less frequently used variable for monitoring flywheel devices) provided more information to distinguish between sets and loads around the angular acceleration peak. Other authors typically use the peak power in flywheel devices, especially to monitor eccentric overload, but we found a higher inter-individual variation in this variable, compared to the acceleration peak. Thus, the acceleration peak may be a more reproducible variable for these purposes.

Finally, the literature provides studies that have used linear and angular speed to quantify the mechanical output using flywheel devices. In our laboratory, we found that the *angular speed*, *angular acceleration* or *power* decreased over 30 repetitions in the leg extension exercise when a cylinder-shaped device was used. During a progressive loading test, a linear relationship has been shown between mean or peak speed and the moment of inertia used, but with a lower coefficient of determination compared to free weights [i.e. $r^2 = 0.66$ (Carroll et al. 2019)], which suggests a limitation in using speed to monitor intensity in such devices.

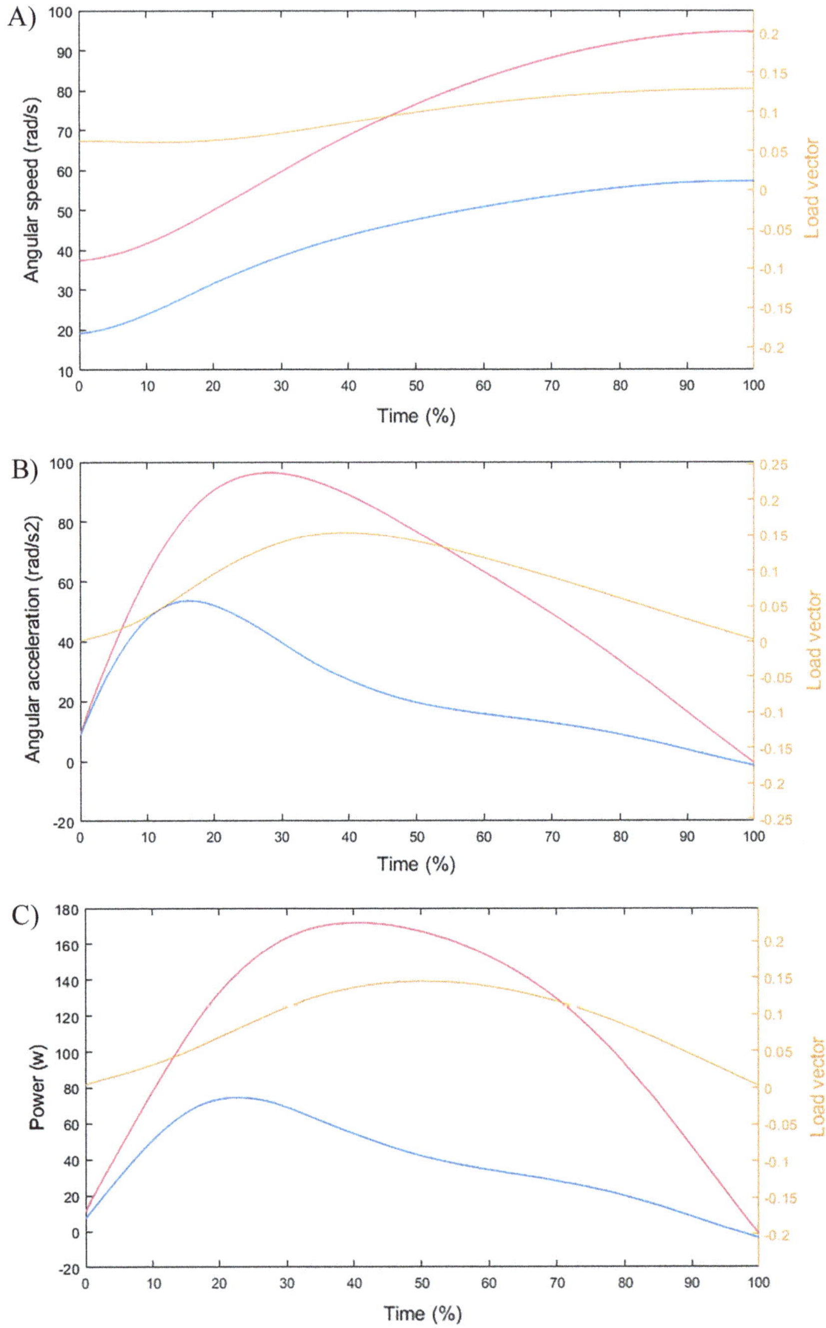

Fig. 1 Raw concentric data for **a** angular speed, **b** angular acceleration and **c** power, during the execution of a unilateral leg extension exercise with a moment of inertia of 0.025 and 0.050 kg m^2 (average). The red line represents the first repetitions, while the blue line represents the last repetitions, out of 30 repetitions. The yellow line (load vector) represents the variation in the curve, analyzed using a Principal Component Analysis test (unpublished private data)

2.4 *Where to Focus Attention*

While traditional equipment has been principally designed to enhance the acceleration of movement or mechanical output during the concentric phase, the flywheel training paradigm has attracted interest because of the eccentric load that can be produced. Several investigations acknowledge that eccentric overload is not always achieved when flywheel devices are used. Indeed, in a recent meta-analysis (pending publication) we show that the eccentric overload depends on the flywheel shaft shape type (and is more likely in a cylinder-shaped device), the moment of inertia used (i.e. lower inertia, ≤ 0.01 kg m^2) and the mechanical output (peak power or peak speed). However, despite those characteristics, the way the eccentric overload is achieved depends on the execution technique.

It is not only the eccentric phase that is of interest in flywheel devices. As previously explained, the flywheel training paradigm is based on a stretch-shortening cycle with a higher overload, compared for example to a countermovement jump using the athlete's bodyweight. Therefore, some authors have quantified the stretch-shortening cycle by considering the mean concentric force during coupled concentric-eccentric phases and the mean concentric force during isolated concentric repetitions. Throws, jumps and sprints could benefit from this kind of quantification during generic preparation in strength and conditioning routines.

3 From Practice

One of the most important aspects of any assessment is data reproducibility and reliability. In flywheel devices, individuals require a good technique when performing exercises. A minimum of two familiarization sessions is required to stabilize the mechanical outputs (i.e., force or power) (Sabido et al. 2018). In addition, when a more complex technique is required (i.e. to eventually achieve eccentric overload), a minimum level of expertise is demanded, even in simple closed-chain exercises such as the flywheel leg curl (Tous-Fajardo et al. 2006).

Flywheel device monitoring can be implemented for several purposes. In the following paragraphs we will cover some of the most important and useful proposals.

3.1 *Progressive Loading Testing*

One of the main differences of progressive loading tests in flywheel devices, compared to free weights, is related to the increment in external load. In free weights, that increment can be higher with light loads and lower with high loads (Pareja-Blanco et al. 2017); using small disks, for example, allows the coach to

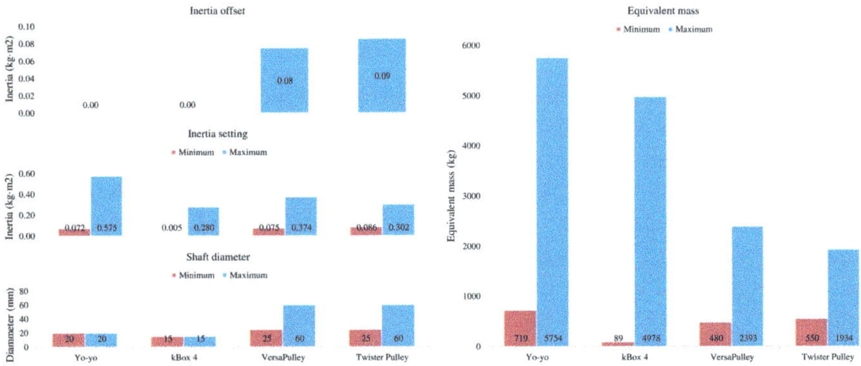

Fig. 2 Mechanical setting of common flywheel resistance training devices. The inertia offset is the inertial load of the base disk included in the flywheel (e.g., the minimum possible inertia to be used). The inertia setting represents the minimum (red) and maximum (blue) available inertial loads for each device. The equivalent mass is calculated as the inertia divided by the squared shaft radius (for more precise information, read the Chap. 1)

gradually increase the relative intensity. However, in flywheel devices, there are fewer options for changing the moment of inertia, especially with a cylinder-shaped flywheel. Hence, sometimes, the rates of increase can be too high, especially when moments of inertia are close to 0.075 kg m^2 in a cylinder-shaped device, or close to 0.2 kg m^2 in the case of a cone pulley. Figure 2 provides information about the available moments of inertia, shaft shape and equivalent mass in some of the most frequently used flywheel devices.

There are several reasons for conducting progressive loading tests using flywheel devices. One of the most common purposes is to determine the load that elicits the highest peak power output (Hoyo et al. 2015). It is also useful to know the amount of *power* or *speed* an individual produces against an absolute moment of inertia, to monitor possible performance enhancements. Finally, the force-velocity profile can also be tested using flywheel devices. For this purpose, a minimum of 6 repetitions averaged over each set and load are required to obtain reliable results, using at least three different moments of inertia (Darjan et al. 2020). In addition, it is important to acknowledge that most studies use two repetitions to accelerate the inertia before starting to monitor the selected variable (Piqueras-Sanchiz et al. 2019), while some use three to four (Sabido et al. 2018). Finally, coaches must consider that, in a flywheel device, arm level modifications and the number of multiplying pulleys used may change the final external load.

3.2 *Eccentric Overload*

To monitor the eccentric phase, several approaches have been proposed. The classic method for monitoring eccentric overload is to use the peak force, typically measured with a strain gauge, and compare the eccentric peak force against the concentric peak force. This allows the calculation of E:C-r. A ratio above 1 means that eccentric overload is produced. However, to monitor this variable using *force*, a force sensor is required, although its setting is sometimes complex. As an alternative, *power*, directly calculated from a rotary encoder, can also be used. Despite this being a common approach, it must be acknowledged that a modification in technique is required to achieve eccentric overload [typically, delaying braking of the kinetic energy, produced during the concentric phase, to the last third of the eccentric phase (Berg and Tesch 1994)]. This technique is not simple and can result in high variability between and within individuals, which may hamper the reliability of measurements for monitoring purposes.

In this regard, the technique for achieving eccentric overload can be modified with training. Recently, a resistance training program, consisting of 9 weeks performing the one-step horizontal acceleration exercise, using a conical pulley device with a moment of inertia that elicited the highest individual mean power output, showed a change in the E:C-r of 0.52–0.61. Despite not producing eccentric overload, the eccentric characteristic of the exercise increased in the group that trained with the flywheel device, but did not change in the control group (Núñez et al. 2019). When monitoring eccentric overload, it is important to account for variation in the reliability of the results, especially due to technique. Accordingly, the E:C-r increased in reliability with training experience, decreasing below a 10% coefficient of variation from the fourth training session. Finally, when eccentric overload is unilaterally tested, it has been shown that the dominant leg showed eccentric overload, but not the non-dominant leg, in a unilateral squat performed with a cylinder-shaped flywheel device with a moment of inertia of 0.010 kg m^2 (Raya-González et al. 2020). This is important because we can easily consider a bilateral eccentric deficit between legs, but this may have a specific sport context (e.g., soccer players).

3.3 *Movement Variability*

One of the main characteristics of flywheel devices is the variable execution of movement, and the possibilities that offers in relation to the degree of uncertainty during resistance training exercises (Gonzalo-Skok et al. 2016). To monitor variability, several indexes can be used. Movement variability is defined as the typical variations that motor performance supports across the execution of the exercise (Stergiou et al. 2006). When using raw data, for example, obtained from a force platform, entropy can be used as a variability index by calculating the SampEn variable (Moras et al. 2019). However, this approach is difficult to conduct in relation to the exercise setting (external sensors) and calculations (raw data processing). A more feasible option may be calculation of the coefficient of variation or, even better, the individual smallest important change (SIC) within the set execution. By definition, the SIC is the minimum change in performance that is related to individual biological variation (Hopkins 2004), in relation to a specific sport task, in our context. It can be calculated by multiplying the within-subject individual standard deviation by a fixed coefficient, chosen for example from the Cohens' effect size thresholds (i.e. 0.2 for a small effect, 0.6 for a moderate effect or above 1.2 for a large effect).

3.4 Real-Time Monitoring Decisions

In recent years, the options for objectively monitoring resistance training mechanical outputs in real time have dramatically increased. In the past, coaches used to program relative intensities (i.e. % of RM), sets, repetitions and resting times for resistance training exercises. Biofeedback options have changed the way those programs are set up, including the option to decide on a suitable training intensity during the exercise (e.g. a given movement speed) and a desired level of fatigue (e.g. number of repetitions or mechanical output losses).

As shown in the previous chapter, velocity-based training supports those ideas. Although its use with free weights is widespread, its scientific implementation in flywheel training is not so common. However, recent work suggests that peak power can be used to determine intra-set fatigue using up to fifteen repetitions (Sabido et al. 2018). Caution should be observed because the authors showed that, when loads from 0.025 to 0.075 kg m^2 in a cylinder-shaped device using the squat exercise were used, a minimum of 10 repetitions were required to achieve significant mechanical fatigue, but only about 6 repetitions were required when a load of 0.1 was used. In addition, the relative mechanical loss is different in the concentric phase than in the eccentric phase. In our laboratory, we have also compared those decrements using the mean or peak speed, acceleration, and power. We observed that every 5 repetitions there was a significant decrement, up to 30 repetitions, using either 0.025 or 0.050 kg m^2 in the leg extension exercise with a cylinder-shaped device, supporting previous data relating to mechanical fatigue. However, we observed that at the 20th repetition, the relative mechanical decrements from the best repetition were different for mean speed (20%), mean angular acceleration (30%) and mean power (40%). The selected variable will depend on the available technology for controlling the exercise in real time but if possible, as explained before, we recommend peak acceleration.

4 Filling Gaps

4.1 In Relation to the Training Intensity

There is little descriptive scientific data, especially correctly classified, relating to exercises performed with flywheel devices, which makes comparisons difficult, even in practical situations. We have shown the importance of providing (1) the moment of inertia, (2) the radius variation of the flywheel shaft, (3) the shaft-shape and (4) the number and type of pulleys used, as a general setting to better reflect the intensity of an exercise performed in a flywheel device. In addition, it is useful to provide the mechanical data on relevant mechanical outputs, such as *speed*, *acceleration* or *power* (either mean or peak).

To establish a proper training intensity, the following recommendations can be followed to perform a progressive loading test using flywheel devices:

1. Avoid using more than four to five loads.
2. Use a minimum of three minutes of rest between sets for light loads and five minutes for medium and high loads.
3. Measure a minimum of six repetitions to be averaged a posteriori, plus two initial repetitions to be discarded.
4. If you are testing a unilateral exercise but using both sides, randomize which side executes each set first, in order to avoid fatigue or post-activation potentiation enhancement effects.

4.2 How to Effectively Achieve Eccentric Overload

In previous chapters it has been noted that the eccentric kinetic energy depends on the concentric energy the individual is able to produce. Hence, the ability to monitor the eccentric phase of the movement completely relies on the concentric phase. There are some options for altering the execution technique to produce eccentric overload, but caution should be exercised when monitoring it because some of them, despite producing a real eccentric overload, will not show it:

1. *Delayed eccentric action*: this refers to the classical approach used in the literature. Concentric kinetic energy is produced by the individual during the eccentric phase, and is braked during the last third of the eccentric phase. Data will reflect the eccentric overload.
2. *Impulse overload*: similar to delayed eccentric action, but the individual goes to the position which corresponds to the last third of the eccentric phase, waiting for braking the kinetic energy at that position. Data will also reflect an eccentric overload.
3. *Overload concentric action*: the individual executes the concentric phase either (1) self-assisted with the use of stronger groups of muscles to support the concentric phase; or (2) spotter-assisted, with the use of external support (i.e. a coach or an implement) to assist in the concentric phase. Data will probably not show eccentric overload due to the likely impossibility of braking kinetic energy much higher than that produced by the individual unaided.

4. *Lateral overload*: the concentric phase is executed with both sides of the body (i.e. bilateral), but the eccentric phase is executed with a single limb. As with the overloaded concentric action, the data will probably not show eccentric overload.

5 Take-Home Messages and Practical Resources

1. To monitor a flywheel device, the shaft shape, shaft radius, number of pulleys used, plane of the movement, and moment of inertia will determine the exercise intensity.
2. When testing eccentric overload in practice, a minimum level of experience is required because the greater the experience, (1) the higher the possibility of achieving eccentric overload, and (2) the higher the reliability of the measurements.
3. Monitoring the mechanical output of the flywheel can help to show intensity and fatigue within a set. Peak acceleration is a good variable for monitoring such outcomes, but coaches and practitioners should be aware of not applying the same relative mechanical losses if different mechanical outputs are monitored.
4. The Smallest Important Change (SIC) can be used to measure movement variability, which occurs in some exercises using flywheel resistance training devices, both within the set and within training sessions.

References

Berg HE, Tesch PA (1994) A gravity-independent ergometer to be used for resistance training in space. Aviat Sp Environ Med 65(8):752–756

Carroll KM, Wagle JP, Sato K, Taber CB, Yoshida N, Bingham GE et al (2019) Characterising overload in inertial flywheel devices for use in exercise training. Sport Biomech 18(4):390–401

Cormie P, McGuigan M, Newton R (2011) Developing maximal neuromuscular power. part 2—training considerations for improving maximal power production. Sport Med 41(2):125–146

Darjan S, Darjan S, Nejc Š (2020) Validity and reliability of force-velocity outcome parameters in flywheel squats. J Biomech 107:109824

De Hoyo M, Pozzo M, Borja S, Carrasco L, Gonzalo-Skok O, Domínguez-Cobo S et al (2015) Effects of a 10-week in-season eccentric-overload training program on muscle-injury prevention and performance in junior elite soccer players. Int J Sports Physiol Perform 10 (1):46–52

González-Badillo JJ, Sánchez-Medina L (2010) Movement velocity as a measure of loading intensity in resistance training. Int J Sports Med 31(5):347–352

Gonzalo-Skok O, Mendez-Villanueva A, Tous-Fajardo J, Berzosa C, Arjol-Serrano JL, Bataller AV et al (2016) Eccentric-overload training in team-sport functional performance: constant bilateral vertical versus variable unilateral multidirectional movements. Int J Sports Physiol Perform 12(7):951–958

Hopkins WG (2004) How to interpret changes in an athletic performance test. Sportscience 8:1–7

Loturco I, Ugrinowitsch C, Roschel H, Tricoli V, González-Badillo JJ (2013) Training at the optimum power zone produces similar performance improvements to traditional strength training. J Sport Sci Med 12(1):109–115

Martinez-Aranda LM, Fernandez-Gonzalo R (2017) Effects of inertial setting on power, force, work, and eccentric overload during flywheel resistance exercise in women and men. J Strength Cond Res 31(6):1653–1661

Moras G, Vázquez-Guerrero J, Fernández-Valdés B, Rosas-Casals M, Weakley J, Jones B et al (2019) Structure of force variability during squats performed with an inertial flywheel device under stable versus unstable surfaces. Hum Mov Sci 66:497–503

Núñez FJ, de Hoyo M, Muñoz López A, Borja S, Otero-Esquina C, Sanchez H et al (2019) Eccentric-concentric ratio: a key factor for defining strength training in soccer. Int J Sports Med 40(12):796–802

Núñez FJ, Galiano C, Muñoz-López A, Floria P (2020) Is possible an eccentric overload in a rotary inertia device? Comparison of force profile in a cylinder-shaped and a cone-shaped axis devices. J Sports Sci 38(14):1624–1628

Pareja-Blanco F, Sanchez-Medina L, Suarez-Arrones L, Gonzalez B (2017) Effects of velocity loss during resistance training on performance in professional soccer players. Int J Sports Physiol Perform 12(4):512–519

Piqueras-Sanchiz F, Martín-Rodríguez S, Martínez-Aranda LM, Lopes TR, Raya-González J, García-García Ó et al (2019) Effects of moderate vs. high iso-inertial loads on power, velocity, work and hamstring contractile function after flywheel resistance exercise. PLoS One 14(2): e0211700

Raya-González J, Castillo D, Domínguez-Díez M, Hernández-Davó JL (2020) Eccentric-overload production during the flywheel squat exercise in young soccer players: implications for injury prevention. Int J Environ Res Public Health 17(10):3671

Sabido R, Hernández-Davó JL, Pereyra-Gerber GT (2018) Influence of different inertial loads on basic training variables during the flywheel squat exercise. Int J Sports Physiol Perform 13 (4):482–489

Sanchez-Medina L, Perez CE, Gonzalez-Badillo JJ (2010) Importance of the propulsive phase in strength assessment. Int J Sports Med 31(2):123–129

Sjöberg M, Berg HE, Norrbrand L, Andersen MS, Gutierrez-Farewik EM, Sundblad P et al (2020) Influence of gravity on biomechanics in flywheel squat and leg press. Sport Biomech

Stergiou N, Harbourne RT, Cavanaugh JT (2006) Optimal movement variability: a new theoretical perspective for neurologic physical therapy. J Neurol Phys Ther 30(3):120–129

Tous-Fajardo J, Maldonado RA, Quintana JM, Pozzo M, Tesch PA (2006) The flywheel leg-curl machine: offering eccentric overload for hamstring development. Int J Sports Physiol Perform 1(3):293–298

Vázquez-Guerrero J, Moras G, Baeza J, Rodríguez-Jiménez S (2016) Force outputs during squats performed using a rotational inertia device under stable versus unstable conditions with different loads. PLoS One 11(4):e0154346

Worcester KS, Baker PA, Bollinger LM (2020) Effects of inertial load on sagittal plane kinematics of the lower extremity during flywheel-based squats. J Strength Cond Res

Alejandro Muñoz-López has a M.Sc. in High Performance in Sports, in soccer performance and injury rehabilitation, and a Ph. D. in Sport Sciences. He is a former strength and conditioning coach of some professional and amateur Spanish soccer teams and of the Latvian Football Federation. His main research line is monitoring the training load of field and strength exercises applying new advances in sports technology. He was the co-founder of Science2improve, a company with the main aim to help coaches and practitioners to apply the new advancements in sports science to their daily routines. He is working as a professor at the University of Seville (Spain) in the sports training subject. Despite being young, the continuous application of the theory to the practice in the strength and conditioning topic over the last ten years realized him about this book idea. He is also an author and co-author of relevant sports science papers. At the moment, he is more focused on the development of a new training methodology to implement the flywheel resistance training devices on resistance training programs. As part of this purpose, he is a sports scientist and consultant of SmartCoach Inc.

Fabio Yuzo Nakamura has M.Sc. and Ph.D. in sports sciences (São Paulo State University). He is a professor at the University Institute of Maia (ISMAI, Portugal) and scientific consultant of AMP: Science-Based Performance. He started his career at State University of Londrina (Brazil) and has worked as the senior physiologist of Nucleus of High Performance in Sport prior to the 2016 Olympic Games. He has more than 200 publications and is currently a member of the editorial boards of relevant scientific journals in the fields of sports training and physiology.

How to Use Force Sensors for Resistance Training in Daily Practice

Alberto Sánchez-Sixto and John J. McMahon

Abstract The testing and assessment of resistance training exercises is a fundamental aspect for coaches and athletes. Through the force-time data measured by force plates, we have the possibility to calculate velocity, displacement, work, and power values of the centre of mass. This chapter has a theory section where we explain why force plate are useful to evaluate isometric and ballistic action during resistance training. Also, we explore how we can obtain velocity, displacement, power and work variables from force-time data through the impulse method. The chapter contains a practice section where we response some key questions when setting up a force plate to assess athletes' physical performances. Then, we describe how to perform an Isometric Mid Tight Pull Test (IMTP) and a Countermovement Jump Test (CMJ). We explain how we can obtain biomechanical variables from both tests and we discuss about the biomechanical variables that provide important information to interpret correctly the IMTP and CMJ tests. Finally, we added a filling the gap section where we provide several recommendations on how to implement the evidence-based theory in real life applied sports environments.

Keywords Force-time data · Physical performance · Isometric mid tight pull test · Countermovement jump test

---

A. Sánchez-Sixto (✉)
Departamento de Deporte, Centro de Estudios Universitarios Cardenal Spinola CEU, Campus Universitario CEU Andalucía. Glorieta de Ángel Herrera Oria, s/n, 41930 Bormujos, Sevilla, España
e-mail: asanchezsixto@ceuandalucia.es

J. J. McMahon
Directorate of Sport, Exercise and Physiotherapy, University of Salford, Salford, UK

A. Muñoz-López et al. (eds.), *Resistance Training Methods*, Lecture Notes in Bioengineering, https://doi.org/10.1007/978-3-030-81989-7_11

1 Introduction

The testing and assessment of resistance training exercises is a fundamental aspect for coaches and athletes. The possibility to measure the effects of loads applied to athletes can be deemed essential to assess the resistance training demands and to determine the effects of the training process. An important aspect that it is necessary to take into consideration is that athletes apply force and the other variables that strength and conditioning professionals like to measure are a consequence of the differences between the force that athletes produced and the force that the resistance (either body weight [BW] or BW plus external load) provided among the time. For this reason, the possibility to measure the force applied by athletes in the resistance training exercise executions is fundamental to understand their performance.

Nowadays, researchers and strength and conditioning professionals have the possibility to obtain information of force, acceleration, velocity, displacement or power from different devices. There are a lot of investigations that analysed the validity and reliability of the devices utilised to control the resistance training process (Balsalobre-Fernández et al. 2015; Ferro et al. 2019). In this chapter we will focus solely on force plates devices, since they are considered the gold standard device to measure force (Bampouras et al. 2013; Crewther et al. 2011). Instruments that measure other variables (i.e. velocity, acceleration…) estimate the force applied, so these devices do not directly measure the force. Force plates can measure the Newtons of force that athletes performed against them over the time, so this device give us a direct measure of the force applied by athletes.

Through the force-time data measured by force plates, we have the possibility to calculate velocity, displacement, and power variables of the centre of mass. Strength and conditioning professionals analyse force, velocity, displacement, and power variables to assess the resistance training process and to evaluate the athlete's performance (Schoenfeld 2010; Kasovic et al. 2019). When professionals evaluate resistance training, like isometric and ballistic actions, discrete and continuous analysis could be performed (Comfort et al. 2015; Suchomel et al. 2020). In that sense, we can analyse discrete values where we obtain information on the key actions that we are evaluating. This is the most common analysis performed by strength and conditioning professionals to assess resistance training, so in this chapter we will explore how we can obtain all of these variables from force–time values obtained by force plates.

2 From Theory

Force plates allow the execution of different actions carried out on it and most systems are able to measure vertical, horizontal and lateral forces that athletes apply against their surface. However, there are modern devices that only include vertical force which is usually evaluated for resistance training actions. When we utilise a force plate, we have to take into account some considerations to ensure that the data registered can be well analysed. First of all, we are going to explain why force plates are an interesting device when we try to assess resistance training exercises.

2.1 Why Force Plates Are a Good Device to Assess Resistance Training?

Force plates are instruments that allow us to obtain information about the force that an athlete is applying when they perform an action on its surface. One of the main advantages that we have when we evaluate resistance training actions with force plates is that we can register force even when the participant remains static. This is very important because we can analyse isometric actions, in which there is no movement, and we can evaluate the ability of athletes to apply force when faced with a static action (Comfort et al. 2015; Wells et al. 2019). Many tests and training exercises use isometric actions to assess and increase the ability of athletes to apply force, so force plates would be a suitable instrument for this purpose (McMaster et al. 2014). Another important aspect is that before a movement occurs in any action, a difference between the force applied by the athlete and the force applied by the load is necessary to generate an acceleration and velocity of the centre of mass. Devices that measure velocity, acceleration or displacement only would not allow us to obtain this information until the athlete starts the movement. So, this means that we will lose important information of how athletes apply force before the movement occurs, thus force data will be useful to understand why the movement start at this way (Badillo and Ribas 2002). This is very important in exercises where the eccentric phase is not performed and we just want to evaluate a concentric action. We have to know that there are some variables where the maximum value will be reached in the static phase of the movement, such as the maximum rate of force development (RFDmax), especially when we use loads that are higher than the 30% of the one maximum repetition (1RM) in the bench press or squat exercise (Badillo and Ribas 2002). Furthermore, during dynamic actions, there can sometimes be a discrepancy between the movement of the athlete's centre of mass and the movement of device that is measuring the velocity, acceleration or displacement, depending on where the device is positioned on the athlete or barbell (Lake et al. 2012). This is another reason why we should use force plates in order to have a complete information of the actions evaluated.

2.2 How to Analyse Vertical Force-Time Data?

As mentioned previously in this chapter, force plates only measures force data during a given time, so it is necessary to calculate other variables commonly used to

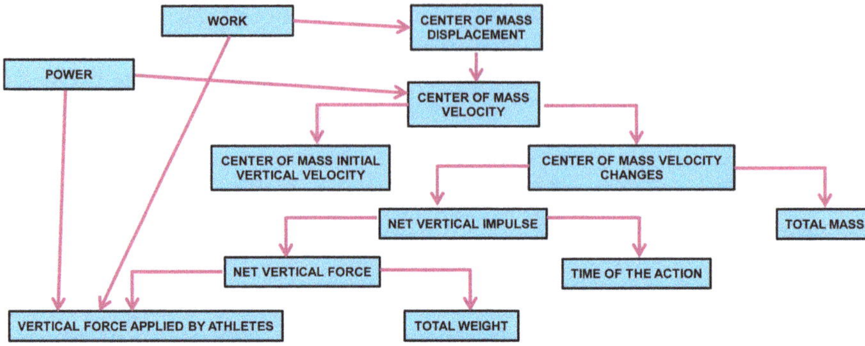

Fig. 1 Diagram of variables from force-time data to centre of mass displacement

evaluate the resistance training performance (velocity, displacement, power…). To perform this analysis, the following steps can be follow. In this section, the impulse method will be explained because it is a method that is frequently used to measure vertical actions, such as jumps (McBride et al. 2010; Kirby et al. 2011). Figure 1 shows how we can obtain biomechanical variables from vertical force-time data.

In order to perform the analysis, we need to establish the weight baseline, including the weight of the load (where applicable) and the BW of the athlete. We will calculate the mean value of the first second of the force data (Owen et al. 2014) or the mean value of the initial 2 s of the force data (Street et al. 2001), while the athlete remains static, to calculate the athlete's BW and, where necessary, the training load. The combination of the athlete BW and training load is often referred to as the system weight or mass.

Once we know this the BW of system weight, it necessary to determine the start of the movement. Following the recommendation proposed by Street et al. (2001) we have to calculate a threshold which is defined as 1.75 times the residual peak found during the first 2 s of the force-time data. The residual peak is the maximum force value registered by the force plate during the 2 initial seconds. After that, when the force values exceed the upper or lower threshold calculated, a backward inspection must be carried out in the force-time data until identifying the instant at which the vertical force crosses the mean force value calculated while the athlete was static (Street et al. 2001). This will be the point of the start of the exercise. Following Owen et al. (2014), there is an alternative to detect the start of the movement. They considered the first second where athletes remained static to establish the BW of athletes. Then, they calculated the standard deviation (SD) during the first second and when the force exceeded BW plus or minus 5 times the SD, the start of movement was considered (Owen et al. 2014).

Once we have detected the start of the movement, we have to calculate the net force value at each instant. The net force is understood as the difference between the reaction forces produced by the athlete and the BW or system weight.

Once we have calculated the vertical net force, we have to calculate the net vertical impulse. The net vertical impulse can be defined as the area under the force-time curve (McBride et al. 2010) and it can be calculated by integrating the net vertical force respect to the time.

Once we have calculated the net vertical impulse of each instant, the next step is calculating the vertical velocity of the centre of mass of the athlete during the resistance training execution. The net vertical impulse divided by the system mass represent the change in velocity of the centre of mass at each instant. Then, knowing the vertical velocity of the centre of mass, we can calculate the vertical displacement of the centre of mass during the exercise.

Another important parameter used to assess the resistance training is power, we can obtain the power that the athlete is performing with the data obtained through the product of the vertical reaction forces and the vertical velocity of the centre of mass at each instant.

Finally, another parameter that has also been used to evaluate strength training is the mechanical work. This parameter can be calculated as the integral of force and displacement.

Figure 2 shows an example of a countermovement jump curves obtained by force-time integration.

3 From Practice

3.1 How to Measure Vertical Resistance Training Exercises with Force Plates?

The procedures that trainers perform when they measure a resistance training action are very important to ensure that the data measured will be useful for the analysis. This section focuses on the steps to follow to analyse exercises in which athletes start on the force plate, perform the exercise and finish on it.

- First of all, we have to be sure that the force plate is placed on a hard and firm surface.
- The data collection duration must be larger than the athlete execution, as we need to obtain information of the entire exercise. This is important in order to avoid losing part of the exercise performance.
- The force plate sampling frequency should be set to at least 1000 Hz, or the highest possible sampling frequency if 1000 Hz is not possible (but ideally no less than 500 Hz).
- Prior to the record of the movement we have to perform a "Zero". This removes any existing force that may registered by the force plate and so helps to reduce signal noise.

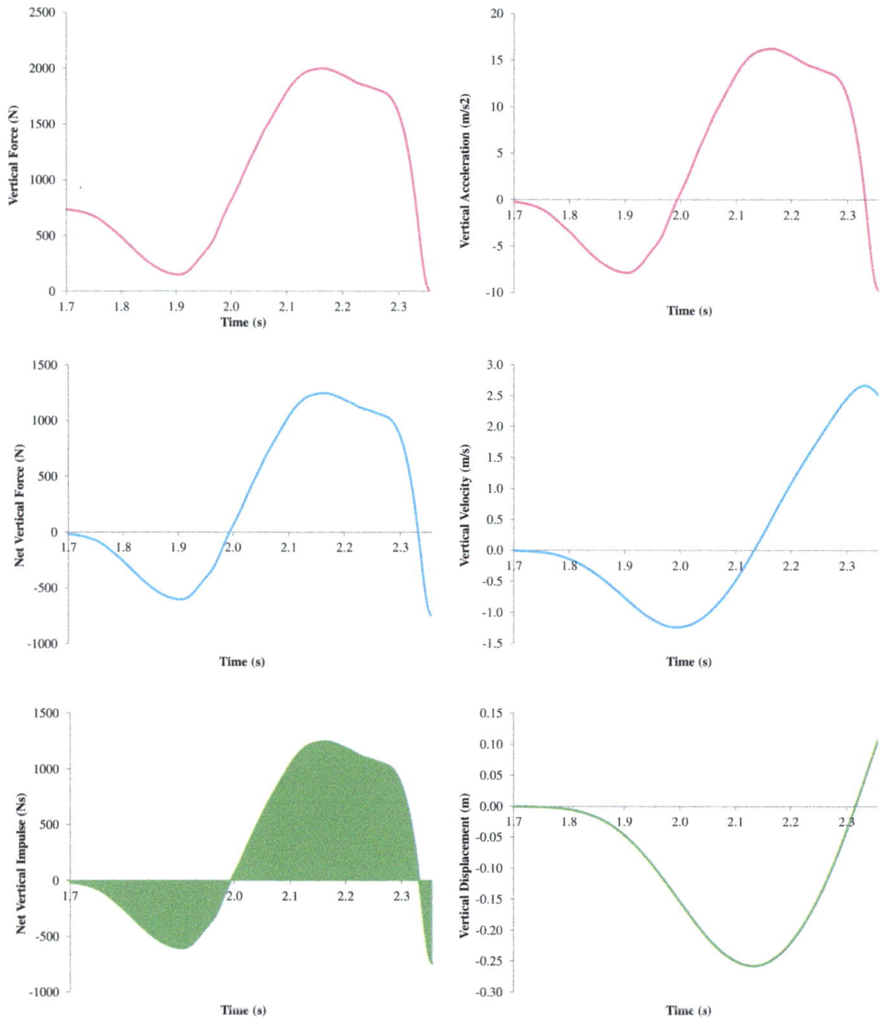

Fig. 2 An example of applying the numerical integration process to a countermovement jump force-time curve, from the onset of movement to take-off

- Athletes have to be static on the force plate prior the start of the movement. They must be static for at least 1 s (Owen et al. 2014) or 2 s (Street et al. 2001) prior to the start of the movement.
- Then they have to perform the exercise and stay on the force plate until the data collection finishes.

Figure 3 shows key questions to answer when setting up a force plate to assess athletes' physical performances.

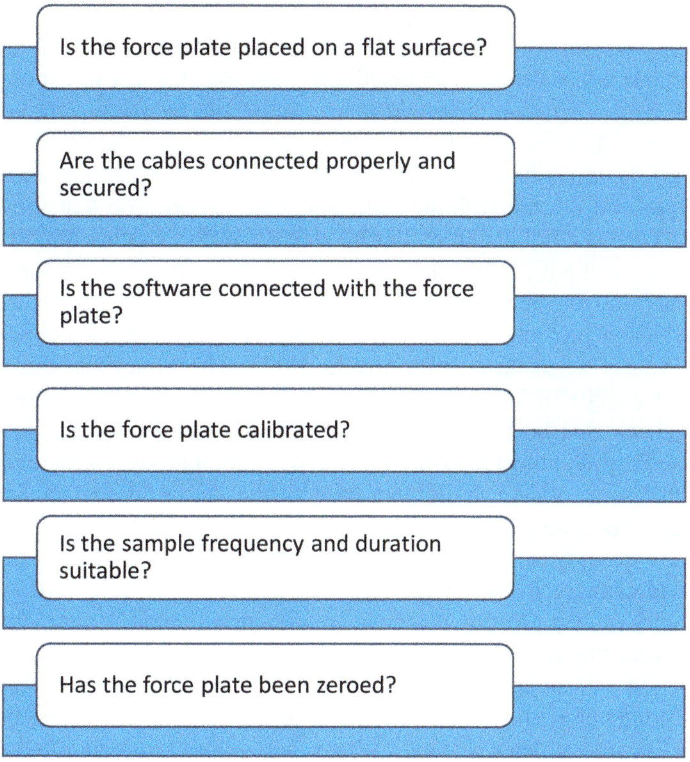

Fig. 3 Key questions to answer when setting up a force plate to assess athletes' physical performances. *Isometric tests with force plates—the isometric mid thigh pull test (IMTP)*

Isometric tests to evaluate the strength capacities of athletes are widely used. The use of these tests is due to the procedures used to conduct this type of test being easier when we have to evaluate a large number of participants. In addition, the test execution is safer in comparison with one maximum repetition dynamic tests (Comfort et al. 2015). There are different isometric tests available to researchers and practitioners, but we focus on the isometric mid-thigh pull (IMTP) in this chapter because is one of the most popular multi-joint isometric actions used to evaluate athletes' force capacities and it has a high test-retest reliability (Comfort et al. 2015; Haff et al. 2015).

The IMTP is a test where athletes adopt a position similar to the second pull of the clean exercise. At this position, athletes can apply the highest force so strength and conditioning professionals have to perform this test with this position in order to measure the maximal force capacities of the athletes (Comfort et al. 2019). In all the isometric tests, the position adopted by the athletes is decisive, since if it is not the same, the possibility of applying force will be different and this will cause the results obtained from the test to be not comparable. Several investigations have analysed the IMTP test, however participants executed the action with different knee and hip angles (Brady et al. 2020). In order to establish a consensus, researchers found that the optimal position to perform this test is with a 130–140 degree knee angle (between thigh and shank) and a 145 degree hip angle (between thigh and torso) (Brady et al. 2020). An important aspect, in order to achieve this hip and knee angle position, is ankle dorsiflexion, where evaluators must instruct athletes to perform the test with the sufficient ankle dorsiflexion that allow athletes to adopt the knee and hip angles explained above. Finally, another important aspect is that athletes must have their trunk upright resulting in an adequate forward trunk oscillation of 5–10 degrees with respect to an upright position (Comfort et al. 2019). This is because a previous investigation found that powerlifters produced higher peak force values when they performed the IMTP with an upright trunk position (Beckham et al. 2012). The bar should be allocated in the correct position after evaluators check that body position is well fitted following the previous recommendations. Then, the most important thing, when the purpose of the evaluation is measuring the same athlete in different parts of the season, is maintaining the bar and the body position equal between the different execution of the IMTP test. In the event that the process is conducted properly, the changes in the force applied will not be due to changes in the execution of the test (Comfort et al. 2019). We just need to modify the bar position when we are measuring young athletes that may be of growing age, this is necessary to maintain the body angles. Once we have the correct position we can start with the test. We will instruct athletes to stay static at least one second before the start the action, as described in the earlier section. Then, we have to instruct athletes to 'pull as hard and as fast as possible' (Brady et al. 2020) or 'push your feet into the ground as fast and as hard as possible' (Comfort et al. 2019). The most important thing again is to instruct every athlete with the same sentence and volume voice. Athletes have to perform an effort of 5 s and then they must stay at least one second in the same initial position.

The next step after measuring an action is the analysis of the data that we obtain. We have to take some considerations to consider that the test was well executed. When analyzing the force–time curve we must check that the force values obtained before the start of the action, remains equal to the athlete BW. In the same way, once we have finished the action, the force values must be similar to the baseline. Another important aspect is that once the gesture starts, the force applied should rise without a countermovement because it will affect RFD values. Then, is necessary to determine the start of the action and there are two possibilities, to establish an absolute or a relative threshold from the baseline (Comfort et al. 2019). In our opinion, setting a relative threshold will be more appropriate to ensure that the weight differences between athletes or within athletes in different moments of the season, do not affect the threshold setting. For that reason, following West et al. (2011) recommendations for the IMTP test, the start of the action could be considered when the force value exceeded the mean plus 5 SDs from the force baseline (West et al. 2011).

Several variables have been used to evaluate the IMTP performance. Among the variables analysed in the literature, we will focus on peak force, the force and impulse obtained at different time instants. The reason for using these variables is because they showed the highest reliability and are considered more suitable for evaluating the IMTP (Brady et al. 2020). The peak force is the maximum force value obtained during the test and gives information about the maximum force capacity of athletes. There are investigations that accounted for the body mass in the force value obtained, in our opinion it is the best option, but as we said previously, the most important thing is to standardize this analysis. The force to different time instants were calculated by dividing the peak force achieved by the time (from 0–50 to 0–250 ms), showing the RFD (Haff et al. 2015). This variable gives information about how athletes are able to achieve an amount of force during a time given (Haff et al. 2015). Some investigations calculate the maximum value of RFD or the RFD average but these variables showed a poor reliability an high coefficient of variation (CV), so it is not recommended for the IMTP interpretations (Brady et al. 2020). Finally, the impulse, defined above, could be calculated over 0–50 to 0–250 ms periods and showed a high interclass correlation coefficient (ICC) and low CV, so it can provide reliable information of the IMTP (Comfort et al. 2015; Brady et al. 2020).

3.2 *Ballistics Tests with Force Plates—The Countermovement Jump (CMJ)*

The ability of athletes to accelerate rapidly is very important in many sports activities. There are several ballistics tests that can be conducted with force plates and, in this section, we will focus on vertical jump assessment. The vertical jump is widely analysed for multiple purposes such as evaluating the stretching-shortening cycle, fatigue, training effects… Many devices are able to evaluate jumping performance but force plates are considered the gold standard and other devices checked their validity and reliability from them (Balsalobre-Fernández et al. 2015; Bampouras et al. 2013; Crewther et al. 2011). It is possible to analyse several jump types but the most popular one used in the literature is the CMJ. For this reason, we focus in this section on the most important aspects when evaluating the CMJ.

When evaluating the vertical jump with a force plate we must check that we follow some considerations. Participants must stand upright with their hands akimbo and remain static for at least 1 s before executing the jump (Owen et al. 2014). During the entire duration of the measurement they must remain with their hands akimbo. A very important aspect is the instruction that evaluators gave to the participants prior the CMJ execution because it will substantially modify the results obtained. In this way, previous investigations showed that instructions which reduce the countermovement depth induced a lower jump height in comparison with instructions where the countermovement depth was self-selected by participants (Kirby et al. 2011; Sánchez-Sixto et al. 2018; Salles et al. 2011). On the other hand, other investigations found higher jump height values when participants performed a deeper countermovement depth (Sánchez-Sixto et al. 2018) or the jump height remains constant (Kirby et al. 2011). Another modifiable variable by instructions is the vertical velocity of the centre of mass during the countermovement phase. Previous investigations showed that instructions that emphasized to jump as fast as possible reduced the jump time, affecting the reactive strength index modified (RSI_{Mod}) (jump height divided by jump time from the start of the movement to take-off) (Pérez-Castilla et al. 2019). Another important aspect to be considered is that resistance training can modify how athletes apply force and the jump execution will suffer modifications (Arabatzi et al. 2010). In that sense, we think that the instructions given should limit athletes as little as possible. For this reason, out of all possible instructions, one of the most appropriate could be the simple one 'jump as high as possible'. Another relevant instruction when jumping ability is used to evaluate the explosiveness of athletes could be "jump as high and as fast as possible" due to this instruction eliciting higher RSI_{Mod} values (Barker et al. 2018; McMahon et al. 2018).

Strength and conditioning professionals utilize several variables to evaluate the vertical jump and the most utilized one is the jump height. However, analyzing other biomechanical variables provide useful information when measuring the effect produced by resistance training programs.

Force variables, such as maximum force and mean force during the propulsive phase are variables used to analyse the vertical jump. When analyzing force during a vertical jump, it is important to know that the centre of mass displacement affects the forces values obtained (Kirby et al. 2011; Salles et al. 2011). Previous investigations showed lower force values and a higher jump height when deeper countermovement depths were achieved (Kirby et al. 2011; Sánchez-Sixto et al. 2018; Salles et al. 2011). In addition, participants reached higher force values and lower jump height values when the countermovement depth decreased (Sánchez-Sixto et al. 2018). This fact is because is harder to apply force when the centre of mass is in a deeper position (Moran and Wallace 2007). Then, higher jump height when lower maximum and mean force values are achieved is explained because the jump height is defined by the net vertical impulse (McBride et al. 2010; Kirby et al. 2011). This information shows how important it is to consider the centre of mass displacement when the force during countermovement jumps is analysed.

Power is another variable widely evaluated when vertical jump is analysed. Previous investigations showed that peak power and mean power have a relationship with the jump height (González-Badillo and Marques 2010). However, recent investigation showed that the centre of mass displacement affects the power output, due to, as was discussed above, the force applied (Markovic et al. 2014; Morin et al. 2019). For that reason, the same peak power was found with different jump height and countermovement depth combinations (Salles et al. 2011). In that sense, considering the centre of mass displacement is important to determine to accurately interpret the power output during vertical jumps.

Finally, velocity of centre of mass variables of the vertical jump is useful for evaluating resistance training programs. For example, the vertical centre of mass velocity is often used to identify different phases of the jump which allows for a more complete analyses of the force-time record between test occasions. Additionally, maximum velocity during the jump propulsive phase seems to be a more sensitive variable to detect between-jumps differences in comparison with the jump height (Jiménez-Reyes et al. 2016). This is due to participants reaching the maximum centre of mass velocity before the take-off, so peak velocity suppresses take-off instant errors. However, the peak velocity should not be used to estimate jump height because of the magnitude of the deceleration that occurs just prior to take-off being dependent on athlete body mass and stature.

4 Filling Gaps

Just because it is possible to objectively evaluate several resistance training exercises with force plates does not mean that we must evaluate all of them. Instead, we must carefully select which resistance training exercises to assess via force plates based on a clear rationale. Otherwise, we risk overassessing our athletes which can result in an overwhelming accumulation of force data that is not used to directly inform the athlete's training priorities (due to the time it takes to process the data) and, possibly, more emphasis being placed on data collection than coaching the athletes in front of us. Indeed, it has been suggested that there is a risk of relying too much on technology generally due to its increased prevalence across sports, but it is possible to strike the right balance between performance assessment and strength and conditioning coaching. To avoid 'tipping the scales' too far towards force plate data collection specifically, there is a requirement to better bridge the gap between some of research presented earlier and applied practice.

From here on in, we provide several recommendations on how to implement the evidence-based theory in real life applied sports environments. Firstly, although a minimum sample frequency of 1000 Hz has been widely suggested in the scientific literature when collecting force-time data for most resistance exercises (Owen et al. 2014; Street et al. 2001), some force plates that are used by practitioners may not be able to acquire data at this rate (e.g. some force plates have a maximum sample frequency of 500 Hz). In instances such as this, we recommend that practitioners use the highest possible frequency that their force plates can be set to, provided that it is a minimum of 500 Hz. The most important thing is that the same sample frequency is used each time the athletes are assessed, as changing the sample frequency will affect many commonly reported variables and thus may mask or inflate any 'true' changes in the athletes' performances.

Secondly, it has been suggested that the number of trials required to minimize the signal-to-noise ratio varies depending on the force-time derived variable being reported. For example, it has been advised that while averaging data across up to eight trials will reduce the coefficient of variation by 1–8%, depending on the variable, it may not be feasible to apply this in practice due to the additional time requirement (Kennedy and Drake 2018). Certain variables, such as jump height, peak velocity and peak concentric power obtained from a countermovement jump, have been shown to exhibit acceptable signal-to-noise ratios when the average of 2–3 trials were taken (Kennedy and Drake 2018). Composite variables, such as reactive strength index modified and flight time to contraction time ratio, displayed poor signal-to-noise ratios, irrespective of the treatment method. Therefore, it is important for practitioners to select force-time (and derived) variables based not only on their suggested importance to the test aims but also on their ability to detect 'true' changes in athletes' performances. For example, if the scientific literature suggests that reactive strength index modified is an important countermovement jump variable to report for a given sport but it does not possess sufficient sensitivity to identify acute changes athletes' performances, then its inclusion is likely to be inappropriate if it is intended to be used as an acute monitoring metric. It should be noted that signal-to-noise ratios should be established for the specific population being tested and it may be the case that reactive strength index modified can be applied to acute monitoring scenarios for certain athletes. Further research is required to determine this.

When focusing on force plate assessments of vertical jumping, most research suggests that athletes keep their arms akimbo throughout the assessment. Whilst this recommendation is to facilitate test standardization, it may not always be the most appropriate advice for jump-based sports. For example, for basketball and volleyball athletes, who perform a large volume of jumping actions that involve arm swing during competition, it may be prudent to conduct their vertical jump assessments with the inclusion of arm swing either in replace of, or in addition to, jumps performed with arms akimbo (Heishman et al. 2020). Per our earlier point, the most important thing here is to standardize whichever approach is adopted between testing sessions involving the same athletes. The decision about which countermovement jump protocol to administer is the practitioner's, however, they should be informed that involving arm swing generally results in higher between-trial coefficients of variation, even if it is more relevant for certain sports.

For certain resistance training exercises assessed on a force plate, it may be suggested in the scientific literature that the involved joint angles should be measured and standardized among all athletes during each test. While the importance of standardizing all methods within subjects is not to be undermined, subtle differences in joint angles (such as hip and knee angles during the isometric mid-thigh pull test) between subjects is unlikely to create a huge problem for the practitioner. For example, 'optimal' hip and knee angles for generating highest peak forces have been reported for the isometric mid-thigh pull test, however, some studies have suggested that the athlete's self-preferred hip and knee angles may yield significantly similar peak forces to the so-called optimal condition (Comfort et al. 2015). Therefore, rather than spending additional time setting athletes up with 'optimal' hip and knee angles, it may be more efficient, yet not detrimental, for practitioners to opt to allow athletes to perform the isometric mid-thigh pull test with their preferred hip and knee angles. One caveat to this suggestion is that practitioners should correct hip and knee angles if some athletes adopt a completely incorrect position (i.e. if they have excessive hip flexion which does not at all resemble the mid-thigh pull position).

5 Take Home Messages

(1) Force plates can measure the Newtons of force that athletes performed against them over the time, so this device give us a direct measure of the force applied by athletes. One of the main advantages that we have when we evaluate resistance training actions with force plates is that we can register force even when the participant remains static. Following the impulse method, we can obtain acceleration, velocity, displacement, power or work variables. We have to be sure that: before the start of a test, we have placed the platform on a flat and firm surface, the data collection duration is larger than the athlete execution, the sampling frequency is set at 1000 Hz (if it is possible), we perform a "Zero" previous to the movement, athletes stay static for at least 1 s prior to the start of the movement and they stay on the force plate until the data collection finished.

(2) The IMTP is one of the most popular multi-joint isometric actions used to evaluate athletes' force capacities and it has a high test-retest reliability. The optimal position to perform this test is with a 130–140 degree knee angle (between thigh and shank) and a 145 degree hip angle (between thigh and torso). The peak force, the force and impulse obtained at different time instants are the variables that showed highest reliability and are considered more suitable for evaluating the IMTP.

(3) The vertical jump is widely analysed for multiple purposes such as evaluating the stretching-shortening cycle, fatigue, training effects. A very important aspect is the instruction that evaluators gave to the participants prior the CMJ execution because it will substantially modify the results obtained. We have to give always the same instruction in order to compare our results and we have to avoid limit the execution as less as we can. Force and power variables must take into consideration the centre of mass displacement because the countermovement depth affect force values and we can have wrong interpretation of the athletes jumping ability.

References

Arabatzi F, Kellis E, De Villarreal ESS (2010) Vertical jump biomechanics after plyometric, weight lifting, and combined (weight lifting + plyometric) training. J Strength Cond Res 24(9): 2440–2448

Badillo J, Ribas J (2002) Bases de la programacion del entrenamiento de fuerza. INDE publicaciones

Balsalobre-Fernández C, Glaister M, Lockey RA (2015) The validity and reliability of an iPhone app for measuring vertical jump performance. J Sports Sci 33(15):1574–1579

Bampouras TM, Relph NS, Orme D, Esformes JI (2013) Validity and reliability of the Myotest Pro wireless accelerometer in squat jumps. Isokinet Exerc Sci 21:101–105

Barker LA, Harry JR, Mercer JA (2018) Relationships between countermovement jump ground reaction forces and jump height, reactive strength index, and jump time. J Strength Cond Res 32(1):248–254

Beckham G, Lamont H, Sato K, Ramsey M, Haff GG, Stone HM (2012) Isometric strength of powerlifters in key positions of the conventional deadlift. J Trainology 1(2):32–35

Brady CJ, Harrison AJ, Comyns TM (2020) A review of the reliability of biomechanical variables produced during the isometric mid-thigh pull and isometric squat and the reporting of normative data. Sports Biomechanics 19:1–25

Comfort P, Jones PA, McMahon JJ, Newton R (2015) Effect of knee and trunk angle on kinetic variables during the isometric midthigh pull: test-retest reliability. Int J Sports Physiol Perform 10(1):58–63

Comfort P, Dos'Santos T, Beckham GK, Stone MH, Guppy SN, Haff GG (2019) Standardization and methodological considerations for the isometric midthigh pull. Strength Cond J 41(2):57–79

Crewther BT, Kilduff LP, Cunningham DJ, Cook C, Owen N, Yang GZ (2011) Validating two systems for estimating force and power. Int J Sports Med 32(4):254–258

Ferro A, Floría P, Villacieros J, Muñoz-López A (2019) Maximum velocity during loaded countermovement jumps obtained with an accelerometer, linear encoder and force platform: a comparison of technologies. J Biomech 95:109281

González-Badillo JJ, Marques MC (2010) Relationship between kinematic factors and countermovement jump height in trained track and field athletes. J Strength Cond Res 24(12):3443–3447

Haff GG, Ruben RP, Lider J, Twine C, Cormie P (2015) A comparison of methods for determining the rate of force development during isometric midthigh clean pulls. J Strength Cond Res 29(2):386–395

Heishman AD, Daub BD, Miller RM, Freitas EDS, Frantz BA, Bemben MG (2020) Countermovement jump reliability performed with and without an arm swing in NCAA Division 1 intercollegiate basketball players. J Strength Cond Res 34(2):546–558

Jiménez-Reyes P, Pareja-Blanco F, Rodríguez-Rosell D, Marques MC, González-Badillo JJ (2016) Maximal velocity as a discriminating factor in the performance of loaded squat jumps. Int J Sports Physiol Perform 11(2):227

Kasovic J, Martin B, Fahs CA (2019) Kinematic differences between the front and back squat and conventional and sumo deadlift. J Strength Cond Res 33(12):3213–3219

Kennedy RA, Drake D (2018) Improving the signal-to-noise ratio when monitoring countermovement jump performance. J Strength Cond Res

Kirby TJ, McBride JM, Haines TL, Dayne AM (2011) Relative net vertical impulse determines jumping performance. J Appl Biomech 27:207–214

Lake JP, Lauder MA, Smith NA (2012) Barbell kinematics should not be used to estimate power output applied to the barbell-and-body system center of mass during lower-body resistance exercise. J Strength Cond Res 26(5):1302–1307

Markovic S, Mirkov DM, Nedeljkovic A, Jaric S (2014) Body size and countermovement depth confound relationship between muscle power output and jumping performance. Hum Mov Sci 33(1):203–210

McBride JM, Kirby TJ, Haines TL, Skinner J (2010) Relationship between relative net vertical impulse and jump height in jump squats performed to various squat depths and with various loads. Int J Sports Physiol Perform 5(4):484–496

McMahon JJ, Jones PA, Suchomel TJ, Lake J, Comfort P (2018) Influence of the reactive strength index modified on force- and power-time curves. Int J Sports Physiol Perform 13:220–227

McMaster DT, Gill N, Cronin J, McGuigan M (2014) A brief review of strength and ballistic assessment methodologies in sport. Sports Med 44:603–623

Moran K, Wallace E (2007) Eccentric loading and range of knee joint motion effects on performance enhancement in vertical jumping. Hum Mov Sci 26:824–840

Morin JB, Jiménez-Reyes P, Brughelli M, Samozino P (2019) When jump height is not a good indicator of lower limb maximal power output: theoretical demonstration, experimental evidence and practical solutions. Sport Med 49(7):999–1006

Owen NJ, Watkins J, Kilduff LP, Bevan HR, Bennett MA (2014) Development of a criterion method to determine peak mechanical power output in a countermovement jump. J Strength Cond Res 28(6):1552–1558
Pérez-Castilla A, Rojas FJ, Gómez-Martínez F, García-Ramos A (2019) Vertical jump performance is affected by the velocity and depth of the countermovement. Sport Biomech 30:1–16
Salles AS, Baltzopoulos V, Rittweger J (2011) Differential effects of countermovement magnitude and volitional effort on vertical jumping. Eur J Appl Physiol 111:441–448
Sánchez-Sixto A, Harrison A, Floría P (2018) Larger countermovement increases the jump height of countermovement jump. Sports 6(131)
Schoenfeld BJ (2010) Squatting kinematics and kinetics and their application to exercise performance. J Strength Cond Res 24:3497–3506
Street G, McMillan S, Board W, Rasmussen M, Heneghan JM (2001) Sources of error in determining countermovement jump height with the impulse method. J Appl Biomech 17:43–54
Suchomel TJ, McKeever SM, McMahon JJ, Comfort P (2020) The effect of training with weightlifting catching or pulling derivatives on squat jump and countermovement jump force-time adaptations. J Funct Morphol Kinesiol 5(2)
Wells JET, Charalambous LH, Mitchell ACS, Coughlan D, Brearley SL, Hawkes RA et al (2019) Relationships between challenge tour golfers' clubhead velocity and force producing capabilities during a countermovement jump and isometric mid-thigh pull. J Sports Sci 37(12): 1381–1386
West DJ, Owen NJ, Jones MR, Bracken RM, Cook CJ, Cunningham DJ et al (2011) Relationships between force-time characteristics of the isometric midthigh pull anddynamic performance in professional rugby league players. J Strength Cond Res 25(11):3070–3075

Alberto Sánchez-Sixto has a M.Sc. in High Performance in Sports and he received his Ph.D. in Sports Sciences form Pablo de Olavide University in 2018 following his research into jump biomechanics. The research lines of Dr. Sánchez-Sixto are jumping and running biomechanics analyzing how training modify athlete's performance. He is also an author and co-author of more than 30 papers and conferences communications in biomechanics and basketball. He was involved in international and national project focus on running mechanics and gait variability. He is part of the International Basketball Research Network. He is working as a lecturer in the CEU Cardenal Spinola University (Seville, Spain) and he is the physical activity instructor for Spanish basketball referee in the Spanish Basketball Federation. In addition, he is an International Basketball Referee since 2019.

John J. McMahon is a lecturer in sports biomechanics and strength and conditioning at the University of Salford (UK). He received is Ph.D. in sports biomechanics from Salford in 2015 following his research into dynamic muscle-tendon stiffness and stretch-shortening cycle function. He has been an accredited strength and conditioning coach with both the National and United Kingdom Strength and Conditioning Associations since 2010 and has worked as a strength and conditioning coach across several team sports. He has also co-authored 65 peer-reviewed journal articles relating to athletic performance assessment and is currently researching alternative methods of analyzing vertical jump force-time characteristics to help better inform athlete's training priorities.

Program Design and Periodization: Combining Strategies

Basics of Programming and Periodization in Resistance Training

Eduardo Sáez de Villarreal

Abstract Athletes can utilize resistance training and the magnitude of strength improvement is influenced by the structure of the training program. Periodization enables systematic, sequential, and integrative scheduling and programming of training sessions to maximize specific physiological adaptations underpinning performance outcomes. There are several ways that a periodized training plan can be implemented. Commonly used forms of periodization are the linear or classic periodization model (LP) (characterized by gradually increases training intensity and decrease volume) and nonlinear or undulating periodization model (UP) (characterized by more frequent alterations in intensity and volume). The complexity of periodized resistance training has evolved to meet the needs of particular sports and guarantee the success of the individual athlete. However, periodization is still based in the concept of training variation, sport specificity, and individualization of the training program. The real periodization is more a combination of various periodization methods, especially if you train for different outcomes. The purpose of the periodization is to achieve planned goals and you can create a million of combinations to get it. Periodization, as a whole, is a planning process that can be used to organize the training process of any athlete, regardless of developmental level or the sport being trained for.

Keywords Linear · Undulating · Intensity · Volume · Training variation · Sport specificity · Individualization · Training program

E. Sáez de Villarreal (✉)
Universidad Pablo de Olavide, Physical Performance and Sports Research Center, Seville, Spain
e-mail: esaesae@upo.es

A. Muñoz-López et al. (eds.), *Resistance Training Methods*, Lecture Notes in Bioengineering, https://doi.org/10.1007/978-3-030-81989-7_12

1 Introduction

While muscular strength is suggested to be a crucial attribute for several athletic disciplines, it's conjointly a necessary part of practicality in daily living (Kraemer and Ratamess 2002). Resistance training (RT) is usually recommended because it might improve muscle mass, strength and bone mass, generate bigger quality, increase health-related quality of life and increase resting rate (Petersen et al. 2017; Yamamoto et al. 2016; Goldfield et al. 2017; Campbell et al. 1994). Athletes will utilize RT and also the magnitude of strength improvement is influenced by the structure of the training program (American College of Sports Medicine 2009). Considering the importance that strength and muscle mass hold in respect to athletic success and also the overall well-being, implementation of a correct RT program is crucial to optimize these attributes. In line with (Haff and Triplett 2016), periodization permits systematic, sequential, and integrative programming and programming of training sessions to maximise specific physiological diversifications underpinning performance outcomes. Coaches and sport scientist's area unit commonly attributable with developing the ideas of programming and periodization of resistance training models. Whereas the tenets of periodization stay constant, there are a unit many ways in which a periodized training set up are often enforced (Rhea and Alderman 2004). Usually used varieties of periodization area unit the linear or classic or traditional periodization model (LP) (characterized by bit by bit will increase training intensity and reduce volume, with these changes being mode close to each four weeks) and nonlinear or undulating periodization model (UP) (characterized by a lot of frequent alterations in intensity and volume) (Rhea and Alderman 2004). It ought to be noted, however, that UP and LP do not have to be mutually exclusive. For example, UP models usually incorporate dimensionality to coincide with coming competitions. Specifically, whereas UP entails frequent variations in loading, repetition schemes will progress from high (hypertrophy-oriented) to low (strength/power-oriented) over the course of many training phases (Poliquin 1988).

LP of resistance follows a general pattern of decreasing training volume and increasing training intensity as a training cycle progress. Sport science analysis incontestable the disc training model will lead to bigger fitness gains than non-varied resistance training will give (Fleck 2011). Though UP may be a comparatively new resistance training model, it will lead to important gains in strength, power, body composition and motor performance (Buford et al. 2007). Information conjointly indicates that UP ends up in considerably bigger changes in fitness variables compared to non-varied and even strength training models. Current analysis indicates that once a nonlinear program is employed, the training zones ought to be alternated on a session-by-session basis. Building on the ideas of training variation used with classic resistance periodization training, each the disc training and also the versatile UP programs have emerged as terribly effective training models (Buford et al. 2007). The current empirical proof provides insights relating to training frequency, volume, rest intervals and repetition ranges

(Schoenfeld et al. 2016). Despite these evidence-based general suggestions for coming up with a training protocol for up strength, muscle hypertrophy, or increase health-related quality of life, there's a dearth of proof relating to completely different periodization ways (Grgic et al. 2018). It appears that the majority RT programs utilize some kind of periodization, however it's still unclear whether or not the effectiveness of periodization is especially associated with the principle of specificity once analysing strength (Mattocks et al. 2016), or a lot of determined by the varied structure of the training programs and also the variations within the volume and intensity of training (Kok et al. 2009).

2 From Theory

2.1 What is Periodization?

Periodization of physical training refers to the manipulation of the method variables of the physical training divided in logical phases and has the aim to perform specific changes for physical performance increase and bar of overtraining (Stone et al. 2000). The employment of training periodization in RT has become significantly well-liked within the last years. Currently, elite athletes, bodybuilders and place of business goers use physical training periodization in RT with the aim to enhance performance (Minozzo et al. 2008). Coaches, athletes and sport scientists from the previous jap coalition countries area unit commonly attributable with developing and researching the ideas of periodized training. The key goal of periodized training, as well as periodized resistance training, was to confirm peaking for major competitions (national and international championships, word championships, or Olympic Games). For the goal of peaking to occur, the training program had to confirm that strength and power were optimally developed, muscular hypertrophy occurred, and there was adequate recovery between training sessions so serial training sessions may well be performed at high intensity (Evans 2019).

Sport scientists and coaches from the previous jap coalition rigorously monitored their athletes' training volume and intensity and came to the conclusion that training volume and intensity of undefeated athletes followed a selected pattern over the course of a training year (Fig. 1) (Fleck and Kraemer 2004). At starting of the competitive year once preparation for competition was simply beginning, training volume was high and training intensity low. Because the competitive year progressed, training volume attenuated, and training intensity inflated. Before major competitions, training intensity was at its highest and training volume was at its lowest (Fleck and Kraemer 2004).

Additionally, training intensity conjointly showed a decrease now before major competition. This decrease in training volume and intensity was thought to be necessary for psychological also as physical recovery now before a significant competition so the simplest doable performance would occur at the key competition

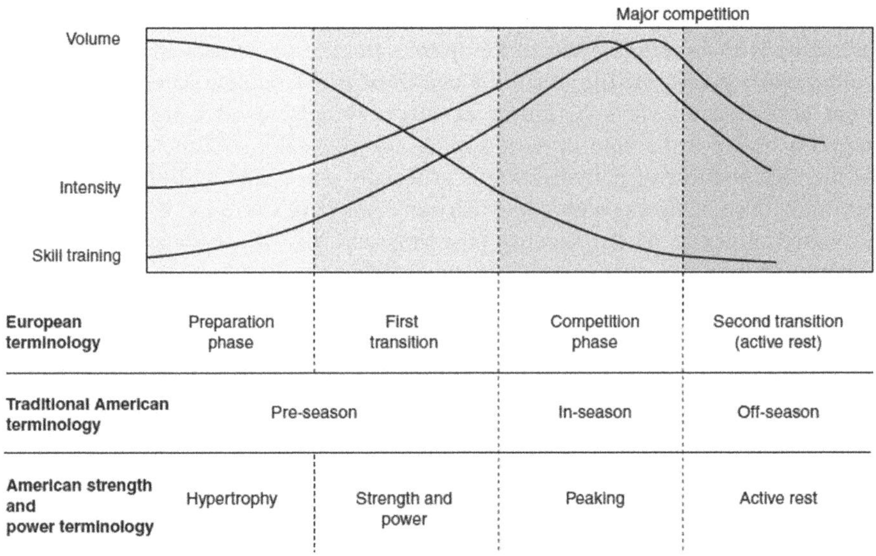

Fig. 1 Training intensity and volume patterns with resistance periodization. Reprinted by permission, from Fleck and Kraemer (2004)

(Fleck and Kraemer 2004). Skill training for the actual sport conjointly showed a pattern almost like that of training intensity. However, ability training peaked slightly nearer to the key competition than training intensity did. However, almost like training intensity, ability training attenuated now before major competitions. This general pattern of ability training, intensity and volume as employed in developing training programs for specific sports and personal training programs for every athlete (Fleck and Kraemer 2004).

Originally, as a result of there have been comparatively few major competitions during a competitive year, the pattern of increased training intensity and decreasing training volume of 1 training cycle occurred over a complete year. Then, as a lot of competitions were another to the competitive year, the timeframe for finishing a complete training cycle was bit by bit shortened. Today, the complete pattern of decreasing training volume and increasing intensity takes place in three to four months. Therefore, the complete pattern is recurrent 3 or fourfold per annum. Because the year progresses, training intensity and volume ideally area unit increasingly higher at the beginning of serial training patterns than they were at the beginning of previous patterns as a result of the contestant is currently in higher wholeness. Likewise, as the pattern is repeated over an athlete's career, training intensity and volume at the start of each year are also higher because physical condition during the athlete's career also increases (Fleck and Kraemer 2004).

2.2 *Physiological Bases of Periodization*

The complexity of periodized resistance training has evolved to fulfil the wants of specific sports and guarantee the success of individual contestant. However, periodization continues to be primarily based within the thought of training variation, sport specificity, and individualisation of the training program.

2.2.1 Training Variation

For a training program to stay effective, it should frequently overload the fibre bundle system (Kraemer and Ratamess 2004). It's been recommended that variations in training stimuli area unit necessary to optimize strength diversifications, since variation might force the fibre bundle system to continually adapt to unaccustomed stress (Rhea and Alderman 2004). Conversely, drawn-out time periods of loading that area unit barren of variation might lead to fatigue and stagnation (Williams et al. 2017). Therefore, one among the needs of periodization is to implement structured variability into training to offset the negative outcomes which will occur through the strain of linear loading. Moreover, the training program had to supply variation within the psychological and physiological stress of physical acquisition and competition (Williams et al. 2017). This was necessary in order to bring about the adaptations needed for long-term increases in the physical condition of an athlete that were critical for success in his or her particular sport. Training variation, like progressing toward bigger training intensities, was conjointly essential for peaking the contestant for major competitions (Williams et al. 2017). additionally, to incorporating variability among the training structure, several periodized training plans implements taper phases before necessary competitions. These phases include reduced training masses with the intent of restoring the contestant from training-induced emotional and physiological stress (Mujika and Padilla 2000).

2.2.2 Sport Specificity

Periodization theory counsel that variation is crucial to maximise fitness diversifications and variation in training stimuli seems to be very important for increasing top strength (Williams et al. 2017). Many authors reveal in their investigations that the periodized teams trained with higher intensities inflated the top strength compared to the non-periodized teams (Williams et al. 2017). This has semiconductor diode some authors to counsel that the superior results knowledgeable by the periodized teams might be because of the principle of specificity instead of the various nature of periodized training (Williams et al. 2017).

2.3 Basic Principles of Periodization

The basic principle of periodization may be a shift from a stress of high volume (exercises × sets × repetitions) and low intensity (% of most effort) training to low volume and high intensity training. The training year is split into distinct phases called mesocycles. Every mesocycle relates to an amendment within the volume and intensity of training and should last for 2–3 months relying upon the contestant. Usually every mesocycle reflects a particular training stress for that part of training. The initial mesocycle is named the preparatory or hypertrophy part and consists of high volume and low intensity training. It's designed to primarily increase muscle mass and muscle endurance, and to organize the contestant for a lot of advanced training throughout the later stages of training. Ensuing 2 mesocycles area unit usually spoken because the strength and strength/power part, severally. In these mesocycles training intensity will increase whereas training volume is reduced. The ultimate mesocycle of the training year is that the peaking part. Throughout this training part the contestant prepares for one contest by more reducing training volume and increasing intensity. Table 1 provides an example of training periods.

It is not uncommon to own short training cycles known as microcycles that facilitate transition from mesocycle to mesocycle. These microcycles area unit usually 1–2 weeks in period and supply an amendment to the conventional training routine permitting the athletes to avoid staleness and aid in recovery. Table 2

Table 1 Training periods (Based on Bompa and Haff 2009, Issurin 2010, Siff 2004, Stone et al. 2007, Haff 2013)

Period	Duration	Description
Multiyear preparation	2–4 years	Also termed a quadrennial plan
Annual training plan	1 year	The overall training plan can contain single or multiple macrocycles. It can be subdivided into preparatory, competitive and transitional phases of training
Macrocycle	Several months to a year	Some authors refer to this as an annual plan. It contains preparatory, competitive and transitional phases of training
Mesocycle	2–6 weeks	This medium-sized training cycle is sometimes called a macrocycle or a block of training. It consists of microcycles that are linked together
Microcycle	Several days to 2 weeks	This small-size training cycle can range from several days to 2 weeks in duration. It is comprised of multiple workouts
Training day	1 day	One training day can contain multiple sessions. Is designed in the context of the microcycle in which it is contained
Training session	Several hours	It generally consists of several hours of training. If the session has >30 min of rest between bouts of training, it should be considered as having multiple workouts

Table 2 Volume and intensity during a periodized strength training program

Mesocycle	Sets	Repetitions	%1-RM
Hypertrophy	3–5	8–12	60–75%
Strength	3–5	6–8	80–85%
Strength/power	3–5	4–6	85–90%
Peaking	3–5	2–4	>90%

provides an example of training manipulations across the assorted mesocycles within the strength/power athlete.

In athletes that participate during a sport putting importance on a complete season (i.e. soccer, basketball, handball), peak condition must be achieved by the onset of the competitive year, and maintained throughout the period of the season. These athletes usually have a brief peaking part preceding training camp or the beginning of the season. However, throughout the season each training intensity and volume area unit manipulated to replicate the importance that's placed on active sport specific skills. Throughout this training part exercise intensity is reduced to levels almost like those used throughout the strength part, whereas training volume is down by reducing the quantity of help exercises. In-season training is usually known as the upkeep part and is mostly performed at a reduced frequency of training also.

2.4 *Objectives of Periodization*

2.4.1 Improvements in Strength

Increases in strength are shown in each periodized and non-periodized RT programs. However, strength enhancements do seem to be bigger as a result of periodized training (Hoffman 2002). Many authors rumored that periodized RT programs seem to be superior that non-periodized training programs in generation enhancements in 1RM bench press, one RM squat and vertical jump performance (Baker et al. 1994; Stowers et al. 1983; Willoughby 1993). These studies offer proof that periodized RT is simpler than non-periodized training in eliciting strength and motor performance enhancements. However, this advantage is also for the most part dependent upon the training standing of the individual (Fleck 1999). The magnitude and rate of strength will increase area unit a lot of bigger in untrained people than in trained people, thus in thought of the fast strength will increase seen in novice athletes, periodized training might not be necessary till a precise strength base has been established (Hoffman 2002). When examining any training program, together with resistance training programs, the primary question that ought to be asked is whether or not the RT program causes the specified physiological diversifications. Consecutive question is whether or not the program leads to bigger will increase in those variables compared to the will increase with alternative training programs. The solution to those queries regarding the classic strength and power periodization model is affirmative. Qualitative investigations (Kraemer and Ratamess 2002; Fleck 2011; Fleck and Kraemer 2004) conclude that the bulk of analysis comes demonstrate that the classic strength and power model brings regarding bigger will increase in supreme strength and power than low-volume (single-set) and higher-volume (3–6 sets), non-varied (same range of set and repetitions per set for the complete training period) training programs. Meta-analysis conjointly concludes that periodized resistance training brings regarding bigger will increase in strength than non-varied training programs (Rhea and Alderman 2004). The notion that a specific periodization model (LP vs. UP) would possibly elicit superior strength enhancements compared to alternative models has been resolved by many investigations. Some authors have urged that UP could manufacture superior strength gains since it incorporates a lot of frequent variations in loading (Poliquin 1988; Rhea et al. 2002). Since the record model usually entails prolonged time periods spent in an exceedingly specific stop, the athlete would possibly quickly adapt to the training stimulant, which can lead to stagnation. Conversely, since the UP model varies the training stimulant a lot of oftentimes, the athlete is also forced to continually adapt to the unaccustomed stress. A meta-analysis by Harries et al. (2015) found no important variations between UP and record for eliciting supreme strength gains, but the authors did show a trend pro UP as being a lot of helpful for leg press strength. A future meta-analysis by Caldas et al. (2016) found UP to provide considerably bigger enhancements in supreme strength than record. The various results are also because of the larger knowledge pool enclosed within the analysis by Caldas et al. (2016). Within the meta-analysis by Williams et al. (2017), it had been found that UP created superior strength gains to record. Considering these findings, it's attainable that UP may be ideal for optimizing supreme strength.

2.4.2 Hypertrophy

Not solely will strength and power periodization leads to bigger strength and power will increase than non-varied models, however the bulk of studies conjointly indicate that this kind of training brings regarding bigger will increase in skim mass, indicating bigger muscle hypertrophy and bigger decreases in share of body fat than non-varied training models (Fleck 2011). Many reviews paid shut attention to reinforce muscle hypertrophy by RT programs. A recent systematic review analysed twelve studies scrutiny periodized RT to non-periodized RT protocols for hypertrophy outcomes (Grgic et al. 2018). The authors complete that each program sorts yield similar results, suggesting that measures of muscle hypertrophy should be created by imaging or ultrasound, to assess potential variations. Some authors (Kraemer and Ratamess 2002, 2004) counsel that a protracted study period is also needed for evaluating variations in muscle hypertrophy. Some studies failed to notice a superior impact of periodized RT over a non-periodized program in muscle hypertrophy, and will ensure to the variations in study period, protocols, volume, intensity and populations (young, older adults, young female). Scrutiny the periodization models (LP and UP), Poliquin (1988) projected that UP model could provide a superior hypertrophic impact compared to the record model, since the record model entail prolonged time spans spent in an exceedingly explicit stop. Four meta-analysis are conducted to match the consequences of record to UP for enhancing muscle hypertrophy (Grgic et al. 2018; Caldas et al. 2016; Krieger 2010; Schoenfeld et al. 2017). Whereas neither analysis found a bonus to either model for hypertrophy outcomes, it ought to be noted that a lot of the studies enclosed in every analysis used strength-oriented program styles with hypertrophy measured as a secondary outcome. It's attainable that totally different results might have occurred if all of the enclosed studies enforced hypertrophy-focused training styles (such as higher repetition schemes used before testing). It'd be that, for muscle gain, training volume is a lot of vital than the employment of record or UP models (Kok et al. 2009), however it remains unclear whether or not these finding area unit generalizable to alternative types of periodization (i.e., block periodization, reverse linear periodization, hybrids between record and UP, etc.). Nonetheless, the consequences of training volume on gains in muscle were confirmed in meta-analysis by Krieger (2010), and later by Schoenfeld et al. (2017). There looks to be a dose-response relationship between resistance training volume and will increase in muscle mass (Grgic et al. 2018; Schoenfeld et al. 2017). A lot of analysis is needed to analyse the consequences of semipermanent periodization programs on muscle hypertrophy.

2.5 *Why Use Periodization?*

The main objective of any fitness enthusiast, amateur or skilled athlete performing arts resistance training is to extend in muscle size, body composition, power or strength. Since athletes started serious training, they and their coaches have created changes in their training programs in an effort to evoke continued fires gains and avoid training plateaus. When a lot of trial and error, coaches and athletes have learned what changes to form in training programs and once to implement those changes. The changes created resulted within the development of planned semipermanent training programs and planned changes in training programs. Terms to explain planned semipermanent training variation area unit program manipulation and periodization. Periodization is that the hottest term for planned training variation. Changes in resistance training in any variable (type of exercise, order of exercise, range of repetitions per set, range of sets, rest periods between sets and exercises, intensity, speed of execution, etc.) are often used as a neighborhood of periodized training set up. Additionally, the amount of training sessions per day, the speed of training, the amount of training sessions per week, the remainder breaks or low-intensity or low-volume training periods will all be incorporated into a periodized training program. Though all of those style of changes are often created, changes in training volume (i.e., range of sets, range of repetitions per set, training sessions per week, training sessions per day) and training intensity (i.e., % of the supreme resistance which will be used for one repetition, the speed of execution) have received the foremost study by sport scientists and area unit usually used because the basis of any periodized training program. Besides continued semipermanent gains in muscle size, body composition, strength and power, there are a unit alternative vital reasons to use a periodized RT program:

- Planned variation in training for several athletes also will facilitate keep the training program psychologically fascinating.
- If a trainee merely goes through training sessions and does not attempt to perform the session at the required intensity and volume due to boredom, fitness gains will stagnate.
- To use a periodized training program is the prevention of overuse injuries.
- Performing an equivalent exercise at an equivalent training intensity associate degreed volume for long periods will eventually lead to an overuse injury.

2.6 Types of Models of Periodization

2.6.1 Linear Periodization

The "classic", "traditional" or linear periodization (LP) model relies on ever-changing exercise volume and cargo across many foreseeable mesocycles (Lorenz et al. 2010). Classical periodization was originally mentioned by Russian scientist Leo Matveyev (1981) and more dilated upon by Stone and O'Bryant (1987) and Bompa and Haff (2009). The program is de-escalated into distinct block that area unit named supported time frames. Designing that spans over a 12-month amount is remarked as a macrocycle, and 2 subdivisions area unit the mesocycle (3–4 months) and therefore the microcycle (1–4 weeks). Coaches follow a scientific progression of strength, power and speed with progression to every section betting on action of specific goals within the previous section. Table 3 shows a one-week sample program of lower extremity strengthening utilizing linear periodization. Table 4 shows the intensity training zones.

2.6.2 Undulating Periodization

The other main model is that the non-linear or "undulating" periodization model (UP), 1st projected by Poliquin (1988). Whereas UP has been used, the term "non-linear" has become a lot of favorable. Non-linear periodization relies on the conception that volume and cargo area unit altered a lot of oftentimes (daily, weekly, biweekly) so as to permit the fibre bundle system longer periods of recovery as lighter hundreds area unit performed a lot of usually (Lorenz et al. 2010). Within the UP model, there are a unit a lot of frequent changes in stimuli. This lot of frequent changes is also extremely contributory to strength gains (Poliquin 1988; Lorenz et al. 2010). Table 5 shows a one-week sample program of lower extremity strengthening utilising UP.

Table 3 Linear periodization in lower extremity

Exercise	Set/rep	Intensity
Hypertrophy/endurance		*Zone 2*
Han clean	4 × 6	55% 1RM
Back squat	3 × 12	70% 1RM
Single-leg deadlift	3 × 12	70% 1RM
Strength		*Zone 3*
Power clean	4 × 3	85% 1RM
Front squat	4 × 6	80% 1RM
Single-leg deadlift	4 × 6	80% 1RM
Max strength/power		*Zone 4*
Hang power clean	6 × 1	90% 1RM
Front squat	3 × 3	90% 1RM
Trap bar deadlift	3 × 5	85% 1RM

Table 4 Intensity training zones

Zones	Strength	Power
Zone 1	General muscle and technical <50%	General neural and technical <25%
Zone 2	Hypertrophy training 50–75%	Ballistic speed training 25–38%
Zone 3	Basic strength training 75–90%	Basic power training 38–45%
Zone 4	Maximal strength training 90–100%	Maximal power training 45–55%

All loads expressed as percentage of 1RM

Table 5 Undulating periodization in lower extremity

Exercise	Set/rep	Intensity
Workout 1		*Zone 3*
Han clean	3 × 3	80% 1RM
Back squat	4 × 5	80% 1RM
Workout 2		*Zone 1/2*
Hang snatch	3 × 5	50% 1RM
Front squat	3 × 12	50% 1RM
Leg press	3 × 12	50% 1RM
Workout 3		*Zone 2*
Deadlift	3 × 8	70% 1RM
Back squat	3 × 8	70% 1RM
Leg press	3 × 5	70% 1RM

2.6.3 Block Periodization

Block periodization (BP) is associate degree approach to the periodization of strength that has experienced a revived interest these days (Issurin 2010). BP involves extremely targeted, specialised workloads. Every step within the training cycle includes a giant volume of exercises cantered on specific, targeted training skills to make sure most adaptation. The principle for BP is that ancient models usually account for under one “peak” annually, whereas several athletes have various competitions throughout the year (basketball, soccer, handball, etc.). The BP system permits to keep up basic qualities throughout the year. This is often called long-lived delayed training impact—retention of changes even when the surcease of training (Issurin 2010). Issurin (2010) has projected that power and strength are often maintained for up to thirty days whereas peak performance are often maintained for 5–8 days. Moreover, record and UP models, have time dedicated to endurance, strength, power and speed, no matter the game. Within the BP approach, if associate degree athlete doesn’t need endurance for his or her sport, it’s not a spotlight of training. Similarly, the BP approach wouldn’t embody balance, strength and lightness in one training block, they would be performed on an individual basis with a particular focus.

Another distinction is that the BP is lessened into 2–4 weeks blocks, whereas phonograph recording, and UP models have a minimum of four-week phases. In alternative words, associate athlete might do strength, power and peaking among four weeks whereas it’s going to be many months before every part is completed within the phonograph recording or UP as a result of, they’re of longer length. The BP approach is split into 3 distinct phases (Dietz and Peterson 2010). the build-up part builds work capability. Compared to the opposite 2 phases, there’s a better volume of exercises performed at 50–70% of 1RM, composed of general movements. Typically, this part might last from 2 to 6 weeks, supported however long the athlete has until the competitive season, additionally as their training history. Primitive athletes would need longer during this part. The second part is that the transmutation part. During this part, specific exercises with larger masses, comprising 75–90% of onc RM arca unit pcrformcd. Finally, thc conclusion part is comprised of even a lot of specific movements than the transmutation part with masses at ninetieth of 1RM or larger. Instead, athletes can perform >90% 1RM squats, deadlift, bench press, cleans, etc. In some cases, there’s every week of reduced loading and volume following the conclusion part to permit recovery thanks to the high intensities used among the conclusion part. Table 6 shows a sample program of lower extremity strengthening utilising block periodization. There are a unit several potential benefits and drawbacks between the various periodization models approach, though no definitive conclusions are often created. Table 7 shows some benefits and drawbacks of the models of periodization.

Table 6 Block periodization in lower extremity

Exercise	Set/rep	Intensity
Accumulation phase week 1–2		
Week 1		
Push press	3 × 10	50% 1RM
Back/front squat	3 × 12	50% 1RM
Less press/hack squat	3 × 12	50% 1RM
Step ups	2 × 12	
Lunges	2 × 8	
Week 2		
Push press	3 × 8	60% 1RM
Back/front squat	3 × 10	60% 1RM
Less press/hack squat	3 × 10	60% 1RM
Trap bar deadlift	3 × 8	60% 1RM
Week 3–4 transmutation phase increasing load		
Week 3		
Hang clean/hang snatch	3 × 4	75% 1RM
Back/front squat	3 × 6	80% 1RM
Deadlift/trap bar deadlift	3 × 6	80% 1RM
Week 4		
Hang clean/hang snatch	4 × 3	85% 1RM
Back/front squat	4 × 6	75% 1RM
Week 5: realization phase. Peak power. Intensity can be based on sport demands		
Hang clean		
Back squat (front or back)	4 × 2	90% 1RM
Alternative lifts: deadlifts, hang snatch	4 × 5	90% 1RM completed as fast as possible
Week 6: Restoration phase*. Reduced loading to follow high intensity work* *Choose several exercises <50% 1RM. Emphasize total body workouts with light loads and high repetitions*		
After the 6-week block, the athlete repeats each phase		

3 From Practice

3.1 Many Coaches, Many Methods

Basically, what we have a tendency to area unit reading within the literature regarding periodization models, are “pure”, “theoretical” varieties of periodization ways, a scenario that isn’t happening therefore typically in reality for the coaches and athletes. The important periodization is a lot of a mixture of assorted periodization ways, particularly if you train your athlete for various outcomes.

Table 7 Advantages and disadvantages of the models of periodization

Model	Advantages	Disadvantages
Linear	• Repetition and loading schemes are predictable for athletes and coaches because they are ultimately determined by what phase the athlete is in • Each phase typically focuses on only one training parameter • Helps ensure that each training parameter (strength, power, speed) is addressed in stepwise progression • It provides the athlete a predictable sequence of loading and repetitions that they can follow • Helps take the "guess-work" out of loading and repetitions schemes	• Originally devised as a training model for preparing for one peak competition per year • For athletes that play several sports or athletes that have multiple competitions in a season, this may not be optimal as an athlete's tolerance to loading may ebb and flow based on injuries or frequency/intensity of competition • Maintenance of specific training parameters is difficult once an athlete transitions to another phase (i.e., strength to power phase)
Undulating	• The weekly fluctuations in training loads may lead to better neuromuscular adaptations compared to the LP approach, as loads are more unpredictable • Accounts for the need for modifications in the training program based on an athlete's recovery from competition or from a previous workout/training session • Several training parameters may be addressed at the same time • An athlete may address power and strength within the same week • Due to the concurrent nature of the training, the detraining effects that occur in a LP approach might be avoided	• In the recovering, the athlete may not be appropriate for lifts focusing on power development, like the clean and snatch, if an appropriate strength base has yet to be achieved or established • UP program may not allow each performance characteristic to be optimally developed due to focus on several parameters at once • Heavy loads are implemented in the first week of workout. Thus, the beginner needs to perform a base program for 4–6 weeks using lighter weights, allowing the individual to gain toleration to the resistance training program
Block	• Is better for athletes who have multiple events per year (cycling, skiing, track, etc.) • Offer several peaks of performance per year, while many team athletes have weekly competitions (soccer, basketball, baseball, handball, etc.) • Allows to maintain quality performances throughout the year	• Untrained athletes would require more experience to adapt to this periodization model • BP requires a high level of technical competency • Developing technical competences while simultaneously establishing special physical preparedness accumulates a large amount of fatigue. High levels of fatigue generally create a scenario where technical skills deteriorate, thus impeding the athlete's overall development

You'll be able to mix totally different periodization ways for various elements of your system, so victimization one periodization technique for strength work and another for speed work. Currently, the utilization of linear periodization model is more and more in decline, thanks to the assorted limitations that have already been exposed antecedent. Actually, there are a unit 3 main teams of periodization ways for resistance training: (1) Sequential periodization method (Undulating periodization), (2) Concurrent periodization method, and (3) Conjugate sequence periodization method (Block model). The purpose of the periodization is to achieve planned goals and you can create a million of combinations to get it.

3.2 *Sequential Periodization Method*

Sequential ways use specific time intervals to develop solely training goal at a time. There are numerous variations of sequential methods, principally classified according the subsequent variables:

- The length of specific time intervals.
- The sequencing of training goals (methods, means, loads).

Below are several examples of sequential methods.

3.2.1 Long Linear Method

May be is that the hottest technique in resistance training. Long linear technique uses longer times intervals (3–4 weeks or microcycles) to develop just one training goal. It yields from high volume-low intensity to low volume-high intensity training. Basically, uses one block (3–4 weeks) to develop strength endurance, one block for hypertrophy, one block for largest strength and one block for power (Fig. 2).

3.2.2 Short Linear Method

Short linear method uses shorter time intervals (1–2 weeks) to develop particular ability. Progress from high-volume low-intensity to low-volume high-intensity training in shorter time (Fig. 3).

STRUCTURAL BLOCK				HYPERTROPHY BLOCK				MAX STRENGTH BLOCK				POWER WORK BLOCK			
W1	W2	W3	W4	W1	W2	W3	W4	W1	W2	W3	W4	W1	W2	W3	W4

Fig. 2 Sequencing of long linear method

CYCLE 1 ONE MONTH				CYCLE 2 ONE MONTH				CYCLE 3 ONE MONTH			
STRUCT WEEK	HYPERT WEEK	MAX STR WEEK	POWER WEEK	S W	H W	MAX S W	P W	S W	H W	MAX S W	P W

Fig. 3 Sequencing of short linear method

3.2.3 Hybrids Between Long and Short Variations

The combinations of methods is used to develop different goals, for example use one month to develop structure and then switch to develop hypertrophy, maximal strength and power, and then repeat (Fig. 4).

3.3 Concurrent Method of Periodization

Concurrent method of periodization develops all abilities in a given time period, mostly one microcycle (week). This does not necessarily means that all the abilities are developed in one training session. This method can be further classified according the emphasis on particular ability: (1) All abilities have same emphasis (volume, training time), (2) One or more abilities are more emphasized than others.

Below are several examples of concurrent methods.

3.3.1 Ordinary Concurrent Method

Ordinary concurrent method of periodization uses the same emphasis to develop all targeted motor abilities in a given time period (one microcycle-week). The problem is that some abilities need more volume to be developed than others (structural and hypertrophy work) (Fig. 5).

3.3.2 Emphasised Concurrent Method

Emphasised variation emphasise one (or more) particular ability within others that are developed concurrently (Fig. 6).

STRUCTURAL BLOCK				CYCLE 1 ONE MONTH		
WEEK 1	WEEK 2	WEEK 3	WEEK 4	HYPERTROPHY WEEK	MAX STRENGTH WEEK	POWER WEEK

Fig. 4 Sequencing of hybrid methods

CYCLE 1 ONE MONTH				CYCLE 2 ONE MONTH			
P WORK	P WORK	P WORK	P WORK	P WORK	P WORK	P WORK	P WORK
M STR WORK	M STR WORK	M STR WORK	M STR WORK	M STR WORK	M STR WORK	M STR WORK	M STR WORK
H WORK	H WORK	H WORK	H WORK	H WORK	H WORK	H WORK	H WORK
STRUCT WORK	STRUCT WORK	STRUCT WORK	STRUCT WORK	STRUCT WORK	STRUCT WORK	STRUCT WORK	STRUCT WORK

Fig. 5 Sequencing of ordinary concurrent method

CYCLE 1 ONE MONTH				CYCLE 2 ONE MONTH			
P WORK	P WORK	P WORK	P WORK	P WORK	P WORK	P WORK	P WORK
M STR WORK	M STR WORK	M STR WORK	M STR WORK	M STR WORK	M STR WORK	M STR WORK	M STR WORK
H WORK	H WORK	H WORK	H WORK	H WORK	H WORK	H WORK	H WORK
STRUCT WORK	STRUCT WORK	STRUCT WORK	STRUCT WORK	STRUCT WORK	STRUCT WORK	STRUCT WORK	STRUCT WORK

Fig. 6 Sequencing of emphasised concurrent method

3.4 Conjugate Sequence Periodization Method

Conjugate sequence method is the most advanced method of periodization. It is based on the advantages and disadvantages of sequential and concurrent methods, trying to apply all the advantages and avoid all the shortcomings. It is based on the premise that the elite athlete is unable to optimally adapt and recover to larger number of stimuli (abilities) in the same time. Elite athlete needs concentrated loading of one particular ability, but this method will be decreased in another developed needful abilities. The solution is to develop one ability while maintaining all others with minimal volume. With this approach, athlete is optimally adapting to one stimulus while maintaining others and avoids stagnation, overtraining and fatigue. After some time, emphasis is switched to another ability. There are numerous variations of conjugate sequence method, mostly classified according the following variables: (1) The duration of emphasis block, and (2) The sequencing of emphasis.

Similar to sequential methods, we can differentiate between long and short emphasis periods, and linear or undulating switching of emphasis.

3.4.1 Short Conjugate Sequence Method

Everything is done with the emphasis/volume varies during one microcycle. This basically means that only one ability is developed while the others are maintained

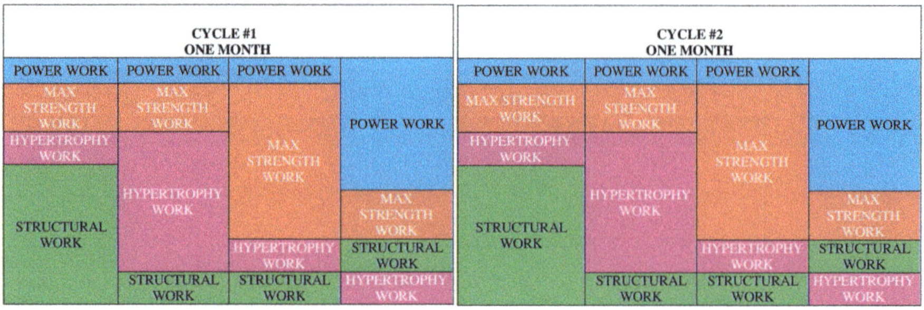

Fig. 7 Sequencing of short conjugate sequence method

(or slightly improved). The sequencing is done on micro level, thus every microcycle (week) there is a switch of emphasis on particular ability (Fig. 7).

3.4.2 Long Conjugate Sequence Method

The only difference is the duration of particular emphasized block. Long conjugated sequence method uses longer periods to develop particular ability (Fig. 8).

4 Filling Gaps

Periodization, as a whole, is a planning process that can be used to organise the training process of any athlete, regardless of developmental level or the sport being trained for. A simplified process that contains seven interrelated steps will highlight, from a very practical and real point of view, how to implement the evidence-based theory and practice previously discussed.

Basic steps in the periodization process [based on Haff (2013)]:

Step 1

- Determine the athlete's long-term goals in order to develop a multi-year periodization. Typically, this is accomplished with a quadrennial periodization.
- Outline the basic structure for the multi-year periodization.

Step 2

- Prioritize the major objectives to be targeted by the annual training periodization (TP).
- Evaluate the previous year's TP, including competitive and performance results, and consult with the athlete or team about the TP.

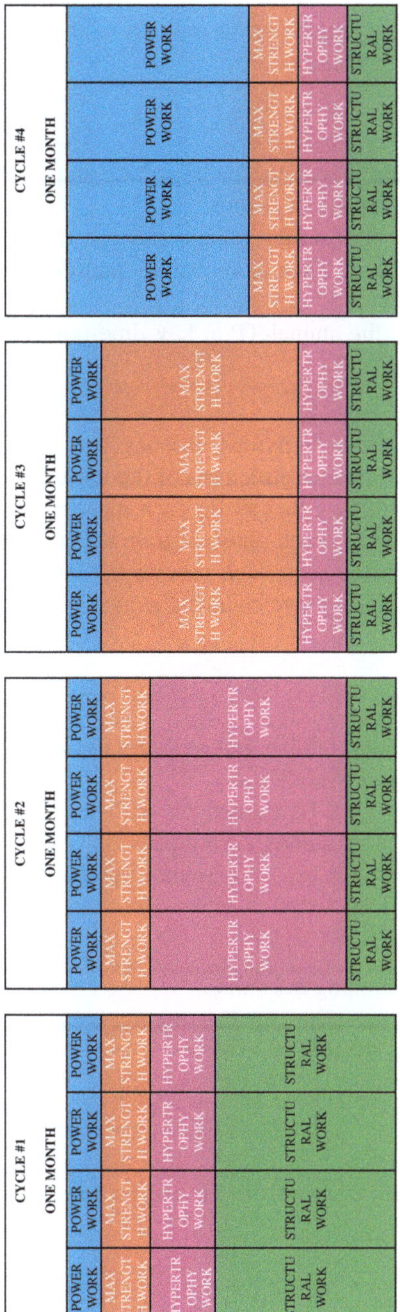

Fig. 8 Sequencing of long conjugate sequence method

- Create a working structure for the next annual TP based on the competitive requirements of the athlete or team.
- Establish the macrocycle lengths in the context of the structure established for the annual TP.

Step 3

- Break the annual TP into preparatory, competitive and transitional phases based on the schedule of the athlete or team.
- Divide the preparatory phases into general and specific subdivisions.
- Create precompetitive and main competitive phases within the competitive phases of the annual TP.
- Insert testing days into the annual TP at key time points.

Step 4

- Determine the lengths of the individual mesocycles.
- Select and sequence the various structures of the mesocycle into the annual TP.
- Prioritize the focus of training factors for each mesocycle, considering how the factors are sequenced across each phase of training in the annual periodization.
- Establish the loading patterns in each mesocycle and determine how loading will progress across the macrocycles in the annual periodization.

Step 5

- Construct each microcycle.
- Divide the microcycle into training and recovery days according to the athlete's level of development and overall goals.
- Establish which factors will be trained on each training day and how many training sessions will be contained during each day.
- Create the loading structures used throughout the microcycle.

Step 6

- Design the individual training sessions.
- Determine the loading structures for the training session.
- Select the activities for the TP.

Step 7

- Implement the TP.
- Continually monitor and evaluate the TP and process.

5 Take-Home Messages

The goal of periodization is to maximize the potential of the athlete to reach peak condition by manipulating both training volume and training intensity. Through proper manipulation of these training variables, not only will the athlete peak at the appropriate time, but also the potential risk for overtraining is reduced.

Prescribing the proper resistance training (RT) program is critical to optimize skeletal muscle hypertrophy and strength. Periodization is a strategy that entails planned manipulations of training variables to maximize fitness adaptations while minimizing the risk of overtraining. Multiple investigations and meta-analyses have shown periodized RT to be superior to non-periodized RT for enhancing muscular strength. These findings are consistent irrespective to training status or training volume. Both the linear model and the undulating model are effective for enhancing strength, although a greater benefit might be achieved through the undulating model. Despite the suggested superiority of periodized RT for strength development, some authors suggest that this might be a consequence of the study designs employed rather than the nature of periodized training. In addition, several limitations exist in the periodization literature, making it difficult to accurately assess the efficacy of periodized RT. With regard to enhancing skeletal muscle hypertrophy, both the undulating model and the linear model appear equally effective; however, this conclusion can only be generated to untrained populations. When comparing periodized RT to non-periodized RT programs, the research is unclear on whether periodized RT is necessary to maximize skeletal muscle hypertrophy.

Summary Points

1. Periodization of training is an essential component of the athlete's long-term development. It should be considered a logical, integrative, and sequential manipulation of training factors in order to optimize training outcomes at predetermined time points.
2. Periodization is a theoretical and practical paradigm that is well established in the scientific literature as a superior method of developing athletes, especially over the long term.
3. Periodization is a planning process that allows the strength and conditioning professional to structure training that targets specific physiological and performance outcomes, while managing the training and life stressors that the athlete is exposed to.
4. Training periods that should be considered and structured when designing an athlete's program include the multiyear training plan, annual training plan, macrocycle, mesocycle, microcycle, and training day and session.
5. Although the linear model of periodization is helpful for novice or beginner athletes, advanced athletes require sequential models of periodization (undulating periodization), concurrent models and conjugate sequence system (block models) of periodization that are more complex.

References

American College of Sports Medicine (2009) American college of sports medicine position stand. Progression models in resistance training for healthy adults. Med Sci Sports Exerc 41:687–708
Baker D, Wilson G, Carlyon R (1994) Periodization: the effect on strength of manipulating volume and intensity. J Strength Cond Res 8:235–242
Bompa TO, Haff GG (2009) Periodization, theory and methodology of training: Human Kinetics
Buford TW, Rossi SJ, Smith DB, Warren AJ (2007) A comparison of periodization models during nine weeks with equated volume and intensity for strength. J Strength Cond Res 21:1245–1250
Caldas LC, Guimaraes-Ferrera L, Duncan MJ, Leopoldo AS, Leopoldo API, Lunz W (2016) Traditional vs. undulating periodization in the context of muscular strength and hypertrophy: a meta-analysis. Int J Sports Sci 6:219–229
Campbell WW, Crim MC, Young VR, Evans WJ (1994) Increased energy requirements and changes in body composition with resistance training in older adults. Am J Clin Nutr 60 (2):167–175
Dietz C, Peterson B (2010) Triphasic training: a systematic approach to elite speed and explosive strength performance. Bye Dietz Sport Enterprise, Hudson, WI
Evans JW (2019) Periodized resistance training for enhancing skeletal muscle hypertrophy and strength: a mini-review. Front Physiol 10:1–7
Fleck SJ (1999) Periodized strength training: a critical review. J Strength Cond Res 13:82–89
Fleck SJ (2011) Non-linear periodization for general fitness & athletes. J Hum Kinet 29A:41–45
Fleck SJ, Kraemer WJ (2004) Designing resistance training programs, 3rd edn. Human Kinetics, Champaign, IL, p 213
Goldfield GS, Kenny GP, Alberga AS, Tulloch HE, Doucette S, Cameron JD, Sigal RJ (2017) Effects of aerobic or resistance training or both on health-related quality of life in youth with obesity: the hearty trial. App Physiol Nut Met 42(4):361–370
Grgic J, Lazinica B, Mikulic P, Schoenfield BJ (2018) Should resistance training programs aimed at muscular hypertrophy be periodized? A systematic review of periodized vs non-periodized approaches. Sci Sports 33:97–104
Haff G (2013) Periodization of training. In: TJB Chandler (ed) Conditioning for strength and human performance, vol 14. Lippincott Williams & Wilkins, pp 326–345
Haff G, Triplett T (2016) Essentials of strength training and conditioning, 4th edn. Human Kinetics, NCSA
Harries SK, Lubans DR, Callister R (2015) Systematic review and meta-analyses of linear and undulating periodized resistance training programs on muscular strength. J Strength Cond Res 29:1113–1125
Hoffman JR (2002) Physiological aspects of sport training and performance. Human Kinetics, Champaign, IL, pp 131–142
Issurin VB (2010) New horizons for the methodology and physiology of training periodization. Sports Med 40(3):189–206
Kok LY, Hamer PW, Bishop DJ (2009) Enhancing muscular qualities in untrained women: linear vs undulating periodization. Med Sci Sports Exerc 41(9):1797–1807
Kraemer WJ, Ratamess NA (2002) Resistance training for health and performance. Curr Sports Med Rep 1:165–171
Kraemer WJ, Ratamess NA (2004) Fundamentals of resistance training: progression and exercise prescription. Med Sci Sports Exerc 36:674–688
Krieger JW (2010) Single vs multiple sets of resistance exercise for muscle hypertrophy: a meta-analysis. J Strength Cond Res 24(4):1150–1159
Lorenz DS, Reiman MP, Walker JC (2010) Periodization: current review and suggested implementation for athletic rehabilitation. Sports Health 2(6):509–518
Mattocks KT, Dankel SJ, Buckner SL, Jessee MB, Counts BR, Mouser JG et al (2016) Periodization: what is it good for? J Train 5:6–12
Matveyev L (1981) Fundamentals of sports training. Progress, Moscow

Minozzo FC, Lira CAB, Vancini RL, Silva AMB, Freitas RJ, Fachina G et al (2008) Periodizacao do treinamento de forca: uma revisao critica. Rev Bras Cien e Mov 16:89–97
Mujika I, Padilla S (2000) Detraining: loss of training-induced physiological and performance adaptations, part I. Sports Med 30:79–87
Petersen BA, Hastings B, Gottschall JS (2017) Low load, high repetition resistance training program increases bone mineral density in untrained adults. J Sports Med Phys Fitness 57(1–2):70–76
Poliquin C (1988) Five steps to increasing the effectiveness of your strength training program. NSCA J 10:34–39
Rhea M, Alderman BL (2004) A meta-analysis of periodized versus non-periodized strength and power training programs. Res Quart Exerc Sport 75:413–422
Rhea MR, Ball SD, Phillips WT, Burkett LN (2002) A comparison of linear and daily undulating periodized programs with equated volume and intensity for strength. J Strength Cond Res 16:250–255
Schoenfeld BJ, Ogborn D, Krieger JW (2016) Effects of resistance training frequency on measures of muscle hypertrophy: a systematic review and meta-analysis. Sports Med 46(11):1689–1697
Schoenfeld BJ, Ogborn D, Krieger JW (2017) Dose-response relationship between weekly resistance training volume and increases in muscle mass: a systematic review and meta-analysis. J Sports Sci 35(11):1073–1082
Siff MC (2004) Supertraining (5° ed.). Supertraining Institute, Denver, USA
Stone MH, O'Bryant HS (1987) Weight training: a scientific approach. Burgess, Minneapolis
Stone MH, Potteiger JA, Pierce KC, Proulx HS, O'Bryant HS, Johnson RL et al (2000) Comparison of the effects of three different weight-training programs on the one repetition maximum squat. J Strength Cond Res 14:332–337
Stone MH, Stone M, Sands WA (2007) Principles and practice of resistance training (Chapter 13). The concept of periodization. Human Kinetics
Stowers T, McMillian J, Scala D, Davis V, Wilson D, Stone M (1983) The short-term effects of three different strength-power training methods. Nat Strength Cond Assoc J 5:24–27
Williams TD, Tolusso DV, Fedewa MV, Esco MR (2017) Comparison of periodized and non-periodized resistance training on maximal strength: a meta-analysis. Sports Med 47:2083–2100
Willoughby DS (1993) The effects of meso-cycle-length weight training programs involving periodization and partially equated volumes on upper and lower body strength. J Strength Cond Res 7:2–8
Yamamoto S, Hotta K, Ota E, Mori R, Matsunaga A (2016) Effects of resistance training on muscle strength, exercise capacity, and mobility in middle-aged and elderly patients with coronary artery disease: a meta-analysis. J Cardiol 68(2):125–134

Eduardo Sáez de Villarreal has a B.P.E., a M.Sc. in High Performance in Sports, and a Ph.D. in Sport Sciences. He is a very passionate strength and conditioning coach that progressively developed into an applied sport scientist, with a main emphasis on strength, power, and velocity development in collective and individual sports. His work (>100 peer-reviewed articles published, >4000 citations; h index 31; h 10 index 45) focuses on intervention strategies and profiling assessments that may improve players' physical and technical potential, using a scientific approach whenever possible. Using his background in strength and conditioning, he has extensive experience with fitness and strength training, injury prevention, recovery, and youth development. Over the past 30 year, he has worked and consulted a bit all over the world and in various sports (soccer, waterpolo, AFL, rugby, ice hockey, individual sports), which allow him to merge theory, experience and pragmatism for a holistic and contextualized approach of high performance.

Programming and Periodisation for Team Sports

Moisés de Hoyo Lora and Luis Suarez Arrones

Abstract Training programming and periodisation is widely acknowledged as crucial for optimising training responses. The achievement of physical aims is impossible when attempted without considering the different possibilities of periodisation, which is understood as the structuring of the training process and competition participation into various phases, periods and cycles. Thus, this chapter introduces different training periodisation concepts and summarises a large body of findings describing their potential benefits and possible limitations. Applying periodised planning to team sports poses unique challenges due to the variety of training goals, volume of concurrent training and practices, and extended season of competition. Therefore, practical suggestions are offered in this chapter to address these challenges and apply programming and periodisation to the design of strength training programmes for different phases of physical preparation for team sport athletes.

Keywords Strength training periodisation · Traditional periodisation · ATR blocks · Microstructure periodisation · Tactical periodisation

1 Introduction

Multiple definitions of the term "periodisation" can be found throughout the sports performance literature. For instance, Isurin (2016) defined periodisation as the main planning strategy for athlete preparation. Lambert et al. (2008) described the term as the purposeful process of systematically planning a short- and long-term training programme by varying training loads and incorporating adequate rest and recovery.

M. de Hoyo Lora (✉)
Physical Education and Sports Department, University of Seville, Pirotécnica s/n, 41013 Seville, Spain
e-mail: dehoyolora@us.es

L. S. Arrones
Sport and Informatics Department, University of Pablo de Olavide, Ctra. de Utrera, 1, 41013 Seville, Spain

A. Muñoz-López et al. (eds.), *Resistance Training Methods*, Lecture Notes in Bioengineering, https://doi.org/10.1007/978-3-030-81989-7_13

According to Manchado et al. (2018), training periodisation is a strategy to promote long-term training and performance improvements with pre-planned systematic variations in training specificity, intensity, and volume organised into periods or cycles within an overall programme. In this sense, in team sports, a typical periodisation plan involves phases or cycles of varying training demands and goals programmed throughout the pre-season, competition, and transition (off-season) phases (Mara et al. 2015; Suarez-Arrones et al. 2019). The transition period in team sports is generally characterised by a substantial reduction in training, even including complete training cessation for a few days. During the off-season, players also participate in other sport activities to retain fitness and/or follow individualised training programmes offered by their clubs to facilitate faster adaptation during the subsequent pre-season phase (Silva et al. 2016). The pre-season is commonly characterised by a high frequency of training sessions and friendly games shortly after returning to training, with rapid increases in training load within a few days (Silva et al. 2016). Regarding the competition period, the playing season in team sports such as football or rugby can span in excess of 35 weeks, particularly in Europe. In this context, the major obstacles for fitness coaches working with these teams are the frequent matches and extended competition period. Given this requirement to continue regular competition over many months, achieving the necessary training periodisation represents a significant challenge that we will attempt to address in this chapter.

Designing periodised training programmes for team sport athletes poses unique challenges and difficulties. This is mainly due to the fact that athletes are required to work on multiple aspects of their individual fitness and physical readiness to perform, while concurrently participating in extensive technical and tactical team training sessions to prepare for upcoming matches, as well as extended periods of competition (Mujika et al. 2018). All these elements must be addressed in the course of the training plan. Therefore, there is a need for planned variations in the training programme to systematically shift the emphasis to promote these different training effects at different phases of the preparation period (Gamble 2006). For fitness coaches, physical periodisation is the systematic planning and variation of training demands, with the aim of optimising physical condition and minimising the risk of injury (Fleck 2011). For team sport players, appropriate strength training (ST) periodisation is needed to achieve these aims.

To achieve these goals, different ST modes with (i) distinct sources of external resistance (traditional free weights, ballistic exercises, plyometrics, weightlifting, flywheel technology, motorised devices and/or sport-specific strength-based actions), (ii) different combinations of the temporal organisation of strength/power training loads (e.g. microcycle and training session variations), (iii) distinct loads, (iv) a wide range of movement velocities, (v) specific biomechanical characteristics, and (vi) different training surfaces, have been adopted with the final end point of achieving an improvement in players' performance in relevant motor tasks (e.g. jumping, sprinting, and changing direction) and reducing the rate of injuries (Silva et al. 2015). In accordance with this, this chapter introduces novel concepts of ST periodisation and summarises a large body of findings describing their potential benefits and possible limitations for team sport players.

2 From Theory

The contemporary theory of training was established in the 1960s and early 1970s by Matveyev (1964) and Ozolin (1970) and became known as the classic or traditional periodisation training model. In this model, training starts with high-volume and low-intensity workloads and shifts to an increase in the intensity and a decrease in the volume as the athlete progresses towards the competition phase (Matveyev 1964). The periodisation model at that time was largely based on individual sports and did not consider team sport scheduling (Manchado et al. 2018). Thus, for example, modern team sports require the achievement of a high level of performance during every match, which the traditional periodisation model may not provide. Team sports also require a high level of performance in many abilities simultaneously (Manchado et al. 2018).

To overcome the limitations of traditional periodisation for team sports, different training models have been developed over the last few decades. In this process of development, the first adaptation of the traditional model for team sports was proposed in 1985 by Issurin (2008). Unlike the traditional concept, this alternative version proposes a high concentration of training means within appropriate preparation cycles. These proposed alternative concepts of athlete preparation were called 'ATR block periodisation' training systems. In this model, the athlete has a higher number of peaks in many games (Issurin 2008). ATR block periodisation is divided into three blocks, which are the accumulation, transformation and realisation blocks. The accumulation block has higher residual training because the coach prescribes aerobic endurance and/or aerobic resistance, maximal strength and basic technical and tactical training (Issurin 2010). The transformation block has medium residual training because the coach prescribes specific training through special and competitive exercises with the aim of improving the physical preparation and the technical and tactical aspects of the athlete (Issurin 2014). The realisation block has a low residual training effect and the objective is for the athlete to achieve their peak during the championship (Issurin 2008). ATR block periodisation is suitable for several sports because the objective is to achieve many peaks during the season (Velásquez 2019).

In the best interest of peak performance at major competitions, these athletes can afford to exhibit subpar performances and even miss competitions that do not fall within the scope of their main goals. By contrast, team sport athletes usually need to perform at a high level week after week if they want to be in contention for the championship at the end of the competitive season (Mujika 2007). Thus, one characteristic which is common to all team sports is that the regular leagues entail competition which takes place over long periods of time. These competitions include at least one competitive match every 7 days, and often two or three competitive matches per week, subjecting the athlete to very high levels of competitive stress for extensive periods of time. This load must be taken as a highly specific load and is considered as such within the planning of the micro-structure (Tarragó et al. 2019). In this context, alternative methods for team sports have been emerging. Seirul-lo (1987) designed microstructure periodisation for team sport games in order for the player to develop the cognitive aspect. Based on this new paradigm, we are able to interpret the athlete as a hyper-complex structure that is made up of interactions and retroactive actions between the following structures: (i) conditioning structure; (ii) coordination structure; (iii) socio-affective structure; (iv) emotional-volitional structure; (v) creative-expressive structure; (vi) mental structure. Each structure must be considered as the expression of underlying processes. Within this context, Seirul-lo Vargas developed a working methodology known as and recognised by the name of "structured training". This method develops a form of organisation and training called the "structured microcycle", which is the smallest structure in the programming and accepts and considers competition as a load that changes and conditions the different structures in the training period between competitive matches (Tarragó et al. 2019). In the "structured microcycle", ST is considered as a part of coadjuvant training (preventive, recovery, structural, and specific-quality), which comprises all the factors that allow athletes to reach and maintain a state of health that enables them to perform the tasks proposed by optimising training on a daily basis (Gómez et al. 2019).

In a similar way, in 1989 Frade designed "tactical periodisation" with the aim of quickly preparing the football team tactically (Delgado-Bordonau and Méndez Villanueva 2018). Within the tactical periodisation training approach, tactical, technical, physiological and psychological aspects are rarely trained in isolation, which is believed to improve specific motor skill acquisition and accelerate tactical learning (Delgado-Bordonau and Méndez Villanueva 2018). In fact, daily training components are not only structured in relation to technical/tactical objectives, but also to the physical capacities to be targeted ("Physiological dimensions provide the biological framework where the football-specific training/recovery continuum lies") (Delgado-Bordonau and Méndez Villanueva 2018). In practice, when playing once a week, the three main training 'acquisition' days allow the successive development/maintenance of the three main physical capacities, i.e., strength, endurance and speed. Focusing more on a given quality on a given day likely allows the training stimulus to be maximised when the other qualities recover, which may decrease physiological interferences (Delgado Bordonau and Méndez Villanueva 2018) and, in turn, lead to greater adaptations (Buchheit et al. 2018). This so-called horizontal alternation in the physical components to be prioritised is often achieved while targeting all intra-session training sequences towards the same quality. For example, a 'strength-conditioned session' would include a strength-oriented warm-up (e.g. light plyometric drills and single-leg horizontal hops), locomotor-based strength work (e.g. accelerations, changes of direction and sled pulling) and game-play sequences including, irrespective of the actual technical/tactical requirements, high and qualitative neuromuscular demands (e.g. high number of players/playing area ratio and maximal intensity of actions with adequate rest periods) (Buchheit et al. 2018). Although this model was designed for use in football, it is now used in other team sports.

2.1 *The Optimal Dose-Response Relationship for Strength Training in Team Sports*

ST has two fundamental aims, which are to optimise performance and reduce the rate of injury. Therefore, in team sports, the capacity to maximise neuromuscular power production is fundamental to success and critical for achieving high levels of performance and greater velocities in sport-specific movements (Cormie et al. 2011). The improvement of high-intensity, explosive actions such as a sprint, change of direction or vertical jump is an important goal for coaches and athletes (Bolger et al. 2015; García-Pinillos et al. 2014). Wisløff (2004) reported that there is a strong correlation between maximal strength, sprinting, and jumping performance in elite football players. Moreover, different authors have shown that when players perform a periodised ST programme, the injury rate is reduced (Hoyo et al. 2015; Askling et al. 2003; Arnason et al. 2007). In this sense, eccentric training for muscle injury prevention has been suggested as a means of decreasing the high prevalence of these injuries (Hoyo et al. 2015; Askling et al. 2003; Arnason et al. 2007) with even more promising results than those observed with concentric training (Mjolsneset al. 2004).

In a meta-analysis, Peterson et al. (Peterson et al. 2004) showed how effect sizes for training frequency (2 and 3 days per week) were similar with no additional benefit to training 3 days per week. Regarding reducing the injury rate, Steib et al. (2017) concluded that ST achieved a reduction in the risk of injury of 42% (and up to 50%) when programmed correctly. For this, it seems that 2–3 sessions per week of about 30 min each are needed, although it is possible to produce similar effects with 10–15 min. It is important to consider the duration of the programme, as it was proven that programmes of more than 6 months in duration were not more beneficial than other shorter ones. What seems clear is that, with less than 30 intervention sessions, the benefits and preventive effects have already been observed (Steib et al. 2017).

In the real-world team sport training environment, there is limited time for ST sessions to take place during the in-season period. The search for time-efficient strategies that concurrently enhance several locomotive-specific actions while preventing injuries is crucial. Furthermore, it is often difficult to develop a science-based ST programme that comprises two or three sessions per week. In this sense, several studies have investigated the effects of frequency of ST and its effects on specific performance (i.e. sprints, jumps, COD, etc.). Hence Peterson et al. (2005) observed that, depending on the training status of the subjects, the best choice in terms of frequency and intensity of training would be three sessions per week for untrained individuals. On the other hand, for recreationally trained nonathletes and athlete populations, maximal strength gains were elicited using two sessions per week (Peterson et al. 2005). In this regard, our group analysed the effects of a combined ST programme (full-back squat, YoYoTM leg curl, plyometrics and sled-towing exercises) on performance among elite young football players when this training programme was performed one or two days per week.

In conclusion, the combined ST programme improved jumping ability, independently of training frequency, though the achievement of two sessions per week also enhanced sprinting abilities (linear and COD) among young football players. In accordance with this, the majority of strength and conditioning coaches in professional leagues typically report to have used ST twice per week in-season (Ebben and Blackard 2001; Ebben et al. 2004; Kraemer 2004). However, the number of sessions in each microcycle could be modified according to the number of competitive games in the period.

2.2 *Strength Training Scheduled During Different Microcycles*

The weekly scheduling of in-season workouts is dictated by the dual need to allow the player to recover from the previous match and avoid excessive residual fatigue at the end of the week in preparation for the next game. Regarding ST scheduled during a typical microcycle consisting of one competitive match, according to the available literature, intense training is not scheduled in the 24–48-h post-match period (Wrigley et al. 2012; Impellizzeri et al. 2004) to facilitate recovery of muscle function and damage. In these microcycles, the greatest volume of on-field training is typically scheduled in the middle of the training week, between 2 and 4 days prior to the match (Malone et al. 2015), with ST programmes most commonly carried out 48–72 h after the match (McCall et al. 2014). Moreover, during congested fixture periods (2 or more matches per week) common to team sports, ST programmes are often sacrificed to prepare for, or compete in, the following match (McCall et al. 2014), for which overtime may result in muscle de-training (Gabbett 2005; Rollo et al. 2014) and render the player more susceptible to injury (Opar et al. 2015). However, Lovell et al. (2018) showed how carrying out eccentric training in the middle of the microcycle (MD+3) increased measures of muscle damage and soreness, which remained elevated on the day prior to the next match (MD+5). Accordingly, the authors indicated that eccentric training should be scheduled early in the microcycle to avoid compromising preparation for the following match. In addition to this, a recent study showed that eccentric exercise of maximal intensity does not per se affect muscle damage biomarkers in an eccentrically accustomed muscle, being evident at the 10th week of training in those training recreationally (Margaritelis et al. 2021). Therefore, muscle damage occurs as a result of the muscles being unaccustomed to responding to a specific muscle contraction pattern, and it is crucial to be adapted to these neuromuscular stimuli.

2.3 The Most Common Strength Training Exercises Used and Optimal Training Load

In team sports it is common to use different ST exercises and loads in accordance with different aims and perspectives. As an example of this, plyometric training is a type of training that involves jumping exercises using the stretch-shortening cycle (Markovic and Mikulic 2010). According to Bedoya et al. (2015), the specific actions performed during plyometric training are similar to team sport demands. Meanwhile, resistance training approaches are based on emphasising the vertical component during the lower body triple extension such as in different squat exercises, due to the fact that these are deemed closer to actions performed at high velocity, such as sprinting and jumping tasks (Kawamori and Haff 2004). Authors have generally tended to use high loads in their studies (70–90% 1 repetition maximum [1RM]) to improve high-intensity actions such as jumping or sprinting (Chelly et al. 2009; Smilios et al. 2013; Styles et al. 2016). In this regard, greater magnitude improvements have been reported in sprinting ability through heavy loads (80% 1RM) in comparison to maximal power output load in the squat jump exercise, though no inter-group differences are presented (Harris et al. 2008). Conversely, other authors have stated that high velocity seems to be crucial to yield positive performance adaptations (McBride et al. 2002). Furthermore, a recent study has shown that a training programme using light loads (40–60% 1RM) at maximal intended velocity may be a preferential stimulus for jumping and sprinting improvements (Hoyo et al. 2016a). Therefore, it seems that the combination of moderate load and maximal intended velocity might also be useful for improving high-intensity actions.

Freitas et al. (2017) published a meta-analysis in which they analysed the short-term adaptations in sprint and vertical jump performance following complex training in team sports. Complex training consists of alternating heavy resistance training exercises with plyometric/power ones, set for set, in the same workout (Freitas et al. 2017). In the aforementioned study, complex training had positive moderate effects on sprint performance and mild effects on vertical jump among team sport athletes. The authors concluded that this training method was a suitable option to include in the season plan. With regards to the intensity of the conditioning activities, intensities below 85% 1RM exhibited greater training effects than maximal loads (>85% 1RM) (Freitas et al. 2017).

On the other hand, resisted sprinting has been implemented as a method for overloading the abilities specific to sprinting acceleration performance (Petrakos et al. 2016). It provides a greater resistance than normal sprint training and may provide a greater stimulus to the working muscles, and optimise training adaptations and crossover to dynamic athletic performance (Hrysomallis 2012). This type of training is commonly used to increase sprinting performance (Spinks et al. 2007). In this regard, the optimal resisted load for sprint training has not been established yet, although it has been suggested that a resistance reducing the athlete's velocity by more than 10% from unloaded sprinting would entail substantial changes in the athlete's sprinting mechanics (Spinks et al. 2007; Lockie et al. 2003).

Until recently, few researchers have exceeded relatively light loading parameters (e.g. approximately ~10% velocity decrement) for fear of creating dissimilar conditions to unresisted sprinting resulting in negative adaptations (e.g. slower running velocity and/or altered running technique) (Spinks et al. 2007; Lockie et al. 2003; Alcaraz et al. 2009). However, there is some preliminary evidence to suggest that training using much heavier loads may be beneficial for accelerative performance (Kawamori et al. 2014). Thus, several studies have proposed that the initial phase of acceleration up to 30 m might be improved using loads ≥20% body mass, while to improve high-speed acceleration phases, loads around 5–12.5% of body mass should be used (Bachero-Mena and González-Badillo 2014; Morin et al. 2017). In this regard, Cross et al. (2018) investigated the effects of resisted sprint training on sprinting performance and underlying mechanical parameters (force-velocity-power profile) based on two different training protocols: (i) loads that represented maximum power output and a 50% decrease in maximum unresisted sprinting velocity and (ii) lighter loads that represented a 10% decrease in maximum unresisted sprinting velocity. Both resisted-sprint training protocols were likely to improve performance after a short training session. However, widely varying individual results indicated that adaptations may be dependent on pre-training force-velocity characteristics (Cross et al. 2018).

As an alternative to these traditional training methods, flywheel inertial devices have appeared increasingly in scientific research and are being incorporated into regular training programmes (Hoyo et al. 2015, 2016a; Tous-Fajardo et al. 2016). The benefits of these devices include both eliciting a greater overall amount of muscle activity than traditional overload exercises (Askling et al. 2003) and a greater eccentric overload (Romero-Rodriguez et al. 2011). Therefore, the introduction of eccentric overload training (EOT) methods, which can overload the eccentric phase, might be appropriate to improve jumping, sprinting and COD abilities in football players as different studies have reported (Hoyo et al. 2015, 2016b). In this sense, several studies have shown that the knee flexors are very likely contributors to sprint acceleration performance, where subjects who produced the greatest amount of horizontal force during sprinting are both able to highly activate their hamstring muscles just before ground contact and present high eccentric hamstring peak torque capability (Edouard et al. 2016). Moreover, it seems that eccentric strengthening exercises for the quadriceps, adductors and hamstrings reduce the thigh muscle strain injury rate in team sports, developing stronger muscles at longer lengths (Hoyo et al. 2015; Askling et al. 2003; Arnason et al. 2007; Brughelli et al. 2010; Núñez et al. 2020).

3 From Practice

In this section, sample microcycles will be provided to illustrate the periodisation strategies proposed for each phase of the training year for a generic team sport, using the example of a football team. The rationale for the approach used for each of the respective training cycles is outlined below. Specific programme variables, such as the length of each phase and exercise selection, will vary depending on the length of the playing season and demands of the particular sport (Suarez-Arrones et al. 2019).

3.1 *Detraining Period (Off-Season)*

Normally, the detraining period in football consists of five-six weeks. During the first two weeks, players are asked to completely rest and avoid any kind of physical activity. Thereafter, for the remaining three weeks, players are instructed to perform an individualised training programme that includes high-intensity running interval training (HIT) and strength training, with the training sessions being carried out 4 days per week. Each training session normally consists of a warm-up, ST in the gymnasium (gym), and HIT at the end of the session or at a different time of day. The warm-up involves joint mobility and active stretching exercises. The ST is structured into three different session types, which are core and sensorimotor exercises, functional exercises (involving different joints and planes during specific movements), and more isolated structural strength exercises (three sessions) for the upper and lower body using different equipment (free-weight, instability, and suspension training). An example of a one-week individual training programme is shown in Table 1.

3.2 *Retraining Period (Pre-Season)*

Usually, in football, the pre-season retraining period lasts six or seven weeks. Our proposal is based on a previous study (Suarez-Arrones et al. 2019) where the players supplemented the football training with an ST programme structured into four different session types, including (i) ST in the gym; (ii) specific ST on the field; (iii) activation training; and (iv) individual training.

ST in the gym is usually organised as circuit training before the football drills are carried out on the field. Players perform one or two laps of a circuit consisting of 10–12 exercises mainly focusing on the lower limbs, combining free weights with non-gravity-dependent flywheel inertial devices and motorised devices, and including some functional exercises for upper-body and core muscles. In addition, complementary ST sessions are prescribed with exercises for upper-body, core, and lumbo-pelvic stability. Strength training sessions in the gym last 30–40 min each, while complementary sessions last 20 min. The main exercises employed in the gym sessions for the lower limbs are as follows: football-specificmovements (side step, cutting, lunge) focusing more on the horizontal force (anterior–posterior/posterior-anterior/lateral and rotational) using Versa Pulley® (VP) (0.19 kg m^2 and 0.26 kg m^2 inertias), several bilateral and unilateral half-squat or lunge exercises

Table 1 An example of a typical training schedule during the off-season (detraining period)

Monday	Tuesday	Wednesday	Thursday	Friday	Saturday	Sunday
Warm-up (5 min) Joint mobility and active stretching **Complementary training** Balance + Core + Functional training (Hip focus, 12 exercises (~20 min): Frontal Bridge TR × 2 × 6 × 6″; side bridge TR × 2 × 6 × 6″ (each side); Bridge 1 leg Fitball 2 × 6 × 6″ (each leg); Russian twists on fitball 2 × 30″; single leg balance (closed eyes) on bosu 2 × 30″; single leg jump on bosu 2 × 30″; band adductor 2 × 10 (each leg); barbell glute bridge 2 × 10; standing adductor on fitball 2 × 10 (each side); push up + side plank 1 leg 2 × 10 (each side); static wall squat 4 × 15″; “the glider” 2 × 10 (each side)	Warm-up (5 min). Joint mobility and active stretching **Complementary training** Balance + Core, 6 exercises (~10 min): Frontal Bridge TR × 2 × 6 × 6″; sidebridge TR × 2 × 6 × 6″ (each side); Bridge 1 leg Fitball 2 × 6 × 6″ (each leg); Russian twists on fitball 2 × 30″; single leg balance (closed eyes) on bosu 2 × 30″; single leg jump on bosu 2 × 30″	Other sport	Warm-up (5 min). Joint mobility and active stretching **Complementary training** Balance + Core + Hip Muscle, 12 exercises (~20 min): Frontal Bridge TR × 2 × 6 × 6″;sidebridge TR × 2 × 6 × 6″ (each side); Bridge 1 leg Fitball 2 × 6 × 6″ (each leg); Russian twists on fitball 2 × 30″; single leg balance (closed eyes) on bosu 2 × 30″; single leg jump on bosu 2 × 30″; band adductor 2 × 10 (each leg); barbell glute bridge 2 × 10; standing adductor on fit- ball 2 × 10 (each side); push up + sideplank 1 leg 2 × 10 (each side); static wall squat 4 × 15″; “the glider” 2 × 10 (each side)	Warm-up (5 min). Joint mobility and active stretching **Complementary training** Balance + Core, 6 exercises (~10 min): Frontal Bridge TR × 2 × 6 × 6″; sidebridge TR × 2 × 6 × 6″ (each side); Bridge 1 leg Fitball 2 × 6 × 6″ (each leg); Russian twists on fitball 2 × 30″; single leg balance (closed eyes) on bosu 2 × 30″; single leg landings on bosu 2 × 30″	Off	Off
	Strength training Upper and Lower body, Circuit training. 3 × 8 exercises (30 min): leg press 8 rep; push up TR × 8 rep; pull TR × 8 rep; single deadlift with dumbbell 6 rep (each leg); standing dumbbell arm swing			**Strength training** Upper and Lower body, Circuit training. 3 × 8 exercises (30 min): half squat 8 rep, bench press 8 rep, seated cable row 8 rep, barbell curl 8 rep, barbell lunge 8 rep (each side), lateral elevation shoulder 8 rep		

(continued)

Table 1 (continued)

Monday	Tuesday	Wednesday	Thursday	Friday	Saturday	Sunday
	30″; high box jump 6 rep; standing dumbbell fly on fitball 8 rep; barbell lunge split jump 6 rep (each leg) **Endurance training** HIT (short intervals). 3 × 6′ [20″ (98 m): 20″ (passive rest)], rest: 3′			(each side), pulley triceps extension 8 rep, high box jump 6 rep		
Endurance training HIT (long intervals). 2 × [3 × 4′ running (800 m)] rest/rep: 2′, rest/set			**Endurance training** HIT (short intervals). 3 × 6′ [10″ (50 m): 10″ (passive rest)], rest/set 4′	**Endurance training** HIT (short intervals). 8 × 30″ running all out straight line, rest: 3′		

using Kbox® (0.10 kg m^2 and 0.05 kg m^2), bilateral and unilateral leg-press and leg-curl using Yo-Yo Technology® (0.11 KG m^2), several exercises focusing on the posterior chain using free weights and inertial devices (i.e. barbell deadlift, barbell hip-thrust, hip-extension or hip-thrust in versa pulley®), anterior cross chain and posterior cross chain using Kine Dimanics®, elastic bands, free weights and/or body weight. The main exercises employed in the gym sessions for upper-body, core, and lumbo-pelvic stability are push-up and pull-up exercises using free weights and Kine Dynamics®, functional bilateral rotational exercises using VP, single leg side, prone and front bridge using Fit ball, cable wood chops using VP and several functional unilateral push and pull exercises using VP + Kine Dynamics®.

Specific ST on the field lasts 20–25 min each session and consists of different combined football drills with goal-shooting (finishing), including high-intensity actions such as plyometric jumps, resisted sprinting, duels, different changes of directions, and high-speed running, among others.

Activation training consists of neuromuscular training exercises in the gym or on the field, as an initial part of the training session and before the specific football drills. Examples of exercises used are those focusing on core, hamstring, groin and abductor muscle activation combined with sensorimotor exercises on stable/unstable surfaces. In addition, some individual activation sessions are also prescribed to some players when the team starts directly on the field ($\sim$10 min). Activation training sessions with the whole group last 20 min each.

Individual training consists of ST sessions in the gym focusing on the player's weak points for injury prevention (i.e. imbalances, posterior chain, groin, abductors, calf, and rectus femoris). The individual training is usually planned after the football training on the field and lasts 10–15 min. An example of the training programme for the team is shown in Tables 2 and 3.

3.3 *In-Season Period*

There is limited time for ST sessions to take place during the in-season period in a professional context. Therefore, an approach with time-efficient strategies that concurrently enhance several locomotive-specific actions while minimising the risk of injuries is essential. The neuromuscular intervention should be carried out both in the gym and on the field, combining different ST methodologies with football-specific strength-based actions, and this type of training session must be delivered to the whole squad with excellent compliance. The criteria of the designated programmes and contents are selected and timed according to current scientific knowledge and the practical and clinical expertise of the staff members involved (coaches, fitness coaches, physiotherapists and medical doctors).

Our proposal, based on the study of Suarez-Arrones et al. (2021), is an approach where ST is usually structured before the football exercises on the field, although there are also some post-training interventions. Players perform one or two sets

Table 2 An example of a typical training schedule during the pre-season (one friendly game/week)

	Monday	Tuesday	Wednesday	Thursday	Friday	Saturday	Sunday
AM	Trip to Training Camp	**Strength training GYM**[2] Individual exercise: individual defensive technique Group Exercise: specific drill pass Group Exercise: game possession 4v4 + 3 jokers	Collective exercise: defensive technique Group Exercise: specific drill pass and finishing Collective exercise: game	Rest	**Strength training GYM**[2] Collective exercise: defensive organisation Group Exercise: offensive evolutions— Collective exercise: game	Group Exercise: generic drill pass Group Exercise: generic rondos	Collective and individual recovery
PM	**Activation**[1] Collective exercise: individual technique—Group Exercise: generic drill pass –Group Exercise: generic rondos –Group Exercise: game possession 4v4 + 3 jokers	Group exercise: individual defensive technique Group Exercise: specific drill pass and finishing Group Exercise: SSGs 6v6 + Gk **Complementary strength training**[3]	Rest	Introduction HIT: medium distance drill Group Exercise: offensive evolutions Collective exercise: conditioned game **Complementary strength training**[3]	Group Exercise: specific drill pass Collective exercise: offensive organisation	**Friendly Game:** (45 min each player)	Rest

[1] 1 × 12 exercises: Russian belt deadlift (8 rep.); Russian belt squat + hip-extension (8 rep.); hip-flexion with band (10 rep./leg); lateral band walk (10 rep./leg); side bridge on fitball (6 rep./side); kneeling on fitball (20 s); single-leg prone bridge on fitball (6 rep/leg); single squat on bosu (8 rep./leg); different landings on bosu (6 rep./leg); lateral lunge on whole body vibration (8 rep./leg); 2-way bunkie side bridge (6 × 6 s./side); skater hops (10 rep.). [2] 2 × 12 exercises: single lateral squat on kbox (6 rep./leg); single-leg Yo-Yo Leg-curl (6 rep./leg); anterior–posterior side step on VP (6 rep./leg); lateral crossover cutting on VP (6 rep./leg); trunk rotation on VP + Kine (10 rep./side); single-leg leg press Yo-Yo multigym (6 rep./leg); single standing pull + trunk rotation with Kine (10 rep/side); lateral step-up box (8 rep./leg); single-leg dumbbell deadlift hop (8 rep/leg); lateral barbell lunge on whole body vibration (6 rep/leg); downward cable wood chops on VP (8 rep/side); hip-extension on VP (6 rep/leg). [3] 2 × 12 exercises: push-up with Kine (12 rep.), pull-up + trunk rotation with Kine (12 rep.), ball pull-over throw (10 rep.), single-arm push + trunk rotation on VP (8 rep./side), single-arm pull on VP (8 rep./side), suspended crunch with Kine (8 rep./side), dumbbell bench press on fitball (12 rep.), cable single-arm row in VP (8 rep./side), seated dumbbell clean on fitball (10 rep.), plank in bench + fitball (20 s.), side plank with fitball (20 s./side), up-down plank (12 rep.)

Table 3 An example of a typical training schedule during the pre-season (two friendly games/week)

	Monday	Tuesday	Wednesday	Thursday	Friday	Saturday	Sunday
AM	**Activation GYM**[1] Group Exercise: generic drill pass Collective exercise: offensive organisation + defensive transition Collective exercise: game possession 9v9 + 3 jokers	**Strength training GYM**[2] Individual exercise: individual technique Group Exercise: specific drill pass + HIT short intervals drill with COD Group Exercise: game possession 4v4 + 3 jokers	Group Exercise: specific drill pass and finishing Collective exercise: offensive organisation + defensive transition Collective exercise: game **Complementary strength training GYM**[4]	Rest	Players with more than 45 min played: Recovery (aerobic training, Foam Roller, and contrast therapy) Players with less than 45 min played: Group Exercise: SSGs5v5 + GK + HIT short intervals drill	Rest	Rest
PM	Rest	**Specific strength training Field**[3] Group Exercise: specific drill pass Collective exercise: game possession 8v8 + 3 jokers	Rest	**Friendly Game**	Collective exercise: individual technique Group exercise: generic rondos **Individual training GYM**[5]	**Individual activation** Group Exercise: specific drill pass Group Exercise: game possession 4v4 + 3 jokers **Complementary strength training GYM**	**Friendly Game**

[1]1 × 10 exercises: Russian belt deadlift (8 rep.); Russian belt squat + hip-extension (8 rep.); adductor slides with kine (8 rep./leg); standing abductor with Kine (10 s./leg); specific side plank (kicking a ball) (6 rep./side); single-leg plank (10 s./leg); side plank with kine opening and closing legs (6 rep/leg); single balance exercises on bosu (15 s./leg); different landings on bosu (10 rep./leg); frontal lunge on whole body vibration (8 rep./leg). [2]1 × 12 exercises: single lateral squat on kbox + bosu (6 rep./leg); hip-extension kick in Exentrix (6 rep./leg); postero-anterior side step on VP (6 rep./leg); lateral crossover cutting on VP (6 rep./leg); trunk rotation on VP + Kine (10 rep./side); soccer kick exercise with VP (10 rep./leg); single standing with trunk rotation with Kine (10 rep/side); step-up box with jump (8 rep./leg); single-leg hamstring bridge on box (5 s. isometric +8 rep/leg); lateral barbell lunge on whole body vibration (6 rep/leg); upward cable wood chops on VP (8 rep/side); landings in bosu with elastic band at the waist (6 rep/leg). [3]Drill 1: 4 × (2 jump hurdle drills + receive and passing the ball + 4 change of directions of 90° + receive the ball + dribbling and finishing. Drill 2: 4 × (20 m sled running with Run Rocket + 30 m of running gradually increasing the speed (>25 km/h) + deceleration + receive the ball + dribbling and finishing. Drill 3:4 × (in pairs, jump up and head the ball + 5 m running pushing between them + 1vs1 and finishing). [4]2 × 12 exercises: push-up + trunk rotation with Kine (12 rep.); pull-up with Kine (12 rep.); push-up + side-plank-leg (12 rep.); single-arm push on VP (8 rep./side); single-arm pull + trunk rotation on VP (8 rep./side); bilateral trunk rotational on VP (8 rep./side); bench press (12 rep.); single-arm dumbbell row (8 rep./side); standing dumbbell fly on fitball (12 rep.), pull up (>6 rep.), bar dip (12 rep.), atomic push-up with Kine on bosu (10 rep.). [5]An example of a player: 2 × 6 exercises: hip-extension kick in VP (8 rep/side), single-leg hip thrust with elastic band (8 rep./leg), deadlift (8 rep), walking dumbbell lunge (8 rep.), single-leg Yo-Yo Leg-curl (6 rep./leg), adductor in VP (8 rep./leg)

(based on weekly matches) of a circuit consisting of 10–12 exercises (between 6 and 10 reps per exercise) mainly focusing on the lower limbs in coordination with the subsequent training that will be carried out on the field (specific ST on the field). This is done by either prioritising strength on the first acquisition day (i.e. Wednesday) using small spaces, and endurance and/or speed on the second and third acquisition days (i.e. Thursday and Friday) in larger spaces (Fig. 1).

In addition, complementary ST sessions are scheduled during the season with exercises for upper-body, core, and lumbo-pelvic stability (usually after training); and activation training sessions consisting of neuromuscular training exercises in the gym or on the field, as an initial part of the training session and before the specific football drills. ST sessions in the gym last approximately 30–40 min, while complementary ST sessions and collective activation training sessions last 15–25 min.

Individual training sessions are also scheduled primarily targeting player weak points such as previous injuries. The individual training is mainly organised as secondary prevention for a specific group of players with a theoretically heightened risk of suffering an injury episode and are mostly carried out in the gym. The sessions' criteria and content are selected and organised according to the player's weak points. As an example, the organisation of the different training interventions with a high-risk player (a biceps femoris long head injury in the previous season) during a typical microcycle is shown in Fig. 2.

This individual work is carried out in combination with the collective ST of the whole team (Fig. 3). Additionally, the same player is subjected to (i) individual activation before an on-field training session in large spaces with a high volume of high-speed running (Fig. 4), (ii) individual ST after the on-field training (Fig. 5), and (iii) individual activation before an official game (Fig. 6).

Additionally, the same player is subjected to (i) individual activation before an on-field training session in large spaces with a high volume of high-speed running (Fig. 4), (ii) individual ST after the on-field training (Fig. 5), and (iii) individual activation before an official game (Fig. 6).

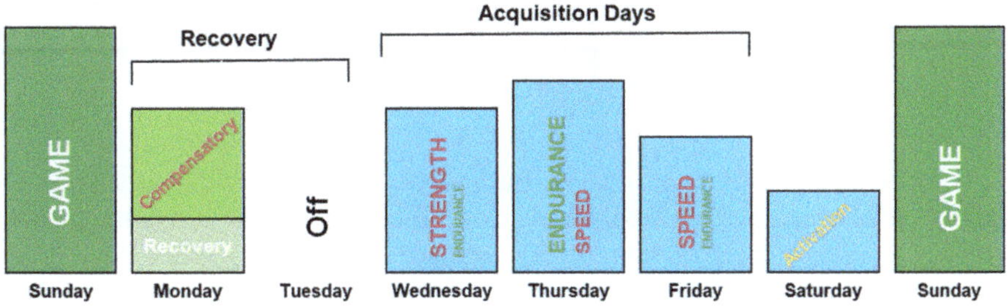

Fig. 1 An example of the organisation of the on-field training contents during a microcycle in football

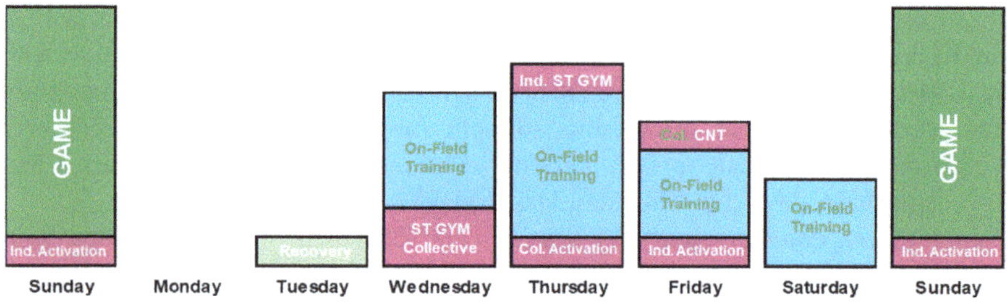

Fig. 2 An example of the organisation of the different training interventions with a high-risk player, who suffered a biceps femoris long head injury in the previous season, during a microcycle in football. *ST* strength training; *Ind* individual; *Col* collective

Fig. 3 An example of collective strength training in the GYM. The training consists of 2 × 10 exercises (between 6 and 10 reps per exercise) mainly focusing on the lower limbs

As reflected in the different images, our proposal combines different sources of external resistance such as free weights and other methodologies with greater eccentric overload such as non-gravity dependent flywheel inertial devices or motorised devices for a stronger eccentric overload. Elastic cords/bands, sliding boards, suspension training, pneumatic cable devices and the player's own body weight are also used. To determine the training intensity (i.e. external load) to be employed during the ST, a velocity-based assessment with incremental loads using free weights and different inertias with flywheel technology and conic pulley should

Fig. 4 An example of different exercises used during individual activation for the posterior chain before on-field training in larger spaces with a high volume of high-speed running

Fig. 5 An example of different exercises used during individual strength training after the on-field training with a player who suffered a biceps femoris long head injury in the previous season

be performed for the most employed exercises. The inertia/load with which the player achieves higher average power is used during the training and individually readjusted every few weeks. For the exercises that employ another source of resistance (e.g. elastic cords, mini-bands and sliding), exercise intensity is regulated subjectively (as usually done by fitness trainers). In addition to overloads related to manipulation of external resistance and speed of execution, overloads in many exercises are also achieved by manipulating motor control/coordination, muscle lengths, and mechanical perturbations. Wide varieties of exercises are employed through the season aimed at providing variability and avoiding stagnation and monotony.

Fig. 6 An example of different exercises used during individual activation before a game

4 Filling Gaps

Team sport athletes need to perform at the highest level week after week if they want to be in contention for the championship at the end of the competitive season. Therefore, there are no unimportant games and there are no performance peaks.

Neuromuscular training in team sports has two main aims, which are to reduce the injury rate and optimise performance. The reality is that there is limited time for ST sessions to be carried out during the in-season period, therefore, time-efficient strategies that concurrently enhance several locomotive- and other high-intensity-specific actions while preventing injuries are crucial.

In the professional field, neuromuscular interventions are commonly carried out both in the gym and on the field combining different ST methodologies with specific strength-based actions, prioritising the greatest volumes of ST in the first acquisition days, several days before competition. A frequently used approach is one where ST is habitually structured before the specific exercises on the field, though there are also different post-training interventions. On these days, the players generally perform ST mainly focusing on the lower limbs in coordination with the subsequent training that will be carried out on the field. In addition to this, individual training sessions are also scheduled primarily targeting player weak points, and complementary ST sessions are planned during the week with exercises for upper-body, core, and lumbo-pelvic stability. As preparation for the joints and muscles for the on-field training, activation sessions are also incorporated combining different neuromuscular exercises in the gym or on the field as an initial part of the subsequent specific training.

5 Take-Home Messages

1. Try to adapt your strength training programme to the field training demands.
2. Time for strength training sessions is reduced in team sports, so try to make your strategies optimise performance and reduce injury risk at the same time.
3. Seek to individualise your strength training strategies as much as possible according to positional profile, imbalances and muscle deficits and history of injuries.
4. If you decide to use strength training programmes with eccentric overload, introduce it progressively so that the athlete can adapt.

References

Alcaraz PE, Palao JM, Elvira JLL (2009) Determining the optimal load for resisted sprint training with sled towing. J Strength Cond Res 23(2):480–485

Arnason A, Andersen TE, Holme I, Engebretsen L, Bahr R (2007) Prevention of hamstring strains in elite soccer: an intervention study. Scand J Med Sci Sports 18(1):40–48

Askling C, Karlsson J, Thorstensson A (2003) Hamstring injury occurrence in elite soccer players after preseason strength training with eccentric overload. Scand J Med Sci Sports 13(4):244–250

Bachero-Mena B, González-Badillo JJ (2014) Effects of resisted sprint training on acceleration with three different loads accounting for 5, 12.5, and 20% of body mass. J Strength Cond Res 28(10):2954–2960

Bedoya AA, Miltenberger MR, Lopez RM (2015) Plyometric training effects on athletic performance in youth soccer athletes. J Strength Cond Res 29(8):2351–2360

Bolger R, Lyons M, Harrison AJ, Kenny IC (2015) Sprinting performance and resistance-based training interventions. J Strength Cond Res 29(4):1146–1156

Brughelli M, Mendiguchia J, Nosaka K, Idoate F, Arcos AL, Cronin J (2010) Effects of eccentric exercise on optimum length of the knee flexors and extensors during the preseason in professional soccer players. Phys Ther Sport 11(2):50–55

Buchheit M, Lacome M, Cholley Y, Simpson BM (2018) Neuromuscular responses to conditioned soccer sessions assessed via GPS-Embedded accelerometers: insights into tactical periodization. Int J Sports Physiol Perform 13(5):577–583

Chelly MS, Fathloun M, Cherif N, Amar M Ben, Tabka Z, Van Praagh E (2009) Effects of a back squat training program on leg power, jump, and sprint performances in junior soccer players. J Strength Cond Res 23(8):2241–2249

Cormie P, McGuigan MR, Newton RU (2011) Developing maximal neuromuscular power. Sport Med 41(2):125–146

Cross MR, Lahti J, Brown SR, Chedati M, Jimenez-Reyes P, Samozino P, Eriksrud O, Morin JB (2018) Training at maximal power in resisted sprinting: optimal load determination methodology and pilot results in team sport athletes. PLoS One 13(4)

de Hoyo M, Pozzo M, Sañudo B et al (2015) Effects of a 10-week in-season eccentric-overload training program on muscle-injury prevention and performance in junior elite soccer players. Int J Sports Physiol Perform 10(1):46–52

De Hoyo M, Gonzalo-Skok O, Sanudo B et al (2016a) Comparative effects of in-season full-back squat, resisted sprint training, and plyometric training on explosive performance in U-19 elite soccer players. J Strength Cond Res 30(2):368–377 de Hoyo M, Sañudo B, Carrasco L, Mateo-Cortes J, Domínguez-Cobo S, Fernandes O, Del Ojo JJ, Gonzalo-Skok O (2016b) Effects of 10-week eccentric overload training on kinetic parameters during change of direction in football players. J Sports Sci 34(14):1380–1387

Delgado-Bordonau JL, Méndez Villanueva A (2018) Tactical periodization—a proven successful training model. SoccerTutor.com

Ebben WP, Blackard DO (2001) Strength and conditioning practices of national football league strength and conditioning coaches. J Strength Cond Res 15(1):48

Ebben WP, Carroll RM, Simenz CJ (2004) Strength and conditioning practices of national hockey league strength and conditioning coaches. J Strength Cond Res 18(4):889–897

Edouard P, Arnal P, Gimenez P, Samozino P, Jimenez-Reyes P, Brughelli M, Mendiguchia J, Morin JB (2016) Athletic injury prevention: determinants of sprint performance. Ann Phys Rehabil Med 59:e22–e23

Fleck SJ (2011) Non-linear periodization for general fitness and athletes. J Hum Kinet

Freitas TT, Martinez-Rodriguez A, Calleja-González J, Alcaraz PE (2017) Short-term adaptations following complex training in team-sports: a meta-analysis. PLoS One 12(6):e0180223

Gabbett TJ (2005) Changes in physiological and anthropometric characteristics of rugby league players during a competitive season. J Strength Cond Res 19(2):400

Gamble P (2006) Periodization of training for team sports athletes. Strength Cond J. 28(5):56

García-Pinillos F, Martínez-Amat A, Hita-Contreras F, Martínez-López EJ, Latorre-Román PA (2014) Effects of a contrast training program without external load on vertical jump, kicking speed, sprint, and agility of young soccer players. J Strength Cond Res 28(9):2452–2460

Gómez A, Roqueta E, Tarragó JR, Seirul·lo F, Cos F (2019) Entrenamiento en deportes de equipo: el entrenamiento coadyuvante en el FCB. Apunt Educ Física y Deport (138):13–25

Harris NK, Cronin JB, Hopkins WG, Hansen KT (2008) Squat jump training at maximal power loads vs. heavy loads: effect on sprint ability. J Strength Cond Res 22(6):1742–1749

Hrysomallis C (2012) The effectiveness of resisted movement training on sprinting and jumping performance. J Strength Cond Res 26(1):299–306

Impellizzeri FM, Rampinini E, Coutts AJ, Sassi A, Marcora SM (2004) Use of RPE-based training load in soccer. Med Sci Sport Exerc 36(6):1042–1047

Issurin VB (2008) Block periodization versus traditional training theory: a review. J Sport Med Phys Fit. 8:65–75

Issurin VB (2010) New horizons for the methodology and physiology of training periodization. Sport Med 40(3):189–206

Issurin VB (2014) Periodization training from ancient precursors to structured block models. Kinesiology 3–9

Issurin VB (2016) Benefits and limitations of block periodized training approaches to athletes' preparation: a review. Sport Med 46(3):329–338

Kawamori N, Haff GG (2004) The optimal training load for the development of muscular power. J Strength Cond Res 18(3):675–684

Kawamori N, Newton R, Nosaka K (2014) Effects of weighted sled towing on ground reaction force during the acceleration phase of sprint running. J Sports Sci 32(12):1139–1145

Kraemer WJ (2004) Strength training for young athletes. In: Fleck SJ (ed) Human Kinetics Publishers, Champaign

Lambert MI, Viljoen W, Bosch A, Pearce AJSM (2008) General principles of training. In: Schwellnus MP (ed) Olympic textbook of medicine in sport, 1st edn. Wiley-Blackwell Publishing, Oxford, pp 1–48

Lockie RG, Murphy AJ, Spinks CD (2003) Effects of resisted sled towing on sprint kinematics in field-sport athletes. J Strength Cond Res 17(4):760–767

Lovell R, Whalan M, Marshall PWM, Sampson JA, Siegler JC, Buchheit M (2018) Scheduling of eccentric lower limb injury prevention exercises during the soccer micro-cycle: which day of the week? Scand J Med Sci Sports 28(10):2216–2225

Malone JJ, Di Michele R, Morgans R, Burgess D, Morton JP, Drust B (2015) Seasonal training-load quantification in elite english premier league soccer players. Int J Sports Physiol Perform 10(4):489–497

Manchado C, Cortell-Tormo JM, Tortosa-Martínez J (2018) Effects of two different training periodization models on physical and physiological aspects of elite female team handball players. J Strength Cond Res 32(1):280–287

Mara JK, Thompson KG, Pumpa KL, Ball NB (2015) Periodization and physical performance in elite female soccer players. Int J Sports Physiol Perform 10(5):664–669

Margaritelis NV, Theodorou AA, Chatzinikolaou PN, Kyparos A, Nikolaidis MG, Paschalis V (2021) Eccentric exercise per se does not affect muscle damage biomarkers: early and late phase adaptations. Eur J Appl Physiol 121(2):549–559

Markovic G, Mikulic P (2010) Neuro-Musculoskeletal and performance adaptations to lower-extremity plyometric training. Sport Med 40(10):859–895

Matveyev LP (1964) Problem of periodization the sport training. Fizkultura i Sport, Moscow

McBride JM, Triplett-McBride T, Davie A, Newton RU (2002) The effect of heavy- vs. light-load jump squats on the development of strength, power, and speed. J strength Cond Res 16(1):75–82

McCall A, Carling C, Nedelec M et al (2014) Risk factors, testing and preventative strategies for non-contact injuries in professional football: current perceptions and practices of 44 teams from various premier leagues. Br J Sports Med 48(18):1352–1357

Mjolsnes R, Arnason A, Osthagen T, Raastad T, Bahr R (2004) A 10-week randomized trial comparing eccentric vs. concentric hamstring strength training in well-trained soccer players. Scand J Med Sci Sport 14(5):311–317

Morin J-B, Petrakos G, Jiménez-Reyes P, Brown SR, Samozino P, Cross MR (2017) Very-heavy sled training for improving horizontal-force output in soccer players. Int J Sports Physiol Perform 12(6):840–844

Mujika I (2007) Thoughts and considerations for team-sport peaking. Olympic Coach 18(4):9–11

Mujika I, Halson S, Burke LM, Balagué G, Farrow D (2018) An integrated, multifactorial approach to periodization for optimal performance in individual and team sports. Int J Sports Physiol Perform 13(5):538–561

Núñez JF, Fernandez I, Torres A, García S, Manzanet P, Casani P, Suarez-Arrones L (2020) Strength conditioning program to prevent adductor muscle strains in football: does it really help professional football players? Int J Environ Res Public Health 17(17):6408

Opar DA, Williams MD, Timmins RG, Hickey J, Duhig SJ, Shield AJ (2015) Eccentric hamstring strength and hamstring injury risk in Australian footballers. Med Sci Sport Exerc 47(4):857–865

Ozolin N (1970) The modern system of sport training. Fizkultura. i Sport, Moscow

Peterson MD, Rhea MR, Alvar BA (2004) Maximizing strength development in athletes: a meta-analysis to determine the dose-response relationship. J Strength Cond Res 18(2):377–382

Peterson MD, Rhea MR, Alvar BA (2005) Applications of the dose-response for muscular strength development: a review of meta-analytic efficacy and reliability for designing training prescription. J Strength Cond Res 19(4):950

Petrakos G, Morin J-B, Egan B (2016) Resisted sled sprint training to improve sprint performance: a systematic review. Sport Med 46(3):381–400

Rollo I, Impellizzeri FM, Zago M, Iaia FM (2014) Effects of 1 versus 2 games a week on physical and subjective scores of subelite soccer players. Int J Sports Physiol Perform 9(3):425–431

Romero-Rodriguez D, Gual G, Tesch PA (2011) Efficacy of an inertial resistance training paradigm in the treatment of patellar tendinopathy in athletes: a case-series study. Phys Ther Sport 12(1):43–48

Seirul-lo F (1987) Opción de planificación en los deportes de largo período de competiciones. Rev Entren Dep 1(3):53–62

Silva JR, Nassis GP, Rebelo A (2015) Strength training in soccer with a specific focus on highly trained players. Sport Med Open 1(1)

Silva JR, Brito J, Akenhead R, Nassis GP (2016) The transition period in soccer: a window of opportunity. Sport Med 46(3):305–313

Smilios I, Sotiropoulos K, Christou M, Douda H, Spaias A, Tokmakidis SP (2013) Maximum power training load determination and its effects on load-power relationship, maximum strength, and vertical jump performance. J Strength Cond Res 27(5):1223–1233

Spinks CD, Murphy AJ, Spinks WL, Lockie RG (2007) The effects of resisted sprint training on acceleration performance and kinematics in soccer, rugby union, and australian football players. J Strength Cond Res 21(1):77–85

Steib S, Rahlf AL, Pfeifer K, Zech A (2017) Dose-Response relationship of neuromuscular training for injury prevention in youth athletes: a meta-analysis. Front Physiol 8(5):311–317

Styles WJ, Matthews MJ, Comfort P (2016) Effects of strength training on squat and sprint performance in soccer players. J Strength Cond Res 30(6):1534–1539

Suarez-Arrones L, Lara-Lopez P, Maldonado R et al (2019) The effects of detraining and retraining periods on fat-mass and fat-free mass in elite male soccer players. Peer J 2019(8)

Suarez-Arrones L, Nakamura FY, Maldonado RA, Torreno N, Di Salvo V, Mendez-Villanueva A (2021) Applying a holistic hamstring injury prevention approach in elite football: 12 seasons, single club study. Scand J Med Sci Sports 31(4):861–874

Tarragó JR, Massafret-Marimón M, Seirul-lo F, Cos F (2019) Entrenamiento en deportes de equipo: el entrenamiento estructurado en el FCB. Apunt Educ Física y Deport 137:103–114

Tous-Fajardo J, Gonzalo-Skok O, Arjol-Serrano JL, Tesch P (2016) Enhancing change-of-direction speed in soccer players by functional inertial eccentric overload and vibration training. Int J Sports Physiol Perform 11(1):66–73

Velásquez C (2019) The ATR model has an alternative system of training and research in sports. Educ Fís 8(1):67–80

Wisloff U (2004) Strong correlation of maximal squat strength with sprint performance and vertical jump height in elite soccer players. Br J Sports Med 38(3):285–288

Wrigley R, Drust B, Stratton G, Scott M, Gregson W (2012) Quantification of the typical weekly in-season training load in elite junior soccer players. J Sports Sci 30(15):1573–1580

Moisés de Hoyo Lora has a M.Sc. in High Performance in Sports and other in soccer performance. He received the Ph.D. degree in sport science from University of Seville, Spain, in 2011. He is a professor in Physical Education and Sports at the University of Seville (Spain). He is a former strength and conditioning coach of some professional Spanish soccer teams (Sevilla FC, Levante UD, Deportivo de la Coruña, and CD Leganés). Currently, he is working as the head of Performance in Watford FC (UK). His major research interests are centered on resistance training programs and training load control in team sports. He is also an author and co-author of more than 40 papers. He is a co-funder of InMuv, a sport clinic, focused on high performance and health (Seville, Spain).

Luis Suarez Arrones has M.Sc. and Ph.D. in sports sciences and is a senior lecturer at the Pablo de Olavide University. He is currently the head Fitness Coach of FC Basel 1893 and a former strength and conditioning coach/sports scientist in several professional soccer teams such as Tianjin Quanjian (Chinese super league), Qatar National Soccer Team (Qatar), ACF Fiorentina (Italy), Aspire Academy (Qatar), Real Betis Balompie or Racing de Santander (Spain). He has published nearly 100 peer-reviewed articles, and currently, his research focus is in training, monitoring, and testing in elite soccer players.

Programing and Periodization for Individual Sports

Filipe Almeida Viana Conceiçao and Helvio Affonso

Abstract Despite the controversy surrounding periodization, it has been assumed over the years as the key tool in training planning for the development and achievement of high-level performance, in individual sports. Given the enormous density of nowadays competitive calendar and the athletes' responsibilities towards their sponsors, the challenge is to train with quality, managing fatigue through specific training programs that correctly handle the load applied to the athlete. The scientific boom experienced today and the availability of information from areas such as physiology, biomechanics, biochemistry and sports training, allowed to overcome myths, improve training prescription/control and the development of new approaches to training periodization in elite athletes. The model presented here has as main characteristics to be timeless and dimensionless. That is, each "momentum" depends exclusively on the athlete's body feedback in relation to the training loads, indicated by the biomarkers used. The duration of "momentums" and "macrocycles" depends on the athlete's performance, ballast and physiological wear (internal biomonitored load). The purpose of this chapter is to present an overview of the components to be considered in individual sports training, their control and how the theory is translated into practice.

Keywords Periodization · Models · Training · Biomarkers · Elite sports

F. A. V. Conceiçao (✉)
Center of Research, Education, Innovation and Intervention in Sports (CIFI2D), Faculty of Sports, LABIOMEP, Porto Biomechanics, Universidade do Porto, Porto, Portugal
e-mail: filipe@fade.up.pt

H. Affonso
APPTO Assessoria Esportiva, APPTO Physiology Laborratory of Excerise, Nutrition and Sports and Tranning, Universidade de Villa Verde, Espirito Santo, Brasil

A. Muñoz-López et al. (eds.), *Resistance Training Methods*, Lecture Notes in Bioengineering, https://doi.org/10.1007/978-3-030-81989-7_14

1 Introduction

Periodization is regarded as a superior method for developing an athlete's peak performance (Fleck 1999; Haff 2004). Despite an apparent lack of scientific rigor to govern its application, periodization is widely practiced and recommended (Fleck 1999; Haff 2004; Durell et al. 2003; Ebben et al. 2004; Cissik et al. 2008).

Conceptually, periodization can be defined as an exercise in stress management, in which coaches aim to achieve peak performance while managing fatigue through specific training programs playing with the volume, intensity and complexity of training loads (Matveyev 1972). The landmark of periodization, is the inverse relationship between volume and intensity progressing from extensive to intensive i.e. from high to low volume or from low to high intensity Several training methods try to combine both variables.

Commonly periodization is applied on a cyclic or periodic basis, structured into macro-, meso-, and microcycles. There is, however, large variability in the time course of each, depending on the approach followed. Macrocycles can overall last from 3 months to 4 years while the mesocycles often grouped into 2–4 week block (Matveyev 1972; Zatsiorsky and Kraemer 2006). Mycrocicles can be structured between 2–2 days and 7–10 days.

Among the most widely implemented/known periodization models, the traditional model introduced in the 1950s by Matveyev (1972) and periodization by blocks stand out (Bondarchuk 1986; Issurin 2008; Verkhoshansky 1985).

Because of the drawbacks and limitations arising from classical theory and the need to optimize periodization and/or training loads with the ultimate goal of preparing top-level athletes for current sports demands, new approaches have emerged. In that sense, sports performance depends not only on controlling training loads, but also on the recovery prescribed for athletes, which is determinant to prevent excessive fatigue and negative effects like overtraining or non-functional overreaching (Grandou et al. 2020; Carfagno 2014; Borresen and Lambert 2009).

This control has been traditionally carried out by using self-perceived scales and metrics like), evolving today with the technological innovations, for example using GPS and accelerometers to control external loads during training and competitive sessions or encoders to measure movement velocity and monitoring intensity during resistance training. The greatest advance in monitoring training comes from integrating exercise biochemistry, more precisely from the use of Point of Care (POCT).

Given the need to train better and best use of the available scientific knowledge, it is important to collect sufficient biomechanical, physiological and biochemical data to establish optimal training sequences and loads avoiding negative transfers to gain insight into how the training should be structured.

Unique challenges involve working with high-level athletes who are often reticent to serve as subjects, coaches who often will not give up training time for measurement, extraordinary demands on ease and speed of measurement and data return, and a different perspective of 'success' in conducting research.

2 From Theory

Variations of training content during periodization is a key premise to promote the neurophysiological, mechanical and psycho-emotional adaptations needed to achieve the best level of adaptation for a given athlete.

In practical terms, this adaptation process results from the imposition of periods of overload, followed by a given recovery period that allows the organism to surpass fatigue reaching a higher level of performance. These principles underpin the development of strength, power as well as all other motor and mental capacities.

Although the fundamentals that support periodization are currently controversial (Kiely 2018), periodization is widely accepted as a powerful tool for training planning (Dick 1997).

The traditional periodization was created in a period when the demands and needs were not as demanding as the current ones and based on fundamentals that have since then been outdated. Therefore, it has limitations such as: (i) inability to allow multiple performance peaks throughout the season; (ii) complex development of many abilities simultaneously with negative overlap between them; (iii) Long periods of monotonous and little motivated training; (iv) high volume with low intensity representing low training stimulus during prolonged periods. This model assumes that a relatively prolonged period of basic training (general preparation) is a prerequisite to a more specific phase (special preparation) (Bompa 1984; Issurin 2013). Despite the more advanced versions of traditional periodization foreseeing the possibility of 2–3 peaks in the year, its limitations had not been resolved and on the other hand nowadays competitive environment in most sports is multi peak.

Strength and conditioning coaches should identify whether basic training is essential for competitive performance in their respective sports disciplines, or whether it is a loss of time (Bohm et al. 2015). This is central when working with elite athletes, since the interaction of non-compatible loads (for instance, a high volume of endurance training together with muscle strength and power) may elicit undesirable responses by shifting the focus from the intended adaptations. Therefore, coaches should carefully choose the contents of the training once endurance loads can inhibit protein synthesis and muscle hypertrophy. In addition, the competitive loads must prevail and be specific respecting the bioenergetic conditions of the modality in view.

According to Issurin (2008), several factors contributed to changes in training periodization. The evolution in training methods and science was linked to a technological revolution that completely changed the sport, developing new equipment to accurately control training in real-time, bridging the gap between science and sport settings. Therefore, considering these factors and change of the competitive schedule news approaches emerged in training periodization and strength training highlighting the block periodization model. This model takes in consideration that:

1. Training cycles with highly concentrated specialized loads should not be directed towards multiple objectives and the number of skills to be worked on reduced.
2. Athletic performance in any sport requires the manifestation of several skills, which, in the case of highly concentrated training, should be developed only consecutively, and not simultaneously.
3. Training with selective concentrated loads, developed consecutively leads to the improvement of targeted skills, while the others do not receive stimulation and therefore decline, so the sequencing of training blocks is extremely important.
4. To achieve morphological, organic and biochemical changes, periods of at least 2–6 weeks are necessary, which corresponds to mesocycle.

This model aims to diminish the total training volume in order to focus more on quality than quantity. Some training routines have been replaced by competitive performances, which contribute to decrease the training volume. For instance, (Bondarchuk 2007) it was reported that former world-class throwers from the track and field used to make 120–150 throws per session, while today such athletes perform only about 30 throws per session.

Two approaches can be found in block periodization namely (i) the concentrated unidirectional training model (Verkhoshansky 1985) and the multi-targeted block approach from Bondarchuk (1986).

The unidirectional approach is based on the phasic changes in training capacity and sports performance. Changes tend to a sharp decline and subsequent improvement in speed/force variables with delayed achievement of peak performance in the target discipline that Verkhoshansky (2006) named it as "the long-term lagging training effect" which corresponds to the delayed training effect phenomenon (Zatsiorsky 1995). The literature argued that the greater the decline in physical fitness, the greater the subsequent increase, however, there are no objective findings that highlight the physiological adaptations associated with these phasic changes (Issurin 2016). The new approaches advocate the concentration of high specialized loads, namely during the "shock microcycles", in order to promote this effect. However, there is some controversy in the literature since the adaptations acquired during the shock microcycles seem to be rapidly lost (Wahl et al. 2014). These findings call into question the delayed training effect, which is not fully supported by the scientific literature (Aubry et al. 2014). Although innovative, the unidirectional model presented by Verkhoshansky (1985) is fundamentally suited to those disciplines whose performance depends on a small number of target skills. Another drawback of this approach concerns to the fact that despite considering sequential stages of high concentration of specialized loads, the residual effect of training was never mentioned.

Despite the fact that current models are increasingly adjusted to reality, sports performance in highly competitive sports is complex and depends on a high number of unpredictable variables, which makes periodization even more difficult.

Traditionally, experts from individual sports believe in a positive force transfer from the early phases of the season to the competitive stages, guiding their routine based on this concept (Zatsiorsky 1995). However, there are conflicting results in the literature (Kurz 2001) with some studies reporting both significant, non-significant improvement or even a decrease in performance related parameters (Wilson et al. 1993; Harris et al. 2000; McBride et al. 2002). Young (2006) suggested that it appears that to maximize transfer to specific sports skills, specificity principle should be a premise, especially with regard to movement pattern and contraction velocity. In this way, for sprinters, the priority must be the production of power, as this way we maximize the adaptations that better simulate the competitive load and not exclusively or maximum force itself. On the other hand, the recruitment capacity can be increased using strategies that maximize strength levels (Trappe et al. 2015).

It is possible that the ability to sustain progress in sporting capability over years will benefit from training strategies, making the sports training process more economical, simple and focused on the specific physical capabilities that really matter to actual competitive outcomes (Loturco and Nakamura 2016).

3 From Practice

Sports training periodization has evolved historically, from a more oriented volume periodization at its beginning. This conception was based on some assumptions such as “the more the better” “no pain no gain” that have been nowadays showed to be wrong and inaccurate, since gains in muscle strength and conditioning are not directly linked to those statements but to scientifically supported training principles.

Nowadays, one of the most used approaches is block periodization where mesocycles play an important role within the competitive macrocycle. In this sense, it is observed that high performance athletes of individual sports who have multiple competitions in the year or who have as a priority the development of strength usually resort to small interrelated blocks particularly to achieve multi peaks. One of the features that characterize contemporary periodization is the fact that they are increasingly less rigid and more flexible in order to allow athletes to respond to the challenges imposed by the need to participate in multiple competitions. Successful outcomes ensure high positions in world rankings and important economic incomes.

The overload, fatigue and overcompensation are the basis of the whole training process capable of producing the adaptations to achieve the athlete’s goal. In dense competitive calendars where athletes have to do their best in all competitions, the application of optimal load is critical. However, how to proceed with the proper handling of loads, fatigue and recovery during the session, microcycle and competition to improve and renew the physical freshness to achieve success throughout the various competitions?

3.1 The Organization of Strength Training in Individual Sports and the Use of Velocity-Based Training for the Prescription and Control of Training

Strength training is an essential component of athlete preparation in the majority of sports, contributing significantly to improving performance.

Reduction of the time of force application with increasing sports performance is one of the most important traces of the high-level sport. Therefore, for a greater speed in a given time interval, the greater the impulse produced should be. This interdependence between time, speed and force poses a complex problem to the athlete, i.e. in more difficult conditions (shorter time interval) apply enough force to run, jump or throw, but simultaneously improve performance. Taking into account the features of high-performance sport and the drawbacks arising of using the traditional 1RM and XRM the velocity based training (VBT), is increasingly used as a way of monitoring and prescribing the strength training allowing a better knowledge of the intensity individually (Verkhoshansky and Charniga 1986; González-Badillo and Sánchez-Medina 2010).

The organization of strength training in individual sports that are part of the speed-strength group in general terms usually obeys the following sequence. The season starts with a cycle dedicated to the development of maximum strength through the hypertrophic path. However, it is necessary to clarify that depending on the level of the athlete and the commitments concerning the competitive calendar, there are athletes who give little importance to hypertrophy or otherwise discard this phase. The next step is the development of maximum neural strength by using few repetitions and high loads (80–100% 1RM). At this stage the promotion of low fatigue and muscle damage is the cornerstone. For instance, the velocities are around 0.3 and 0.5 m/s for the squat. After this phase, the focus will be the strength-velocity that is the ability to quickly execute the movement against a relatively large external resistance. It is assessed in terms of loads where the velocities are between 0.7 and 1 m/s. After the conclusion of this period, speed-strength follows, whose aim is to perform fast movements against light external loads, with velocities between 1.0 and 1.3 m/s. Although this phase is dedicated to speed-strength, the athlete also executes sub maximum or maximum strength punctually. The idea associated with this sequence is to go from the general to the specific, combining the traditional strength training with special and specific exercises. In this way, an extensive use of the so-called special exercises, exercises that use the same muscles, same systems, and parts of the competitive movement is necessary; In this group we also find the assisted and the resisted exercises whose essential feature is the introduction of minor changes in the specific exercises, whether overload or lightening. The use of assisted or resisted exercises as well as many other special exercises allows the athlete to learn how to apply more force or speed or execute a given action that is being trained which is decisive for performance. This sort of exercises can provide a training stimulus outside the traditional force-velocity spectrum, like overspeed. Below we present the evolution of an elite

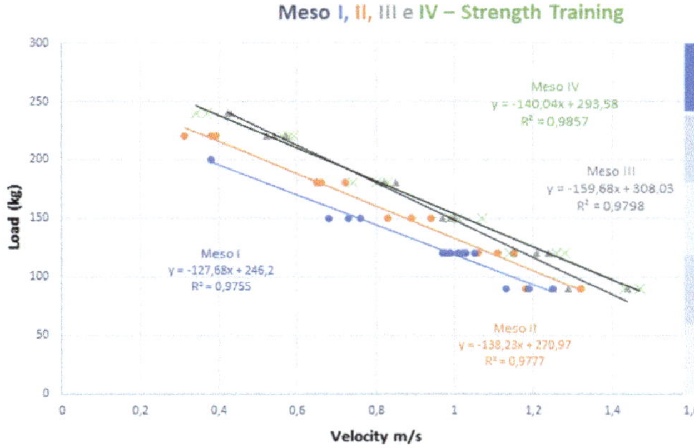

	0,30m/s	1 m/s
Meso I	207,9	118,5
Meso II	229,5	132,7
Meso III	260,2	148,4
Meso IV	251,6	153,5

Fig. 1 Evolution of a Long jumper during four mesocycles of strength training using VBT. The meso I (blue (•)) was dedicated to hypertrophy; meso II (orange (•)) to maximum strength I (black (▲); meso III maximum strength II and meso IV (gray (×)) power. The table on the right shows the evolution of the maximum force (0.3 m/s) and maximum power (1 m/s) in kilograms over the four mesocycles

male long jumper along the indoor season during four mesocycles with load dynamic of 2–1 (Fig. 1).

It can be seen as the season progressed the straight lines moved up and to the right in accordance with coach prescription. During training sessions with VBT the coach can control the training, monitoring the intensity adjusting the loads when the athlete does not reach or exceed the previously established values as can be seen in the figure below.

As you can see in Fig. 2, on 11/18/2019 the training was planned to move loads around 0.5 m/s and 0.40 m/s in some number of sets. Because the athlete moved the loads at speeds below those prescribed, it should be noted that in the last series the coach adjusted the load. A week later for identical training, the athlete was able to achieve higher speeds than the previous week with higher loads, showing an improvement in strength in the various zones. It is important to mention that although the mechanical component is highly important in VBT, it should be combined with physiological parameters, always keeping recruitment and muscle activation at high level.

When working with the VBT methodology it is noticed that as the load increases the speed decreases which is in agreement with force velocity relationship. Studies carried out so far have shown that each exercise has its own strength-velocity profile, which means that the same velocity or percentage represents very different loads or zones (Gonzalez-Badillo 2000). This occurs because the velocity at which the 1RM is reached is exercise-dependent (each exercise has its specific 1RM-velocity).

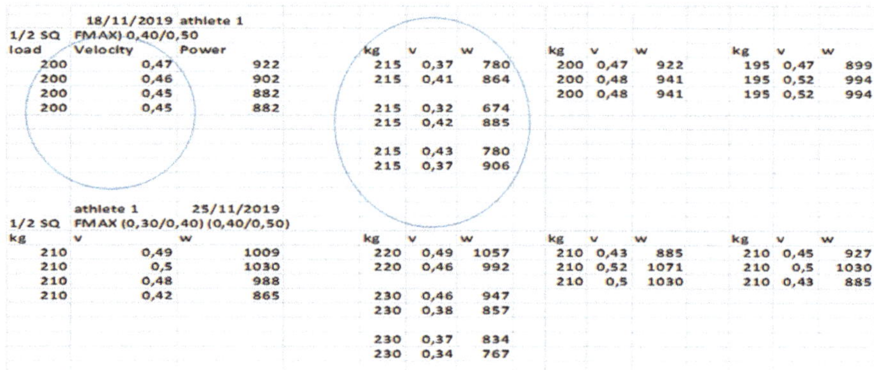

Fig. 2 Training control through velocity-based training method

Another point to be highlighted is that during the strength session athletes do not necessarily perform all the possible repetitions (Fig. 2) and may interrupt the set or the session when the degree of fatigue intended by the coach is achieved (González-Badill et al. 2016).

3.2 *Strength Training Assessment*

One of the essential tasks of any training process is the evaluation of its effectiveness. Usually, tests to evaluate muscular strength are applied to the most important exercises performed, to establish an index of general and specific strength throughout a training period (Brown and Weir 2001). The procedure followed over many decades for strength evolution assessment is to isolate a microcycle, or worse at the beginning and end of the season to carry out the control.

In addition to these problems, there is the fact that a poor selection of exercises can lead to wrong conclusions. According to the literature, strength can undergo variations of 18% over the course of a week (Jovanović and Flanagan 2014). Current training trends suggest that training control is part of the daily routine so that at any moment the coach and athlete can know the athlete's condition and prescribe the training according to the athlete's metabolic, biochemical and mechanical response. For the evaluation and control of strength training there are numerous tests, variables and technologies available. Describing them all would require an effort incompatible with this chapter. In this sense, we chose to describe those that in our view best describes the performance status of a given athlete or else are more appropriate for a given moment and can be performed in the laboratory or on the ground. Improvements in performance can include increased muscular strength, power and both low and high intensity exercise endurance (Paavolainen et al. 1999; Robinson et al. 1995). Changes in these variables (strength, power) as a result of resisted training may be related to improved results on athletic performance, such as in vertical jumping, sprint times, distance running times and agility measures (Paavolainen et al. 1999). These observations indicate that resistance training can have a transfer of training effect that results in a change in specific functional skills and capacities.

3.2.1 Maximum Strength

The ability to produce the maximum strength is considered the basic quality that influences the speed-strength performance (Schmidtbleicher and Komi 1992). The best way to determine it is by an isometric squat contraction with the knee flexed at 120° (Kulig et al. 1984). Therefore knowing the so-called maximum dynamical strength (Young 1995) expressed as a percentage of the maximum strength that a subject can generate isometrically it can be used to switch the emphasis of training from maximum strength to speed-strength and vice-versa.

3.2.2 Force-time, Power and Force-velocity Profile as a Tool for Periodization in Individual Sports

Some of the most important parameters to be considered are, total work, rate of force development (RFD), power and peak power, the highest instantaneous power over a range of motion. Power output is probably the most important factor in separating sports performance. While average power may be more associated with performance in endurance events, for activities such as jumping, sprinting and weightlift movements, peak power is typically strongly related to success (McBride et al. 2002; Garhammer 1993).

3.2.3 Rate of Force Development and Force-velocity Profile

In most sports, the time available for strength production is insufficient to develop maximum strength and therefore the RFD, seems to be the decisive factor with respect to maximum strength (Zatsiorsky and Komi 2003). This parameter is commonly used to measure the ability to generate high rates of muscle strength, and consequently the ability to accelerate objects including body mass (Cormie et al. 2011). For RFD evaluation, the slope of the force-time curve in isometric or in dynamic condition is used (Zatsiorsky and Kraemer 2006; Aagaard et al. 2002).

Maximum RFD values are reached between 80 and 120 ms (Aagaard et al. 2002). In most sports actions and daily life, the movements last between 50 and 150 ms wherein the RFD may be divided into two phases, early (<100 ms) and late stage (>100 ms) (Andersen and Aagaard 2006; Samozino et al. 2008). Several studies have reported that strength training can induce different adaptations with regard to early and late phases as well as the increase of strength production

(Aagaard et al. 2002). Every sport has its critical temporal expression. For sprinting, long jump and high jump it corresponds to support times that vary between 80–120 ms, 110–120 ms and 180–220 ms respectively (Zatsiorsky and Kraemer 2006; Kuitunen et al. 2002). The most relevant data to be extracted from the force-time curve are:maximum isometric strength; time to reach the different percentages of the maximum isometric or dynamic force; peak force developed and time; impulse and different RFD intervals, 50, 100, 150, 200 and 250 ms.

In sports movements, the underpinning features of mechanical output are force and velocity. Accordingly with Samozino et al. (2012) high ballistic performances are determined both by maximizing power capacities and optimizing the mechanical force-velocity profile (F-V) of the lower limb neuromuscular system. Therefore, for each individual exists an optimal F-V profile that maximizes ballistic performance and represents the optimal balance between the strength and speed qualities for these movements (Samozino et al. 2012, 2014). Thus, more and more coaches in individual sports are using the force-velocity profile, whether vertical or horizontal, to assess the athlete's performance status or adjust their training.

3.2.4 Stretch Shortening Cycle

Stretch Shortening Cycle (SSC) are present in our daily action, running, jumping, throwing distinguishing two types according to the duration, slow ≥ 250 ms and fast $\leq$ 250 ms (Schmidtbleicher and Komi 1992) which suggest that different mechanisms of action. The SSC exercises play with (a) the pre-activation, (b) the variable activation of the muscles that precede the functional phase (c) the myotactic reflex that influences muscle-tendon stiffness (d) the elasticity of the tendon, and (e) changes of muscles length versus tendon, making them the best exercises to develop the ability to apply a high level of force in a short time interval (Komi 2000; Conceição 2004).

Speed and jumps events require highly specialized needs of strength and power as they fall within the fast SSC. The time available to apply force in these events is very short and its magnitude high. Although the explosive force is intrinsically associated to a high level of maximum force, the isolated development of the maximum force or the speed component does not translate into the improvement of the force-velocity curve as a whole, mainly in the specific area where these sports work. With regard to training, the focus should be on reactive components (rapid SSC) through the so-called plyometric and ballistic exercises associated with maximum strength (Komi 2000). There are several tests located in the area of the slow SSC recommended assessing the speed and strength qualities.

- Squat jump (SJ) from a mat or force platform. The athlete placed in a force platform, with the knees at 120º, in a squat position jump vertically as high as possible by extending he legs after an instruction from the coach. No countermovement is allowed. The data of the force-time curve will be analyzed particularly those related to the speed-strength qualities, and the height of the jump.

- Countermovement Jump (CMJ). This test is performed in a similar condition as the squat jump, starting the athlete in an upright position. A fast countermovement is allowed prior to the extension of the legs. The force-time curve is analyzed and the main parameters and the jump height are registered.
- Standing Long Jump (SLJ). This test assesses an athlete's leg power. The athlete is in the upright position and after rapid flexion and extension of the legs jump forward looking for the maximum possible horizontal distance. Performance in the standing long jump is evaluated by the total jump distance, which is the horizontal distance from the takeoff line to the mark made by the heels at landing.

For the fast SSC the Reactive Strength Index (RSI) test is recommended. It is measured through depth or drop jumps from different drop heights by using a contact platform or force plate to measure the jump height as well as other parameters of the force-time curve. The subject should be instructed to jump as high as possible with minimum contact time. This exercise is appealing for:

1. The muscle's ability to decelerate the downward velocity of the body that falls from a high height
2. Promote high stretch in the muscle tendon unity and quickly accelerate the body during the shortening phase, which is similar to many dynamic actions, and
3. To be successful in many sports the ability to tolerate a high level of load stretch is fundamental (Young 1987).

The optimum drop height according to Schmidtbleicher (1993) can be over 100 cm for elite athletes and is used to prescribe drop jump training. Young (1995), suggests that a drop jump below "optimum" provides an insufficient overload and training stimulus, whereas a stretch load above the optimum may cause a neuromuscular inhibition resulting in a weakened contraction and training effect.

4 Filling Gaps

4.1 Biomarkers and Sports Performance: Physiology and Biochemistry Applied to Training Control and Prescription (New Periodization Models)

Modern high-performance bioanalytical platforms, in combination with computational resources for data analysis and interpretation, make it possible to quantify hundreds of metabolites in a single analysis. An important highlight here is to "look" at the metabolism involved in each planned training action, as well as correlate it with the competitive load.

There are currently many biomarkers available to assess metabolic fitness and performance. Due to its easy analysis and accessibility, besides being one of the most important signatures of cellular metabolism, lactate has been a preferred biomarker to assess physical and metabolic fitness. The lactate test is probably the best method for, (i) predicting performance in endurance events, (ii) discriminating different levels of athletic performance, and iii) being useful in prescribing training zones for athletes (D'Alessandro 2019).

When there is mitochondrial dysfunction or, simply, the mitochondria are not enough efficient to remove lactate, due to the low capacity of the oxidative pathway (aerobic capacity), it accumulates in the cytosol and is exported to the blood for later oxidation in most of the body tissues becoming a substitute for mitochondrial function (D'Alessandro 2019). Before being exported to the blood, lactate is oxidized in the muscle fibers with some mitochondrial capacity, i.e., muscle fibers that are not completely mobilized for work.

Now, let's reflect on sports with a predominance on muscle mechanic power production, such as sprinting, jumping, throwing. A point that deserves to be highlighted in strength training based on VBT is the paradox between the speed zone to be worked on, the role of muscle recruitment and the decision of the coach. What would be the best strategy to maximize performance? focus on metabolism or on the ability to produce strength? The answer seems to be considering the two approaches in a more complex analysis? To achieve the maximum recruitment of muscle fibers for a given exercise, loads between 90 and 100% of the maximum load must be selected. This type of load induces a more successful neural than metabolic adaptation.

This inevitably generates gains in movement economy and maximizes performance, however it transiently decreases the capacity to generate power or to be fast in training athletes (Gandevia 2001; Allen et al. 2008). It is important to know fatigue mechanisms, if they are central, peripheral, derived from the "lack of substrates", or even in the difficulty for resynthesis of ATP, prescribing pauses that allow the elimination or minimization of those mechanisms. This control can be performed by monitoring lactatemia and glycemia during training or testing (Ament and Verkerke 2009). For lactatemia and glycemia control is necessary to guarantee the training happening in the high energy pathway and with intervals that allow the full recovery of ATP resynthesis as well as the removal of the produced lactate that tends to be high in athletes with biopositive adaptation for muscle recruitment (Zwarts et al. 2008).

So we have two scenarios that deserve full clarification about lactate, strength, the velocity of execution and muscle recruitment: (a) increasing the recruitment capacity that generates an increase in lactate production, almost instantly at the central command of the contraction (b) pay close attention to the intervals to allow maintenance of strength production with adequate ATP resynthesis as well as lactate removal. It is important to avoid that the training with neural emphasis becomes metabolic due to the excessive volume and very short intervals generating greater magnitudes and accumulation of metabolites including hydrogen ions. The result of this would be a greater unnecessary acidosis at this point of the periodization (Sundberg et al. 2018).

It is noteworthy that if the training loads remain for a long time, being applied above the optimal level the probability of injuries, immunosuppression and diseases becomes greater, impairing the entire process. However, if the loads are below a level of stress that allows important physiological adaptations, we will be on the path of "detraining" the athlete, which brings damage to performance.

4.2 Adjustments and Monitoring of Training Sessions with Control by Lactate and Glycemia

Another interesting point is to consider the competitive load taking into account the athlete's metabolism and recovery windows. In this sense, it is only possible to adopt decisions that effectively contribute to improve performance, after the accurate identification of the hydration status, tissue damage, inflammatory processes, among other aspects, through biomarkers.

Sometimes the exclusive use of lactate threshold protocols or parameters established in the literature such as CK, AST, ALT, among others, portray tissue injuries, which can be and most of the time are very far from the competitive burden. That said, we can reflect on the drop of the performance or loss of possibilities for optimal adaptations. Figure 3 shows an example of competitive load monitoring to identify the threshold of an athlete in the training environment.

This can generate a false impression of adjustments based on minimum productions (below 5 Mmol) which will inevitably incur less assertive decisions since the competitive load is much higher than that depending on the athlete close to 20/25 Mmol. The "lactate threshold" is an indicator of high importance in the training control of the middle and bottom of athletics and in all modalities of prolonged effort. Putting the lactatemia evolution curves in progressive effort in the analysis of the sprinter's training is not correct, but unfortunately it happens a lot in the indications of coaches and physical trainers due to the lack of specific data from sprinters.

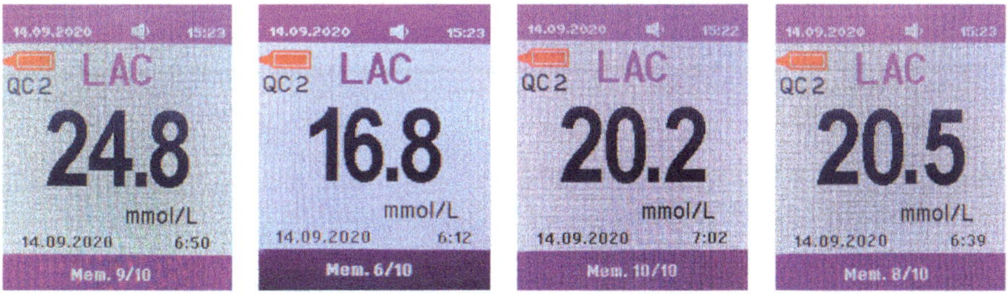

Fig. 3 Example lactate by training of elite sprinter

4.3 New Proposal About Periodization

Currently, new proposals for periodization modeling are emerging in order to meet the new demands of the sport, with a high number of competitions and even older athletes acting in high performance. The main bias or failure of these proposals is the lack of consideration of the internal training load, that is, the wear or depletion of all biomotor capacities and systems in general. Then the concepts of flexible or Bioflexible (Affonso and Silva 2018) arise, within this context we highlight the best control of volume and intensity of training, respecting the scientific principles, however adapting to the new realities and demands (Fig. 4). The integration of continuous monitoring reaffirms the foundation of periodization in GAS—General Adaptation Syndrome, to model individual responses to training, thus allowing coaches to validate and optimize the training process (Cunanan et al. 2018).

The central axis of the model is always the competitive load. Thus, it must be dimensionless and timeless. Specific production (competitive specific load) always guides momentum variations, volume, duration, and intensity depends entirely on monitoring and adjustment based on the internal and external loads.

4.4 Monitoring and Control Indications for Each Momentum of the Bioflexible Periodization

4.4.1 Strength Momentum

Before starting this *momentum* it is essential to ensure that the athlete has low levels of inflammatory processes and oxidative stress as well as great mobility to enable him to carry out ballistic movements and complex actions without limitations. From this, understanding the relationship between load and speed becomes the key point. As already described in the literature for different age groups, this relationship differs in the adaptation of strength gains. However, in this proposal as important as the increase in strength is the capacity for muscle recruitment aiming to enhance the activation of specific metabolism for energy production. It is important to ensure an optimal adaptation for recruitment gains, thus combining strength production with an active metabolism.

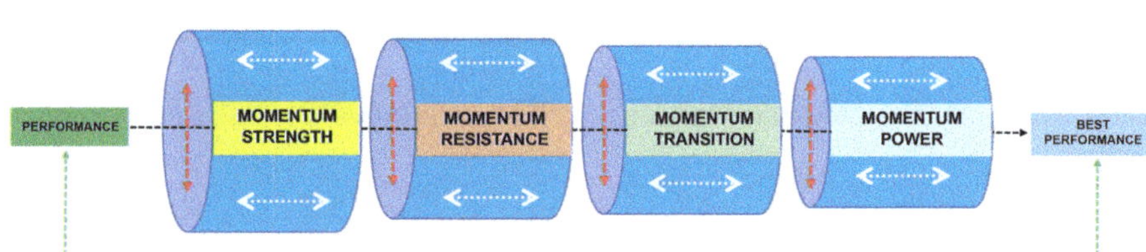

Fig. 4 Model design proposed by Affonso and Silva (2018)

One way to induce this adaptation during the sessions involves establishing complete intervals, high activation and use of energy and low accumulation of metabolites. In this sense, biomarkers have a fundamental role in allowing the control of the internal load in addition to the mechanical strength or power developed by the athlete. In this way, it allows the crossing of the external and internal load, helping in the team's decision making regarding the importance of the "physiological price" and not just the production of strength and/or power.

Another fundamental point is to monitor the force plateau (external load), so that the physiological wear (internal load) does not accumulate without transformation into strength and recruitment gains. With the stabilization of the strength and recruitment within the planning expectations (It is suggested here the lactate and blood glucose monitoring, acute phase proteins and tissue injury markers), the sooner we move on to the new periodization momentum the better. This is intended to avoid non-functional overreaching from accumulated central fatigue, as it is a high demand for this momentum strength.

It is worth mentioning that in this training momentum additional loads are indicated, however, they can cause a transient drop in speed and strength. This effect should be compensated in the next momentums, in addition to sharply boosting the recruitment capacity and not "only" the force production.

4.4.2 Resistance Momentum

At this momentum we notice an increase in muscle wear, the adaptations mainly neuro muscular mastering new levels of strength production and recruitment. Therefore, the loads will be higher to increase the special resistance. This is justified by the maintenance of residual effects of maximum strength throughout the periodization. Here it is also important to ensure a complete recovery so that at the beginning of the week or microcycle you can do more training focused on central (neural) adaptations and, throughout the week, progressively move to resistance (metabolic emphasis).

Accumulations of wear are allowed here with respect to systemic inflammatory processes, through incomplete breaks, and moderate increases in oxidative stress always from the second session of the microcycle, since the first aim is to ensure the maintenance of the strength momentum.

For a better understanding of this momentum, we explain in more detail. An important indication here is the choice of smaller series with an emphasis on quality (high percentages of 1RM where the athlete produces the expected speeds concomitant with the activation of metabolism). The ideal volume of resistance, still focusing on quality, must be guaranteed by the number of blocks and not sequential series. For instance, instead of performing 2 sets of 6 repetitions, one would choose 4 sets of 3 repetitions divided into 2×2 sets of 3 repetitions. Always prioritizing incomplete breaks between small sets and complete breaks between blocks. A relevant note to retain here is that the priority is to train the resistance specifically without using complementary training in the power room or as little as possible.

Therefore, this can guarantee less wear and more adaptations aiming at the competitive load. It is noteworthy that this method is applied either in complementary (in the power room) or specific (in the modality) training. The power room can only be used to control imbalances and / or stabilizing muscle work.

4.4.3 Transition Momentum

Here the main objective is to "bridge" the gap between high power outputs achieved through high loads with high power outputs by using loads that allow maximum power outputs located at the confluence of the strength-speed and speed-strength zone. That is, the education of the ability to present a high mobilization level of the energetic system through high recruitment capacity under light loads similar to high loads, thus, establishing a bridge between the ability to produce strength/power and mobilization of the specific energy system. For example, (i) activation of the glycolytic pathway without the need of high load in complementary training or (ii) activation of it in very short bouts on the track, indicating that an anticipation mechanism was built for high demands and power outputs. The main objective here is to enhance adaptations, aiming at maximum speed and power. Here a new window of opportunity exists, to understand the specific demand and take care of optimal intervals for competitions, that is, how long before we must stop and recover the athlete for the best performance. Another fundamental point to be considered is the elimination of the accumulations coming from the resistance *momentum*, by promoting recovery and almost a return to the *momentum* strength, with the best condition to transition to the next *momentum* that aims exclusively at power. The monitoring here should be to ensure that the recruitment previously only possible at high loads now takes place against small loads and that is more "lasting" by increasing the athlete's resistance. The complex training here seem to be more suitable, guaranteeing a pre-activation and moving to power. The main objective here is to improve the adaptations, aiming at maximum speed and power. With regard to specific training, the indication here is to use special exercises with additional regressive loads, for example using sled races: 15% BW 1 race, 10% BW 2 races, 5% BW 3 races. Another example using vertical jumps: 2 jumps with 30 kg + 2 jumps with 20 kg + 2 jumps with 10 kg + 2 jumps with body weight. These strategies aim to maintain active metabolism even with small overloads but with quick movements.

4.4.4 Power Momentum

In the power momentum the most important consideration is that the athlete can specifically develop actions with the highest power even without the need for high loads promoting less wear and movement savings (energy). It is important to be "fresh" for optimal recruitment and "active metabolism" according to the high specific demands of the sport.

Additional charges are not welcome at this time, except for two reasons:

1. Failure to adapt in recruitment resulting from any interruption in the training process, for example, any injury during the specific recruitment period (Moment of strength).
2. Use for pre-activation of training for maximum speed production (post-activation performance enhancement).

5 Take-Home Messages

1. Periodization is widely accepted as a powerful tool for effective training planning and improving the athlete's performance. Contemporary training is characterized by a large concentration of specific loads aimed at a reduced number of goals and capacities, performed in small consecutive cycles.
2. In the search for training with higher quality, VBT is increasingly the most adopted method in strength training for both prescription and control as well as a means to identify the intensity of training in individual and individual sports.
3. Training control is currently part of the daily routine, with several tools available to be used on the ground, such as photoelectric cells, encoders, lactate analyzers, contact platforms, load cells, video, mobile applications etc. The most important parameters to be considered for assessment and currently present in the training of individual sports are, work, rate of force development, power, peak power and the relationship between force-velocity. Increasingly, coaches seek to optimize training according to the individual needs of each athlete linking strength, velocity and technique.
4. Plyometric exercises are critical and indispensable for improving strength, speed and power, as they appeal to the SSC, which is present in most sports actions and should be included in any training routine.
5. High lactate in elite sprinters after very short efforts is at odds with common knowledge in bioenergetics. This can characterize a physiological signature of these elite sprinters. Thus, lactate can be used to monitor changes in physiological adaptations, especially muscle recruitment, over the season, and for talent identification. In addition, we highlight glycemia as a complementary point to lactate and simple to use to monitor and control training routines. Objectively to glycemic kinetics, during training, allows inferring if glycogen stores are preserved and thus the maximum production capacity for the energetic-glycolytic pathway. Our focus is on the integration of mechanical production (for example, stride or stroke frequency and speed with the energy / metabolism triggered. Finally, we quote Brooks (2020), who describes lactate as the focus of metabolic regulation.

References

Aagaard P, Simonsen EB, Andersen JL, Magnusson P, Dyhre-Poulsen P (2002) Increase rate of force development and neural drive of human skeletal muscle following resistance training. J Appl Physiol 93(4):1318–1326

Affonso H, Silva A (2018) Um novo conceito de periodização do treinamento desportivo. Personal Communication

Allen DG, Lamb GD, Westerblad H (2008) Skeletal muscle fatigue: cellular mechanisms. Physiol Rev 88(1):287–332

Ament W, Verkerke GJ (2009) Exercise and fatigue. Sports Med 39(5):389–422

Andersen L, Aagaard P (2006) Influence of maximal muscle strength and intrinsic muscle contractile properties on contractile rate of force development. Eur J Appl Physiol 96(1):46–52

Aubry A, Hausswirth C, Louis J, Coutts AJ, Meur YL (2014) Functional overreaching: the key to peak performance during the taper? Med Sci Sports Exerc 46(9):1769–1777

Bohm S, Mersmann F, Arampatzis A (2015) Human tendon adaptation in response to mechanical loading: a systematic review and meta-analysis of exercise intervention studies on healthy adults. Sports Med Open 1(1):1–18

Bompa TO (1984) Theory and methodology of training: the key to athletic performance. Kendall Hunt, Boca Raton, Florida

Bondarchuk AP (1986) Training of track and field athletes. Health Publisher, Kiev

Bondarchuk AP (2007) Transfer of training in sports: ultimate athlete concepts

Borresen J, Lambert MI (2009) The quantification of training load, the training response and the effect on performance. Sports Med 39(9):779–795

Brooks GA (2020) Lactate as a fulcrum of metabolism. Redox Biol 35

Brown LE, Weir JP (2001) ASEP procedures recommendation I: accurate assessment of muscular strength. Jeponline 4(3):1–21

Carfagno DG (2014) 3rd JCH. Overtraining syndrome in the athlete: current clinical practice. Curr Sports Med Rep 13(1):45–51

Cissik J, Hedrick A, Barnes M (2008) Challenges applying research on periodization. Strength Conditioning J 30:45–51

Conceição F (2004) Estudo biomecânico do salto em comprimento: modelação, simulação e optimização da chamada. Universidade Porto, Porto, Portugal

Cormie P, McGuigan MR, Newton RU (2011) Developing maximal neuromuscular power: part 2—training considerations for improving maximal power production. Sports med 41(2):125–146

Cunanan AJ, DeWeese BH, Wagle JP, Carroll KM, Sausaman R, Hornsby WG et al (2018) The general adaptation syndrome: a foundation for the concept of periodization. Sports Med 48 (4):787–797

D'Alessandro A (2019) High-throughput metabolomics: methods and protocols, methods in molecular biology. Springer Nature, New York, p 459

Dick FW (1997) Sport training principles. A&C Black, London

Durell D, Puyol T, Barnes J (2003) A survey of the scientific data and training methods utilized by collegiate strength and conditioning coaches. J Strength Conditioning Res 17:368–373

Ebben W, Carroll R, Simenz C (2004) Strength and conditioning practices of National Hockey League strength and conditioning coaches. J Strength Conditioning Res 18:889–897

Fleck S (1999) Periodized strength training: a critical review. J Strength Conditioning Res 13:82–89

Gandevia SC (2001) Spinal and supraspinal factors in human muscle fatigue. Physiol Rev 81 (4):1725–1789

Garhammer J (1993) A review of power output studies of Olympic and power lifting: methodology, performance prediction, and evaluation tests. J Strength Cond Res 7(2):76–89

Gonzalez-Badillo JJ (2000) Bases teóricas y experimentales para la aplicación del entrenamiento de fuerza al entrenamiento deportivo. Infocoes. 5(2):3–14

González-Badillo JJ, Sánchez-Medina L (2010) Movement velocity as a measure of loading intensity in resistance training. Int J Sports Med 31:347–352

González-Badill JJ, Rodríguez-Rosell D, Sánchez-Medina L, Ribas J, López-López C, Mora-Custodio R et al (2016) Short-term recovery following resistance exercise leading or not to failure. Int J Sports Med 37(4):295–304

Grandou C, Wallace L, Impellizzeri FM, Allen NG, Coutts AJ (2020) An exploratory systematic review and methodological appraisal of the literature. Sports Med 50(4):815–828

Haff G (2004) Roundtable discussion: periodization of training—Part 1. Strength Conditioning J 26:50–69

Harris GR, Stone MH, O'Bryant HS, Proulx CM, Johnson RL (2000) Short-term performance effects of high power, high force, or combined weight-training methods. J Strength Cond Res 14(1):14–20

Issurin VB (2008) Block periodization versus traditional training theory: a review. J Sports Med Phys Fitness 48(1):65–75

Issurin VB (2013) Training transfer: scientific background and insights for practical application. Sports Med 43:675–694

Issurin VB (2016) Block periodization versus traditional training theory: a review. Sports Med 46 (3):329–338

Jovanović M, Flanagan E (2014) Researched applications of velocity based strength training. J Aust Strength Cond. 22(2):58–69

Kiely J (2018) Periodization theory: confronting an inconvenient truth. Sports Med 48:753–764

Komi PV (2000) Stretch-shortening cycle: a powerful model to study normal and fatigue muscle. J Biomech 33(10):1197–1206

Kuitunen S, Komi PV, Kyröläinen H (2002) Knee and ankle joint stiffness in sprint running. Med Sci Sports Exer 34(1):166–173

Kulig K, Andrews JG, Hay JG (1984) Human strength curves. Exerc Sport Sci Rev 12:417–466

Kurz T (2001) Science of sports training: how to plan and control training for peak performance. Stadion Publishing Company, Island Pond, Vermont

Loturco I, Nakamura FY (2016) Training periodisation: an obsolete methodology. Aspetar Sports Med J. 5:110–115

Matveyev LP (1972) Periodisierang des SprotichenTraining. Berles and Wernitz, Berlin, Germany

McBride JM, Triplett-McBride T, Davie A, Newton RU (2002) The effect of heavy vs. lightload jump squats on the development of strength, power, and speed. J Strength Cond Res 16(1):75–82

Paavolainen L, Hakkinen K, Hamalainen I, Nummela A, Rusko H (1999) Explosive-strength training improves 5-km running time by improving running economy and muscle power. J Appl Physiol 86(5):1527–1533

Robinson JM, Stone MH, Johnson RL, Penland CM, Warren BJ, Lewis RD (1995) Effects of different weight training exercise/rest intervals on strength, power, and high intensity exercise endurance. J Sports Sci Med 9(4):216–221

Samozino P, Morin JB, Hintzy F, Belli A (2008) A simple method for measuring force, velocity and power output during squat jump. J Biomech 41(14):2940–2945

Samozino P, Rejc E, Prampero PED, Belli A, Morin JB (2012) Optimal force–velocity profile in ballistic movements—Altius. Med Sci Sports Exer 44(2):313–322

Samozino P, Edouard P, Sangnier S, Brughelli M, Jimenez P, Morin J-B (2014) Force-velocity profile: imbalance determination and effect on lower limb ballistic performance. Int J Sports Med 35(6):505–510

Schmidtbleicher D (1992) Training for power events. In: Komi PV (ed) Strength and power in sport. Blackwell, Oxford, England, pp 381–385

Schmidtbleicher D (1993) Personal communication

Sundberg CW, Hunter SK, Trappe SW, Smith CS, Fitts RH (2018) Effects of elevated H+ and Pi on the contractile mechanics of skeletal muscle fibres from young and old men: implications for muscle fatigue in humans. J Physiol 596(17):3993–4015

Trappe S, Luden N, Minchev K, Raue U, Jemiolo B, Trappe TA (2015) Skeletal muscle signature of a champion sprint runner. J Appl Physiol 118(12):1460–1466

Verkhoshansky YV (1985) Programming and organization of training process. FiS Publisher, Moscow

Verkhoshansky YV (2006) Special strength training. A practical manual for coaches. Ultimate Athlete Concepts, Muskegon
Verkhoshansky YV, Charniga A (1986) Fundamentals of special strength-training in sport (A. Charniaga, Trans). Sportivny Press, Livonia, MI
Wahl P, Güldner M, Mester J (2014) Effects and sustainability of a 13-day high-intensity shock microcycle in soccer. J Sports Sci Med 13(2)
Wilson GJ, Newton RU, Murphy AJ, Humphries BJ (1993) The optimum training load for the development of dynamic athletic performance. Med Sci Sports Exer 25(11):1279–1286
Young WB (1987) The triple jump and plyometrics. Strength Conditioning J 9(2):22–25
Young WB (1995) Laboratory strength assessment of athletes. New Stud Athletics 10:89–96
Young WB (2006) Transfer of strength and power training to sports performance. Int J Sports Physiol Perform 1(1):74–83
Zatsiorsky VM (1995) Science and practice of strength training. Human Kinetics, Champaign, Illinois
Zatsiorsky VM (2003) Biomechanics of strength and strength training. In: Komi PV (ed) Strength and power in sport. Blackwell Science, Oxford, pp 439–487
Zatsiorsky V, Kraemer W (2006) Science and practice of strength training, 2nd edn. Human Kinetics
Zwarts MJ, Bleijenberg G, Engelen BV (2008) Clinical neurophysiology of fatigue. Clin Neurophysiol 119(1):2–10

Filipe Almeida Viana Conceiçao received his Ph.D. in 2005 from the Faculty of Sports, University of Porto. Since 2005, he has been an auxiliary professor in the Faculty of Sports and a member of Porto Laboratory Biomechanics of the University of Porto (Portugal) and the Center for Research, Training, Innovation, and Intervention in Sports. His current research interests are biomechanics and sports training as well as coach of athletes of international level in track and field jumps.

Helvio Affonso has Ph.D. and MSD in Pharmaceutical Sciences and post-graduated in High Performance Sports Training and in Exercise Physiology. Main titles as a Physiologist: World Champion in 2015 and OLYMPIC GOLD Medal at Rio 2016 Games—Beach Volleyball; World Record 50 m butterfly (2018) and fifth place in the Rio 2016 Games; he is responsible for monitoring and controlling the training of the fastest sprinter in Brazil and second best/100 m; 4x100 m relay world champion in 2018; technical responsible for APPTO Physiology, Laboratory of Exercise, Nutrition and Sports Training, Vitória Brasil; Currently, he is a post-graduate professor at the University of Vila Velha.

The Role of Resistance Training in Strategies to Reduce Injury Risk

Oliver Gonzalo-Skok

Abstract The main aim of this chapter is to review the role of strength training in reducing the risk of sports injuries, which are a major concern in any sport. As strength training has been recognized as one of the best strategies to minimize the risk of injury, we will focus on the current evidence-based information and how we can directly apply this to our practices. This chapter will also explain an alternative neuromuscular training approach based on dynamic correspondence, using the most common actions players employ throughout a soccer game (our examples are taken from soccer but can be extrapolated to other sports). Traditionally, soccer players train by lifting weights bilaterally in the vertical axis, focused on the concentric phase of the movement and with no making-decision actions. However, it is helpful to train several force-vectors (three-dimensional movements) applying force unilaterally (instead of bilaterally) with emphasis on the eccentric phase, and including stochastic or unexpected situations to mimic real game situations. After reviewing the theoretical concepts, a practical viewpoint is presented, allowing coaches and trainers to adapt their practices to the specific requirements of their sport.

Keywords Eccentric · Unilateral · Injury prevention · Unexpected · Neuromuscular · Ecological validity

1 Introduction

One of the worst things that can occur during any sport is injury. When one player is injured, it can affect both the player's and the team's performance and have financial consequences. The current chapter is based on strategies and practical applications primarily focused on sports involving intermittent activity (which can be extrapolated to individual sports), and we use specific soccer strategies to provide practical examples. Firstly, having all players available may be an advantage in achieving better performance. Teams that have higher levels of player availability

O. Gonzalo-Skok (✉)
Sevilla FC, Seville, Spain

A. Muñoz-López et al. (eds.), *Resistance Training Methods*, Lecture Notes in Bioengineering, https://doi.org/10.1007/978-3-030-81989-7_15

during games, as a result of lower injury incidence and injury burden, have achieved more points per league match and better final league rankings (Hägglund et al. 2013). Secondly, not selecting a player as part of the squad for the next game is deemed an economic loss. Premier League clearly shows the injury costs per season. Specifically, one team (Newcastle United) lost 25.6 million pounds during the 2014–2015 soccer season. Thus, it is important to find practical solutions to minimize the risk of injury and thus avoid effects on both performance and economics. Strength training has been recognized as one of the best strategies to achieve this goal (Lauersen et al. 2014). Thus, we will focus on the current evidence-based information and how we can directly apply this to our practices.

2 From Theory

2.1 An Alternative Neuromuscular Training Approach Based on Dynamic Correspondence

In the real-world soccer training environment, limited time is available for strength training and conditioning sessions during the in-season period. Developing time-efficient strategies that enhance locomotive-specific actions while preventing injuries is seen as crucial. Given the highly dynamic and stochastic nature of soccer movements, there is a need to introduce more challenging training methods that allow players to perceive affordances (possibilities of action) that may induce emergent behaviors for generating optimal movement synergies. Therefore, differential training and/or variability of practice principles must be introduced to enhance complex-skill efficiency in real game situations. The following sections describe the specific contents of our neuromuscular training approach.

2.2 Mono-Axial/Vertical Versus Rotational/ Three-Dimensional

Traditionally, team-sports players have performed vertical movements to improve specific abilities. This approach has been based on bodybuilding training, where all movements involve lifting weights in the frontal plane. One of the main changes related to specific game actions training is the involvement of several force vectors. While this theory has only recently appeared in the scientific strength and conditioning area, it has been applied in practice for several years.

In recent years, there has been an emphasis on the importance of applying force in the desired direction to achieve optimal movement performance (Rabita et al. 2015) and avoid overloading when the movement is not properly performed. Specifically, in linear sprinting, the production of a higher ratio of the horizontal force component seems to have greater relevance compared to total force production; that is, a more horizontal force resultant in every step. Indeed, elite sprinters are able to develop a greater horizontal force per body mass unit at any velocity than sub-elite sprinters (Rabita et al. 2015). Hence, the main question is how we can apply more horizontal force as well as a better orientation of the resultant force. A recent study assessed the force-vector influence in specific abilities such as vertical jumping (axial force-vector) and linear sprinting (postero-anterior force-vector) in a group of young athletes (Contreras et al. 2017). The main aim was to compare the effects of performing the front squat exercise (vertical) or the hip thrust exercise (horizontal) over a period of 6 weeks (biweekly × 4 sets × 6–12 1RM). The results confirmed the previous hypothesis that powerful hip extensors might result in better orientation of the resultant force (Rabita et al. 2015). While the hip thrust group substantially improved their performance in the 20-m linear sprint (ES = 0.46), the front squat group achieved a greater vertical jumping performance (ES = 0.45). Therefore, this study supported the force-vector training theory for enhancing specific abilities in team-sports players.

Our research group developed the first longitudinal experimental study of team-sports complexity. The aim of this study was to compare a traditional paradigm (6 sets of the squat exercise performed bilaterally in the vertical axis) with a novel paradigm (1 set of 6 movements executed unilaterally and multidirectionally), using a conical pulley or Versapulley™ in both paradigms (Gonzalo-Skok et al. 2016). As in Contrera's study (Contreras et al. 2017), the main findings were that the traditional group showed more robust adaptations in those actions where the force application was mainly vertical, whereas the multidirectional group achieved greater improvements in those actions that required the force to be applied horizontally/laterally. These findings indicate that the force-vector application may play a key role in producing optimal movement.

If we focus on medial-lateral and rotational displacements, it is normal to use exercises that involve the external rotators, adductors or gluteus medius. We attempted to justify why we should use these types of exercises. The effect of performing a pre-planned or unexpected landing or change of direction (COD) on neuromuscular mechanisms was analysed in NCAA I players (Meinerz et al. 2015). Unexpected events showed a greater torque for both the external rotators and the abductors. However, the most interesting findings were that the gluteus medius and gluteus maximus displayed significantly ($p < 0.05$) greater electrical activity (measured through EMGs) in comparison to the pre-planned condition. Team-sports players are continuously dealing with uncertainty; consequently, an appropriate activation of both gluteus maximus and medius is key to hip control. Hip internal rotation, which is connected to knee valgus, is one of the main anterior cruciate ligament (ACL) injury mechanisms, so improved hip control can reduce the risk of ACL injury.

Another study involving female NCAA I soccer players examined the possible relationship between both hip adductor and hip external rotator strength and unexpected one-legged COD kinetics and kinematics (Malloy et al. 2016). Players were on a platform connected to an interface that showed a signal on a screen just in front of them: red (single-leg landing), an arrow to the right (landing and side-step) or a forward arrow (landing and front side-step). All landings were registered through a 3D system and a force platform. The main conclusion was that those players who had greater external rotator strength showed greater motor control (less external rotation of the hip and knee valgus) in both unexpected landings and COD.

Finally, a prospective study developed with team-sports players investigated whether isometric hip strength (external rotation and abduction) measured before the beginning of the competitive season could predict the likelihood of ACL injury (Khayambashi et al. 2016). Fifteen non-contact ACL injuries were registered during the competitive season. The baseline hip strength values for those players who had suffered an injury were significantly lower ($p = 0.003$ and $p < 0.001$) than those who did not suffer an injury. Greater probability of suffering an injury was defined as a cut-off point of <20.3% or <35.4% of body mass for hip external rotator and hip abductor strength, respectively.

2.3 *Bilateral Versus Unilateral*

2.3.1 Effects of Between-Limbs Asymmetry in Injury Prevention and Physical Performance

Specificity is a key factor in the development of training-induced adaptations for specific performance (Villarreal et al. 2012). Most high-intensity actions engaged in by team-sports players, such as sprinting, changing direction or landing require the production of force from a single leg in a very high percentage of cases (Gonzalo-Skok et al. 2016). However, some of the most career-threatening injuries (anterior cruciate ligament injuries) occur primarily in a one-legged loaded leg, and when both limbs are loaded the main load is typically still on one leg (i.e., the leg that sustains the injury) (Koga et al. 2010; Waldén et al. 2015). Hence, the nature of team sports requires players to apply force with each limb independently. One of the most important aims throughout the return to play process is recovering the muscle volume and muscle strength in the injured limb to levels comparable to those in the uninjured limb. From a functional point of view, the most significant thing is that both limbs have similar strength levels, which will be discussed below.

A between-limbs strength difference is known as functional asymmetry, limb symmetry index or between-limbs imbalance. Such asymmetries started to be noticed by the scientific community through the use of hop tests. These tests require the subject to perform different jumps unilaterally and they analyse the between-legs asymmetry. Functional asymmetry is calculated as follows (Gustavsson et al. 2006):

$$Functional\,asymmetry = (worst\,leg/best\,leg) \times 100$$

Thus, the result is presented as a percentage from 0 to 100%, where the further the value is from 100%, the greater is the asymmetry.

This test has been deemed a valid and useful tool to detect functional deficits in players who have suffered a previous ACL injury (Gustavsson et al. 2006). In the same study, only 30–50% of those players who suffered an ACL injury or reconstruction were considered as symmetrical (i.e. had an asymmetry value lower than 10%). The rest of the players (50–70%) were deemed to have a high injury risk as they had a fourfold greater probability of suffering another lower-body injury (Gustavsson et al. 2006). In addition, the success of return to play after an ACL injury was analysed in a sample of 503 soccer players (Ardern et al. 2012). The conclusions were dramatic: those players who had a functional hop test asymmetry $\geq 15\%$ had a relative risk or probability of returning to play 2.5 times lower than those who showed normal values (i.e., <15%). Therefore, including a hop test assessment in any return-to-play process may be one of the most useful indicators of successful return to sport.

2.3.2 Effects of Unilateral Training to Improve Functional Asymmetries

Given the influence of functional asymmetries on the risk of injury, likelihood of return to play and performance, we now offer ideas about the most appropriate training to decrease functional asymmetries in players returning from injury.

A recent case-study assessed the horizontal force application with each limb on a non-motorized treadmill, determining the weak and strong leg in reference to the horizontal force (Brown et al. 2017). After a 6-week control period, a unilateral training program to strengthen the weaker limb (hip extension exercises) was added to the bilateral training usually performed during the previous 6-week training program. The current results showed a substantial improvement in horizontal force in the weak leg and a horizontal force asymmetry decrement, as well as both maximum velocity and power improvements during sprinting. These results suggest that unilateral training might help to decrease the horizontal force asymmetry and, consequently, to improve the linear displacement performance.

In a study developed by our research group, 22 young male basketball players were randomly assigned to either unilateral or bilateral combined strength training (Gonzalo-Skok et al. 2017). Each group performed 3 90°-squat sets of a post-determined number of repetitions (stopping the set when power output dropped to <10% of maximum power established at the beginning of each set), 2 sets of 4 drop jumps and 2 sets of 4 CMJ biweekly during a 6-week period. The only difference was that all exercises were performed either unilaterally (starting with the weakest leg) or bilaterally. After the training program, the unilateral group substantially decreased the unilateral power asymmetry in the squat exercise and also showed a substantially greater improvement than the bilateral group.

2.4 Concentric Emphasis (Acceleration) Versus Eccentric Overload (Deceleration)

It's surprising to observe how many S&C coaches continue to ignore and even reject the benefits that eccentric overload can provide. It's true that until the appearance of inertial resistance machines (i.e., YoYo™, Versapulley…), it was very difficult to avoid isokinetic training to provide a "real" eccentric overload (greater peak force in the eccentric phase than in the concentric phase). However, using inertial resistance machines, we can achieve this aim, and incorporate it in any analytic or specific movement. Reaching an eccentric overload is easier with the upper body compared to the lower body. Previously, we have described the need for a minimum level of experience to develop an eccentric overload (Tous-Fajardo et al. 2006). Many team-sports players have significant difficulty learning to decelerate in small spaces in a real-world context. Given that maximum decelerations can be registered throughout a game using tri-axial accelerometery, it has been determined that such decelerations are key in team-sports players. Specifically, a greater number of high-intensity decelerations have been reported (24 vs. 19 in the 1st and 2nd half, respectively) in comparison to accelerations (14 vs. 12) in elite soccer (Russell et al. 2016). In addition, performing either a maximum acceleration or deceleration had a negative impact on performance (10.4 and 11.4%) 5 min post-execution, though recovery had occurred by 10 min after such maximal activity (Akenhead et al. 2013).

It is well known that braking forces determine both COD performance and stability. Jindrich et al. (2006) showed that such forces prevented excess rotation during CODs, while Spiteri et al. (2015) reported that the fastest players were those who applied greater vertical braking forces in the last step during sprinting with CODs of 180°. However, the most interesting finding was that the most effective deceleration strategy consisted of applying greater horizontal braking forces in the penultimate step instead of the last one (Jones et al. 2017). Given that braking earlier allows reduction of high horizontal torque, a more stable support can then be provided, allowing the application of a greater propulsive force.

The adaptations developed through eccentric overload exercises are multiple and can be found in several specific reviews of this topic (Maroto-Izquierdo et al. 2017; Nuñez Sanchez et al. 2017; Tesch et al. 2017; Vicens-Bordas et al. 2018). In this regard, there are several peer-reviewed papers that have shown beneficial effects in minimizing injury risks in soccer players (Hoyo et al. 2015; Askling et al. 2003). The inclusion of eccentric overload exercises (e.g., YoYo squat and leg curl) seems to be effective in decreasing both the incidence and severity of injuries.

2.5 Preplanned Versus Stochastic/Unexpected

One of the more difficult traditional methodologies to remove is the stimulus predictability that is offered to team-sports players. In individual sports it is very common to act from a 'fine detail' viewpoint, trying to both predict and control every variable, when aiming to design successful training programs. However, because of the variability inherent in team sports it is impossible to predict exactly what will happen. Even though a stochastic (i.e., unpredictable) event can be studied or modelled, it should be analysed using a broad overview that offers the necessary perspective. The studies conducted by Besier et al. (2001) and Lloyd (2001) were the first step in this process, as they introduced the unexpected COD concept and, consequently, the increased threat to the knee joint. Later, similar results were found in other movements such as stop-jump tasks (Sell et al. 2006), landings (Brown et al. 2009) and adding such tasks when players are fatigued (Borotikar et al. 2008). By systematically introducing such unexpected events over time, players can decrease the level of threat to the joints after either several years of practice (Kipp et al. 2013) or a sensorimotor training period. In our opinion, such adaptations may be achieved using feedforward mechanisms, defined as the player's ability to anticipate without producing a sensory registration, based on the identification of a situation that the player can relate to previous experiences, allowing adequate muscle pre-activation (Romero and Tous 2010). Task variability allows the proprioceptive-vestibular-visual system to be trained to interpret situations that occur during a game or training, and reduce the stimulus perception time. Here we are talking about previous experience of specific movements, such as supporting, landing, throwing, kicking, and fighting. Unless there is an appropriate pre-activation before such actions, there will be an increase in both execution difficulty and risk of injury, as found by Zebis et al. (2016). A very simple neuromuscular training that included some unexpected events and fighting exercises showed improved pre-activation as a result of a decrease in the vastus lateralis-semitendinous relationship (Zebis et al. 2016).

An interesting neuroscience study (Dux et al. 2009) assessed the effect of three types of training over a period of 2 weeks in a task that interrelated visual, auditory and motor aspects: auditory stimulus (2 possibilities) with a vocal response depending on the stimulus; visual stimulus (2 possibilities) with a motor response depending on the stimulus; or a combination of the previous proposals. After the training sessions, a decrease in time required for central processing of the task was observed, without there being a modification of either the perception time or the execution time. This was explained by an increase in information processing speed in the pre-frontal cortex of the brain. The main practical application of the study and its integration in team-sports is that with the strength-training proposed, involving constant decision-making with several possibilities, there is a modification of the movement patterns in both the eccentric and concentric phases, which will allow players to respond faster to the stimulus.

Hernández-Davo investigated the training effect of unknown loads (Hernández Davó et al. 2015). One of the most interesting findings was that in the first part of a bench press movement, in the concentric phase, ignorance of the magnitude of the moved load elicited a greater rate of force development and power compared to situations in which the subject was informed of the moved load. The electromyographic activity of the main muscles that produce a bench press movement was similar in both situations (i.e., known and unknown loads), though the main mobilizer (i.e., deltoid) was activated 50 ms before starting such movements (Hernández Davó et al. 2015). These small details can help us to understand how the inclusion of unexpected elements in basic movements may configure several areas in the long term. The group that trained with unknown loads showed substantially improved throwing velocity in handball compared to the group that used the same load, with knowledge of its magnitude, over a 4-week period (Sabido et al. 2016).

One of the main limitations to measuring the effect of introducing decision-making actions (i.e., unexpected events) during neuromuscular training is how such measurements can be made. In this regard, Martín et al. (2018) developed a repeated sprint ability test with a decision-making COD in every sprint. This test consisted of 6 × 20 m (10 m + 10 m) sprints separated by 20 s of active recovery while jogging back to the initial position. Each sprint was compounded by a 10 m linear sprint plus a 10 m sprint with three potential directional options according to a traffic light fixed at the first 10-m linear sprint: (a) linear sprint, (b) a linear sprint after performing a 45º COD to the right, and (c) a linear sprint after executing a 45º COD to the left. This method has been shown to be reliable, useful, responsive to detecting training-induced changes, able to detect differences between age categories, and valid (showing significant correlations with match-related performance) in young soccer players (Martin et al. 2018).

A longitudinal study was subsequently carried out with 24 young soccer players (U-17) (Martín 2019). Our idea was based on performing four movements (lunge, defensive shuffling, side-step/crossover step and lateral crossover step; only 1 set per exercise) using a conical pulley. The main difference was that one group executed such exercises with decision-making included, while the other group performed them with no decision-making, over an 8-week period. After the training period, the decision-making group achieved substantially greater improvements in mean times (over 6 sprints) for the RSA-RANDOM test compared to the non-decision-making group. These results support the inclusion of unexpected events in neuromuscular training to improve both performance and injury prevention.

3 From Practice

Having described the most important points of our training philosophy, it is very important to implement such ideas on a daily basis to minimize the risk of injury.

3.1 Mono-Axial/Vertical Versus Rotational/Three-Dimensional

To optimize any movement pattern and minimize the risk of injury, players need to apply force in the desired direction. Therefore, the inclusion of movements in different force-vectors seems a key factor. However, before starting to directly train such movements, strength in the hip external rotators, abductors, gluteus medius and gluteus maximus is of great importance in hip control and can therefore help to minimize the risk of severe injuries such as those to the ACL. Therefore, the progression from anteroposterior (e.g., lunges), posteroanterior (e.g., front side-step), latero-medial (e.g., lateral squat), rotational (e.g., lateral crossover step) and axial (e.g., vertical squat) force-vectors to more complex movements (e.g., from one to two steps, changing the trained leg) is crucial to bring players close to the injury mechanism.

3.2 Bilateral Versus Unilateral

Those players who present with a functional asymmetry may have a greater probability of suffering a lower-limb injury, a lower probability of returning to play successfully, and poorer performance in displacement activities. This fact justifies the importance of performing training workouts aimed to minimize the presence of functional asymmetries in both return-to-play processes and training optimization. Although scientific information is scarce, it seems that unilateral training may help to reduce functional asymmetry. Given the dynamic correspondence of the current actions observed in team-sports players and the impact of training to reduce functional asymmetries, unilateral training should be emphasized in training sessions designed for performance optimization (i.e., improving performance and minimizing the risk of injury).

3.3 Concentric Emphasis (Acceleration) Versus Eccentric Overload (Deceleration)

As eccentric overload exercises seem to be closer to those actions where muscle injuries appear (e.g., muscle lengthening), such exercises should be included in injury prevention programs. In this regard, micro-doses should be used to avoid muscle soreness as limited time is available in the real-world professional context.

3.4 Preplanned Versus Stochastic/Unexpected

When preparing any player for the requirements of play on a soccer pitch, the program should progress from preplanned to unexpected events. The inclusion of perturbations or decision-making drills may help to improve the feed-forward mechanism; that is, the ability to anticipate a new situation. Therefore, any training program that integrates unexpected events will have more likelihood of minimizing the risk of injury.

4 Filling Gaps

The new approach based on performing only one set per exercise instead of a conventional multiset scheme seems to be time-effective for minimizing the risk of injury in team-sports players. S&C coaches should choose the required specific force-vector for the selected injury mechanisms. We advocate a multi-exercise setup where different movement families are interspersed with short recovery periods (i.e., start with longer rest periods and progressively decrease them). In this way, instead of stressing the same structures and functions several times, our aim is to concatenate movements to boost the post-activation potentiation effect while searching for optimal movement sequences (e.g., a lateral side step followed by a lateral cross-over step, or a front single step followed by a front double step with a forward jump). Furthermore, unilateral exercises may help to balance both limbs and achieve the two-fold aim in a time-efficient manner: improve performance and minimize the risk of injury. When the aim is to decrease inter-limb asymmetry, both the specific force-vector, starting with the weaker leg, and task-specificity seem to be the appropriate strategy when the volume is equalized between both limbs. In the real-world context, our program aims for permanent fine-tuning (i.e., inter- and intra-set and session) and not choosing one option or another. The idea is to evolve the program by fine-tuning vectors, external elements like balls or perturbations, and unilateral/bilateral exercises depending on what coaches and trainers expect to get based on specific player demands.

5 Take-Home Messages

(1) Team-sports players should focus on training movements instead of muscles. However, to be ready to perform such movements properly, muscles should be optimally activated.
(2) Fine-tuning inter- and intra-set and sessions is a must to increase players' ability to respond to new stimuli appropriately and, consequently, to minimize the risk of injury.

(3) Specific movement patterns should be progressively included through unilateral force application, perturbations and/or decision-making, trying to mimic the most common injury mechanisms.

References

Akenhead R, Hayes PR, Thompson KG, French D (2013) Diminutions of acceleration and deceleration output during professional football match play. J Sci Med Sport 16(6):556–561

Ardern CL, Taylor NF, Feller JA, Webster KE (2012) Return-to-sport outcomes at 2 to 7 years after anterior cruciate ligament reconstruction surgery. Am J Sports Med 40(1):41–48

Askling C, Karlsson J, Thorstensson A (2003) Hamstring injury occurrence in elite soccer players after preseason strength training with eccentric overload. Scand J Med Sci Sport 13(4):244–250

Besier TF, Lloyd DG, Ackland TR, Cochrane JL (2001) Anticipatory effects on knee joint loading during running and cutting maneuvers. Med Sci Sports Exerc 33(7):1176–1181

Borotikar BS, Newcomer R, Koppes R, McLean SG (2008) Combined effects of fatigue and decision making on female lower limb landing postures: central and peripheral contributions to ACL injury risk. Clin Biomech 23(1):81–92

Brown TN, Palmieri-Smith RM, McLean SG (2009) Sex and limb differences in hip and knee kinematics and kinetics during anticipated and unanticipated jump landings: Implications for anterior cruciate ligament injury. Br J Sports Med 43(13):1049–1056

Brown SR, Feldman ER, Cross MR, Helms ER, Marrier B, Samozino P et al (2017) The potential for a targeted strength-training program to decrease asymmetry and increase performance: a proof of concept in sprinting. Int J Sports Physiol Perform 12(10):1392–1395

Contreras B, Vigotsky AD, Schoenfeld BJ, Beardsley C, McMaster DT, Reyneke JHT et al (2017) Effects of a six-week hip thrust vs. front squat resistance training program on performance in adolescent males: a randomized controlled trial. J Strength Cond Res 31(4):999–1008

De Hoyo M, Pozzo M, Sañudo B, Carrasco L, Gonzalo-Skok O, Domínguez-Cobo S et al (2015) Effects of a 10-week in-season eccentric-overload training program on muscle-injury prevention and performance in junior elite soccer players. Int J Sports Physiol Perform 10(1): 46–52

De Villarreal ES, Requena B, Cronin JB (2012) The effects of plyometric training on sprint performance: a meta-analysis. J Strength Cond Res 26(2):575–584

Dux PE, Tombu MN, Harrison S, Rogers BP, Tong F, Marois R (2009) Training improves multitasking performance by increasing the speed of information processing in human prefrontal cortex. Neuron 63(1):127–138

Gonzalo-Skok O, Tous-Fajardo J, Valero-Campo C, Berzosa C, Bataller AV, Arjol-Serrano JL et al (2016) Eccentric-overload training in team-sport functional performance: constant bilateral vertical versus variable unilateral multidirectional movements. Int J Sports Physiol Perform 12(7):951–958

Gonzalo-Skok O, Tous-Fajardo J, Suarez-Arrones L, Arjol-Serrano JL, Casajús JA, Mendez-Villanueva A (2017) Single-leg power output and between-limbs imbalances in team-sport players: unilateral versus bilateral combined resistance training. Int J Sports Physiol Perform 12(1):106–114

Gustavsson A, Neeter C, Thomeé P, Grävare Silbernagel K, Augustsson J, Thomeé R et al (2006) A test battery for evaluating hop performance in patients with an ACL injury and patients who have undergone ACL reconstruction. Knee Surgery, Sport Traumatol Arthrosc 14(8):778–788

Hägglund M, Waldén M, Magnusson H, Kristenson K, Bengtsson H, Ekstrand J (2013) Injuries affect team performance negatively in professional football: an 11-year follow-up of the UEFA Champions League injury study. Br J Sports Med 47(12):738–742

Hernández Davó JL, Sabido Solana R, Sarabia Marín JM, Sánchez Martos M, Moya Ramón M (2015) Unknown loads affect force production capacity in early phases of bench press throws. J Sports Med Phys Fitness 55(10):1122–1128
Jindrich DL, Besier TF, Lloyd DG (2006) A hypothesis for the function of braking forces during running turns. J Biomech 39(9):1611–1620
Jones P, Thomas C, Dos'Santos T, McMahon J, Graham-Smith P (2017) The role of eccentric strength in 180° turns in female soccer players. Sports 5(2):42
Khayambashi K, Ghoddosi N, Straub RK, Powers CM (2016) Hip muscle strength predicts noncontact anterior cruciate ligament injury in male and female athletes: a prospective study. Am J Sports Med 44(2):355–361
Kipp K, Brown TN, McLean SG, Palmieri-Smith RM (2013) Decision making and experience level influence frontal plane knee joint biomechanics during a cutting maneuver. J Appl Biomech 29(6):756–762
Koga H, Nakamae A, Shima Y, Iwasa J, Myklebust G, Engebretsen L et al (2010) Mechanisms for noncontact anterior cruciate ligament injuries: knee joint kinematics in 10 injury situations from female team handball and basketball. Am J Sports Med 38(11):2218–2225
Lauersen JB, Bertelsen DM, Andersen LB (2014) The effectiveness of exercise interventions to prevent sports injuries: a systematic review and meta-analysis of randomised controlled trials. British J Sports Med 48(11):871–877
Lloyd DG (2001) Rationale for training programs to reduce anterior cruciate ligament injuries in Australian football. J Orthop Sports Phys Ther 31(11):645–654
Malloy PJ, Morgan AM, Meinerz CM, Geiser CF, Kipp K (2016) Hip external rotator strength is associated with better dynamic control of the lower extremity during landing tasks. J Strength Cond Res 30(1):282–291
Maroto-Izquierdo S, García-López D, Fernandez-Gonzalo R, Moreira OC, González-Gallego J, de Paz JA (2017) Skeletal muscle functional and structural adaptations after eccentric overload flywheel resistance training: a systematic review and meta-analysis. J Sci Med Sport 20(10): 943–951
Martín V (2019) La capacidad de repetir acciones de alta intensidad con componente cognitivo. Universidad Pontífica de Salamanca
Martin V, Sanchez-Sanchez J, Ramírez-Campillo R, Nakamura FY, Gonzalo-Skok O (2018) Validity of the RSA-RANDOM test for young soccer players. Int J Sports Med 39(11): 813–821
Meinerz CM, Malloy P, Geiser CF, Kipp K (2015) Anticipatory effects on lower extremity neuromechanics during a cutting task. J Athl Train 50(9):905–913
Nuñez Sanchez FJ, Sáez de Villarreal E, De Villarreal ES (2017) Does flywheel paradigm training improve muscle volume and force? A meta-analysis. J Strength Cond Res 31(11):3177–3186
Rabita G, Dorel S, Slawinski J, Sàez-de-Villarreal E, Couturier A, Samozino P et al (2015) Sprint mechanics in world-class athletes: a new insight into the limits of human locomotion. Scand J Med Sci Sport 25(5):583–594
Romero D, Tous J (2010) Prevención de lesiones en el deporte. Madrid Médica Panam
Russell M, Sparkes W, Northeast J, Cook CJ, Love TD, Bracken RM et al (2016) Changes in acceleration and deceleration capacity throughout professional soccer match-play. J Strength Cond Res 30(10):2839–2844
Sabido R, Hernández-Davó JL, Botella J, Moya M (2016) Effects of 4-week training intervention with unknown loads on power output performance and throwing velocity in junior team handball players. PLoS One 11(6)
Sell TC, Ferris CM, Abt JP, Tsai YS, Myers JB, Fu FH et al (2006) The effect of direction and reaction on the neuromuscular and biomechanical characteristics of the knee during tasks that simulate the noncontact anterior cruciate ligament injury mechanism. Am J Sports Med 34(1): 43–54
Spiteri T, Newton RU, Binetti M, Hart NH, Sheppard JM, Nimphius S (2015) Mechanical determinants of faster change of direction and agility performance in female basketball athletes. J Strength Cond Res 29(8):2205–2214

Tesch PA, Fernandez-Gonzalo R, Lundberg TR (2017) Clinical applications of iso-inertial, eccentric-overload (YoYo^{TM}) resistance exercise. Front Physiol 8:241

Tous-Fajardo J, Maldonado RA, Quintana JM, Pozzo M, Tesch PA (2006) The flywheel leg-curl machine: offering eccentric overload for hamstring development. Int J Sports Physiol Perform 1(3):293–298

Vicens-Bordas J, Esteve E, Fort-Vanmeerhaeghe A, Bandholm T, Thorborg K (2018) Is inertial flywheel resistance training superior to gravity-dependent resistance training in improving muscle strength? A systematic review with meta-analyses. J Sci Med Sport 21(1):75–83

Waldén M, Krosshaug T, Bjørneboe J, Andersen TE, Faul O, Hägglund M (2015) Three distinct mechanisms predominate in noncontact anterior cruciate ligament injuries in male professional football players: a systematic video analysis of 39 cases. Br J Sports Med 49(22):1452–1460

Zebis MK, Andersen LL, Brandt M, Myklebust G, Bencke J, Lauridsen HB et al (2016) Effects of evidence-based prevention training on neuromuscular and biomechanical risk factors for ACL injury in adolescent female athletes: a randomised controlled trial. Br J Sports Med 50(9): 552–557

Oliver Gonzalo-Skok has a M.Sc. in high performance in team sports and a Ph.D. in sport sciences. He is currently working as the head of Rehab in Sevilla FC of the Top Spanish soccer league. Also, he is a professor in several M.Sc. about high performance. He has more than 50 JCR publications about strength training, injury prevention and testing. Previously, he was working 11 years as the head of Strength and Conditioning Coach in Basket Zaragoza (Top Spanish basketball league). Furthermore, he participated as a Strength and Conditioning Coach in Spanish Basketball Federation (U-16 to U-20) from 2014 to 2019. Finally, he was a lecturer for six years at San Jorge University (Zaragoza).

Special Considerations in Resistance Training

Resistance Training in Older Adults

Borja Sañudo and Michael E. Rogers

Abstract Interventions based on resistance training have been shown to counteract muscle disuse and therefore combat muscle strength and muscle mass loss, with positive effects on physical functioning, mobility, independence, psychological well-being, and quality of life. Unfortunately, a low percentage of older adults meet international recommendations on resistance training probably due to fear, health concerns, pain, fatigue, or lack of social support. There is a need for evidence-based guidelines and recommendations for resistance exercise for older adults to safely and gradually introduce this type of training into their routines. However, there is no "average" older adult, and so it is impossible to provide a single recommendation that is fully representative, especially across age groups. All individuals respond differently to resistance training, and progression should be closely monitored to be able to individually adjust the training program; consequently, different methods available for assessing muscle strength and physical function that serve to analyze the effectiveness of interventions are discussed in this chapter. New strategies used in combination with resistance training in older adults are also addressed, in order to provide novel insights regarding the resistance strength training in this population group.

Keywords Strength training · Frailty · Elderly · Physical function · Force-velocity

---

B. Sañudo (✉)
Departamento de Educación Física y Deporte. Universidad de Sevilla, Calle Pirotécnia s/n, 41013 Sevilla, Spain
e-mail: bsancor@us.es

M. E. Rogers
106G Heskett Center, Wichita State University, Wichita, KS 67260-0016, USA

A. Muñoz-López et al. (eds.), *Resistance Training Methods*, Lecture Notes in Bioengineering, https://doi.org/10.1007/978-3-030-81989-7_16

1 Introduction

Societies are progressively ageing and the number of people over 85 years is expected to triple by 2050. Inevitably, ageing is associated with the accumulation of undesirable changes, including the loss of skeletal muscle mass and function, often resulting in an increased risk of co-morbid diseases and premature mortality (Bernardo-Filho et al. 2020). The inevitable impairment of the neuromuscular system (Borzuola et al. 2020) leads to a muscle atrophy and loss of maximal strength and muscle power output with subsequent loss of independence (Valenzuela et al. 2019a).

Sarcopenia (i.e., loss of muscle mass) leads to poor physical performance in older adults (Larsson et al. 2019) and physical exercise, specifically muscle strength, is considered a main determinant in the performance of everyday tasks. Together with sarcopenia, frailty is defined as a multidimensional geriatric syndrome leading to a decline in muscle mass and strength as well as in multiple physiological systems (e.g., endocrine or immune systems). The figures of these geriatric syndromes are exponentially growing and consequently, the development of strategies for improving physical capacities of an individual should be considered a priority to promote healthy ageing (Beard et al. 2016). Being physically active is a key factor in maintaining normal functioning across the life-course; therefore, exercise appears as a potential solution to combat frailty and disability at an advanced age (Valenzuela et al. 2019b).

Despite the association 'exercise-older adults' is commonly used, there is no "average" older adult, and so it is impossible to provide a single recommendation that is fully representative, especially across age groups. In this chapter we provide novel insights regarding the resistance strength training in this population group. Relatively simple and pragmatically implementable approaches are presented that aim to promote health and independence for their quality of like.

2 From Theory

Muscle mass and strength are progressively compromised with ageing with decrements ranging 1–2% per year above the age of 50 (Haehling et al. 2010). Further, the decline in muscle power is even more rapid (Reid and Fielding 2012). These changes are attributed to numerous factors which include quantitative and qualitative changes in muscle mass and neuromuscular function. A decrease in the number of α-motoneurons, voluntary activation, and conduction velocity or the motor unit firing rate have usually been reported in older adults (see for (Valenzuela et al. 2019a) a review). Older adults experience a decrease in force production capacity related to the quantity of muscle mass usually attributed to muscle fibre atrophy, especially in fast-twitch fibres (Nilwik et al. 2013), and changes in some muscle architectural properties (e.g., decreased muscle thickness and pennation angle with ageing (Kubo et al. 2003) or a reduction in fascicle length). Endocrine function can also play a role in lower power production capacity, decreased levels of growth hormone (GH) and insulin-growth factor (IGF)-1 (Hameed et al. 2002) and decreases in adrenal androgens (i.e., testosterone) (Feldman et al. 2002). Changes in the immune system have also been described as decreases in anti-inflammatory cytokines and increases in pro-inflammatory cytokines (e.g., IL-6 or TNF-α) can increase muscle catabolism and thus reduce muscle mass and strength in older adults (Schaap et al. 2006). All these changes contribute to decreased maximal force and power production capacity and consequently the loss of muscle function.

The incidence of geriatric syndromes worldwide (e.g., frailty, sarcopenia, cognitive impairment) lead to the need for cost-effective strategies that can be implemented in a large number of settings (Bernardo-Filho et al. 2020). Among the public health strategies, lifestyle interventions to mitigate the effects of these syndromes are, therefore, of great interest. Unfortunately, a low percentage of older adults meet international physical activity recommendations (Keadle et al. 2016) and most of them are reluctant to even initiate a physical exercise program. Consequently, in order to prevent functional decline, different strategies should be used (Valenzuela et al. 2019a).

Interventions based on resistance training have been shown to counteract muscle disuse and therefore combat muscle strength and muscle mass loss, with positive effects on physical functioning, mobility, independence, psychological well-being, and quality of life (Fragala et al. 2019).

2.1 Exercise Programs for Muscle Mass, Muscle Strength and Physical Performance in Older Adults

The health benefits with physical activity in older adults are widely described in the literature. Numerous studies highlight the influence of physical exercise in reducing the risk of numerous conditions associated with advanced age, including frailty (Silva et al. 2017), falls (Souto et al. 2019) and cognitive decline (Gujral and Oberlin 2021). However, and despite physical exercise being considered a core therapeutic element, the number of older adults meeting international physical activity recommendations is low (Keadle et al. 2016). Physical inactivity is associated with numerous conditions including cardiovascular disease, metabolic dysfunction, some types of cancer and sarcopenia. As displayed in Fig. 1, despite ageing being associated with multisystem deterioration including changes in the metabolic, skeletal, cardiovascular and immune systems, multiple physiological systems can be improved by increasing physical activity.

Although loss of muscle mass is generally gradual with ageing, these age-related declines (i.e., sarcopenia) are very prevalent in adults older than 60 years with an estimated prevalence of 10% that can rise to 50% in adults older than 80 years

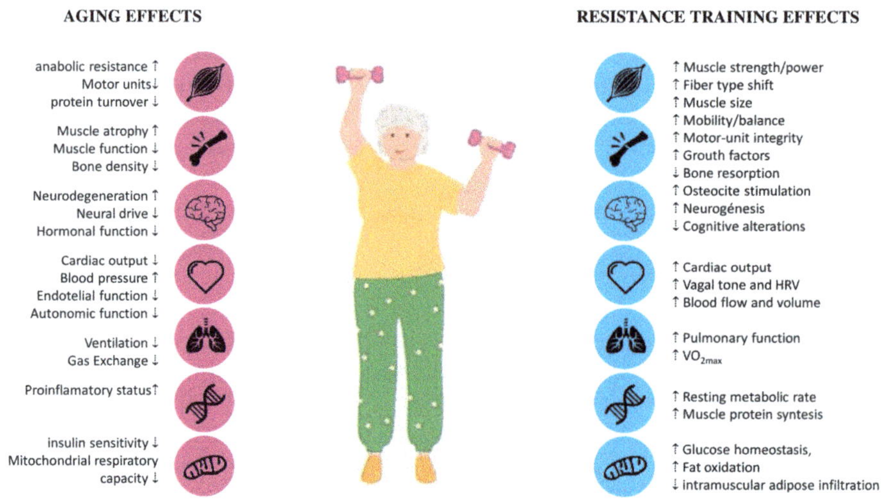

Fig. 1 Summary of the main multi-systemic effects of ageing and the anti-ageing effects of resistance training exercise (VO_{2max}: maximal oxygen consumption)

(Shafiee et al. 2017). There is a high association between muscle weakness and physical disability; consequently, not just muscle mass loss but also strength loss are important contributors to the age-related functional decline (Papa et al. 2017a). Declines in muscle power have also been shown to be even more important than muscle strength in the ability to perform daily activities. Considering the aforementioned consequences of ageing, strategies to prevent muscle disuse are warranted since this is a preventable and reversible factor. Further, a recent report by the World Health Organization on ageing and health (Beard et al. 2016) prioritizes the importance of functional ability and the maintenance of autonomy and independence as the starting point of healthy ageing. Bearing in mind the importance of muscle strength preservation in older adults, and especially muscle power as a stronger predictor of functional disability in this population, undertaking resistance exercise training programs should be considered a key strategy in improving the health of older adults. There are several physiological mechanisms that could explain, at least partly, the benefits of resistance exercise training for older adults (Lavin et al. 2019). From the activation of anabolic signalling pathways (e.g., Akt/mTOR pathway) to hypertrophy across all fibre types (e.g., fast-twitch fibres) (Bickel et al. 2011). The modulation of the rate of motor unit activation or the increases in both muscle volume and neural drive have also been described (Reeves et al. 2004b).

The American College of Sports Medicine (ACSM), recommends the inclusion of resistance training in any multi-component exercise regimen for older adults (Chodzko-Zajko et al. 2009) as this paradigm may improve mobility, frailty, and loss of independence in ageing. The effectiveness of this type of intervention also leads to improvements in strength, muscle power, muscle hypertrophy, as well as functional outcomes that will be discussed below. Although the beneficial effects of resistance exercise training on these outcomes in the general older population are well established, some peculiarities and new findings will be briefly described below.

2.2 *Training Characteristics and Adaptations*

2.2.1 Maximal Strength

Healthy older adults experience a decrease in strength (~0.8–3.6% per year), especially in very old adults (Frontera et al. 2000). Resistance training has been shown to be an effective mode of physical activity to mitigate the dynapenia (i.e., the age-related loss of muscle strength) with improvements ranging from 9 to 147% (see (Fragala et al. 2019) for a review). However, these strength adaptations are influenced by numerous factors. General recommendations suggest using 60–80% of the 1RM to induce improvements in maximal strength (American College of Sports Medicine 2009); however, a recent study tried to establish a dose-response relationship of resistance training in healthy old adults (Borde et al. 2015). Results from this meta-regression revealed that *training period, intensity*, or *total time under tension* had significant effects on muscle strength and the authors suggested using, more specifically, 70–79% of the 1RM and an extended time under tension of 6 s per repetition. Other factors might also influence these changes with duration (a training frequency of two sessions per week), training volume (2–3 sets per exercise), number of repetitions (7–9 repetitions per set) and even rest between sets (60 s) showing the largest effect on muscle strength.

Recent reviews have reported that training volume is a key training variable to induce substantial gains in muscle strength in older adults (Csapo and Alegre 2016). Authors have suggested that when comparing heavy (~80% of 1RM) with light-moderate loads (~45% of 1RM), strength gains tended to be larger following resistance training with higher intensities of load. However, if a sufficient number of repetitions is performed, lower intensities than traditionally recommended might be enough to induce these gains in muscle strength in this population (Csapo and Alegre 2016). Finally, older adults may experience strength gains in response to resistance training that are dependent on the movement velocity. Studies comparing the effects of fast velocity versus low to moderate velocities (Orssatto et al. 2019) suggest that using similar loads, training at fast velocity induces greater improvements in maximal strength when compared with slow velocity.

2.2.2 Muscle Power Output

Muscular power is defined as the product of the force of a muscular contraction and its velocity. In comparison with maximal strength, this outcome declines at a greater rate in older adults (McNeil et al. 2007). Further, numerous studies have suggested a high association between skeletal muscle power and physical function in this group (Bean et al. 2003). There is high heterogeneity in the improvements in muscle power induced with resistance training (i.e., from 14 to 97%) and the main explanation to these discrepancies has been attributed to the movement velocity (Straight et al. 2016). It appears that faster velocity of contraction may be more effective at improving muscle power when compared with low to moderate velocities. Fast velocity of training (focusing on contracting as fast as possible) with a wide range of intensities (30–85% of 1RM) should be recommended to induce power improvements.

2.2.3 Explosive Force (Rate of Force Development)

Together with the declines in skeletal muscle power reported above, the Rate of Force Development (the rate of rise in contractile force at the onset of contraction, a measure of explosive or rapid strength, -RFD- has also been reported to decrease at a greater extent than maximal strength (Gerstner et al. 2017; Thompson et al. 2014). Ageing is associated with peak force reductions and, in addition to the importance of this outcome in physical function (Palmer et al. 2016), recent studies have highlighted its contribution to falls in older adults (Palmer et al. 2015). Further, RFD is considered an important predictor of mobility in older adults (Hester et al. 2020). In a recent meta-analysis (Guizelini et al. 2018), moderate beneficial effects on RFD were reported after a resistance training program ($\sim$27% faster). Again, movement velocity seems to be a key parameter when performing resistance training and participants should be encouraged to train with the intention of contracting muscles as fast as possible. The effects of training on RFD were assessed according to age, training type, sessions per week and training duration but significant influences on the changes in RFD were found.

Improvements in the neural drive, as well as in the fibre type and myosin heavy chain (MHC) composition, seem to influence the changes in RFD (Andersen and Aagaard 2006); therefore, significant effects on RFD are most likely to be observed during the first weeks of training (e.g., 2–8 weeks) due to the resistance training-induced changes in the neural system (Maffiuletti et al. 2016).

2.2.4 Muscle Hypertrophy

Age-related muscle atrophy is mainly determined by a reduction in total myofiber size, especially the atrophy of fast (type IIa and IIx) myofibers (Nilwik et al. 2013; Roberts et al. 2018). Numerous studies have assessed the effect of resistance training on measures of lean body mass, cross-sectional area (CSA) or muscle fibre composition (Borde et al. 2015; Peterson et al. 2011). Increases of $\sim$20% in MHC I and $\sim$30% in MHC II muscle fibre size could be expected after a traditional resistance training intervention (Straight et al. 2020). Increases of $\sim$1 kg in lean body mass were observed after long term resistance programs (an average of 20.5 weeks). Increases between 4 and 33% in CSA have also been reported (Rosa Orssatto et al. 2019; Reeves et al. 2004a) even in the oldest old (Kryger and Andersen 2007). Despite some authors reporting adaptations in muscles and tendons (e.g., increasing stiffness) (Reeves et al. 2006), these changes are mainly dependent on myofiber hypertrophy, particularly the type II fibres typically compromised by ageing (Lavin et al. 2019).

The effectiveness of resistance training on muscle size in older adults may be influenced by sex (Ivey et al. 2000), age (Straight et al. 2020), decreased muscle fibre capillarization (Snijders et al. 2017), chronic inflammation (Toth et al. 2005) or altered hormonal responses to resistance training (Negaresh et al. 2017); however, increases in lean body mass and CSA are highly determined by training volume. To date, the effects of resistance training on morphological adaptations at the cellular level (e.g., skeletal muscle fibre size) in older adults have been inconsistent (Straight et al. 2020). Light load and high repetition resistance training has been recommended in older people; further, ACSM recommendations for muscle hypertrophy in older adults include moderate loading (60–85% of 1RM) and slow to moderate velocities (American College of Sports Medicine 2009). Intensity has been suggested to be inversely associated with the magnitude of hypertrophy in MHC II fibres (Straight et al. 2020). A recent meta-analysis found that moderate-training intensities (51–69% of 1RM) elicited the greatest increases in muscle hypertrophy following resistance training in older adults (Borde et al. 2015); however, in young people, higher training intensities (3–5RM) were suggested to be more effective for inducing hypertrophy than lower intensities of training (Campos et al. 2002). One possible explanation for these inconsistencies was reported in a recent systematic review and meta-analysis (Straight et al. 2020) showing that resistance training increases MHC I and II muscle fibre size and that a higher training intensity was associated with a reduced improvement in MHC II fibre size.

2.2.5 Functional Capacity

Age-associated decreases in neuromuscular function can negatively impact older adults' ability to perform activities of daily living (ADLs) required for independent living (2016). Research has demonstrated that resistance training is related to delayed disability and an important tool to restore independent functioning. As reported above, physical functioning is related to muscular strength and power (Foldvari et al. 2000), while muscle mass increases would not be the major determinant for increased function in older adults. Consequently, interventions to increase strength and muscle power are necessary for maintenance of physical functioning. Despite multimodal exercises having been reported to have a broad effect on physical functioning (Liu et al. 2017), isolated resistance training has been described to result in the most consistent gains in functional tasks (Papa et al.

2017b) with benefits to gait speed, dynamic balance, fall risk reduction, overall physical performance, and ADL even in frail individuals (Giné-Garriga et al. 2014).

Resistance training (2 times per week for 10 weeks at an intensity ~10RM) resulted in improvements in the ability to perform ADL by 21% (Manini et al. 2007). The use of large muscle groups was recommended with sessions lasting for 30–60 min. Recent studies suggest greater frequencies (i.e., 3 times per week) to achieve functional benefits (Nakamura et al. 2007). Regarding intensity, starting at moderate intensities (e.g., 55% of 1RM) and progressing to higher intensities (e.g., 80% of 1RM) as tolerated by the individual can maximize functional gains (Papa et al. 2017b). Recent evidence also suggests that focusing on eccentric muscle contractions improves physical function more than concentric contraction-focused resistance training in older adults (Katsura et al. 2019). Those authors emphasized the use of eccentric manual resistance exercise training to improve mobility, and postural stability of older adults to a greater extent than with concentric training.

3 From Practice

We have just described the many positive effects of strength training in older people; however, these recommendations would not make any sense if they do not actually perform this type of training. Burton et al. (2017) recently described the barriers to participation in resistance exercise for older adults and included safety, fear, health concerns, pain, fatigue, and lack of social support. They suggested the need for evidence-based guidelines and recommendations for resistance exercise for older adults to safely and gradually introduce this type of training into their routines.

3.1 Typical Configurations Used for Strength Training

Training dose is dependent on several factors, including number of sets and repetitions, frequency, and intensity. Previous recommendations for resistance training in older adults included progressive weight training program (8–10 exercises involving the major muscle groups of 8–12 repetitions each) and a frequency of at least 2 days per week at moderate intensity (i.e., 5–8 on a scale of 0–10) (American College of Sports Medicine 2009). Recent studies suggest that interventions should be performed with 3 sets of 8–12 repetitions and an intensity starting at 20–30% and progressing to 80% of 1RM, 3 times a week to induce muscle gains. Finally, the recent Position Statement from the National Strength and Conditioning Association (NSCA) on resistance training for older adults (Fragala et al. 2019) recommend an individualized training program consisting of 2–3 sets with 10–12 exercises (i.e., involving 1–2 multi-joint exercises per major muscle group) at intensities of 70–85% 1RM on 2–3 days per week. The NSCA also suggests including power exercises performed with concentric movements at higher velocities using moderate intensities (i.e., 40–60% of 1RM).

There is considerable variance across studies with regard to exercise programs and these differences are often dependent on the outcome of the training (e.g., hypertrophy, strength, functional mobility). No differences across training frequencies (i.e., 1–3 sessions per week) were found in different studies Ribeiro et al. (2015); however, higher volumes seem to be more effective with longer programs (e.g., >20 weeks) (Lavin et al. 2019). The number of sets per exercise does not seem to be the primary variable responsible for muscle strength increases in older adults. Rather, training volume (i.e., total amount of weight lifted during a training session) has been recognized as an important parameter as moderate volume (i.e., 24 repetitions) increased muscle power more than low (i.e., <24 repetitions) or high volume of resistance training (i.e., >24 repetitions) (Borde et al. 2015).

Moreover, greater effects have also been found after high-intensity resistance training compared with moderate- and low-intensity resistance training (Steib et al. 2010). Authors suggested that intensities higher than 75% of 1RM achieved greater effect on maximal strength and functional performance than moderate (55–75% of 1RM) or lower intensities (less than 55% of 1RM). Similarly, a meta-analysis by Borde et al. (2015) also found that high intensities (70–79% of 1RM) induced larger effects on muscle strength than lower intensities in sedentary older adults. In close association with intensity, some authors have shown greater functional enhancements comparing explosive resistance training and traditional resistance training in older adults (Ramírez-Campillo et al. 2014). Greater improvements in chair rise or stair climbing were observed, although the results in walking speed and timed up and go (TUG) test are controversial (Steib et al. 2010).

Putting together all these recommendations one could simplify and state that, for the enhancement of maximal strength, slow-velocity repetitions for a maximum of 3 sets per exercise, 8–12 repetitions per exercise (10–12 exercises of the major muscle groups) at an intensity of 60–80% of 1RM, should be prescribed and, for increasing power, guidelines should include 1–3 sets per exercise using high velocity at 30–60% of 1RM for 6–10 repetitions. For an adequate recovery, rest intervals between sets ranging from 60 to 180 s are advised. However, response heterogeneity is evident in the older population; thus, the optimal exercise dose is variable by individual.

3.1.1 Resistance Training for Older Adults from General to Specific Recommendations

In order for exercise to be safe for older adults, familiarization to proper exercise technique and structured progression is strongly advised. It cannot be forgotten that there are risks with strength training in this group (e.g., the potential risk of dangerous elevations in blood pressure). Therefore, unstable coronary heart disease, decompensated heart failure, uncontrolled arrhythmias or hypertension, severe and symptomatic aortic stenosis, acute myocarditis, endocarditis, or pericarditis, or severe pulmonary hypertension, should be considered absolute contraindications to resistance training. Individuals who have implanted pacemakers or defibrillators, risk factors for coronary heart disease (e.g., diabetes, hypertension), or musculoskeletal limitations should consult a physician before participating. Finally, those with low functional capacity should be exercised with caution.

Recently, the Copenhagen Consensus statement 2019 (Bangsbo et al. 2019) provided general recommendations to prevent or minimize fears and other barriers to implementation of training programs for older adults. Resistance training programs should begin with a proper and specific warm-up that includes dynamic mobility exercises with the aim of reducing the risk of musculoskeletal injuries (Fragala et al. 2019). Training volume and intensity, and especially power training sessions, should be increased progressively and participants should be fully supervised (Fragala et al. 2019). All individuals respond differently to resistance training, and progression should be closely monitored to be able to individually adjust the training program. It is important to note that the term'older adults' represents a highly heterogeneous population so when designing a strength training program, it is not appropriate to consider a relatively healthy community-dwelling older adult (with a high intrinsic capacity) to be the same as a frail older adult (with low intrinsic capacity) or even those with age-associated neurodegenerative diseases. The findings reported in this chapter are frequently found in studies that involve older subjects with varying health problems and use various assessment protocols (Marques et al. 2013). Consequently, it is important to specify, as far as possible, the guidelines for the prescription of resistance training in different subgroups of older individuals (Fig. 2).

The heterogeneity among older people means that tailored strategies and evidence-based recommendations for resistance training are required (Bangsbo et al. 2019). A highly functional person aged 65 years typically cannot follow the same session for a nursing home resident aged 90 years; however, the subtitle of an editorial recently published in BMJ suggested that "*Advanced age is no barrier to the benefits of tailored exercise*" (Izquierdo et al. 2020). Consequently, we have to just modify these recommendations. An interesting review by Valenzuela et al. (2019c) provided evidence on how physical exercise can help to attenuate the loss of many of the systems affected by ageing, showing some examples of adaptations in the oldest old people (aged >80–85 years). As an example, a recent randomized controlled trial showed that a short-term intervention (i.e., eight week) based on low to moderate intensity leg press exercises increased muscle strength and decreased risk of falls in nursing home residents aged 90–97 years (Serra-Rexach et al. 2011).

The guidelines for resistance training for 'healthy' older adults are (Fragala et al. 2019): 1–3 sets per exercise per muscle group (8–12 or 10–15 repetitions) progressing to 70–85% of 1RM and including 8–10 exercises of the major muscle groups with a frequency of 2–3 days per week per muscle group. If tolerated, the modality of exercise should be free-weights (e.g., barbells, dumbbells, kettlebells, and medicine balls) with exercises trying to mimic tasks of daily living. Beginners or those with frailty should start with 1 set and progress to multiple sets (2–3) per exercise and perform 10–15 repetitions at a lower relative resistance. Major muscle

	Highly functional person aged 65 years	Beginner older adult aged 80 years	Frail nursing home resident aged 90 years
INTENSITY	70–85%1RM (power 40-60%)	50–85% of 1RM	20-30% to 80% of 1RM
SETS	2-3 Sets	2-3 sets	Start with 1 and then 3 sets
REPETITIONS	6-12 reps	8-12 or 10-15 reps	8-15 reps
MODALITY	Free-weight RT	Functional Movements	machine-based RT
FREQUENCY	3 times/week	2-3 times/week	2-3 times/week
VELOCITY	Combine High/mod MPV	Combine High/mod MPV	high speed at low to moderate intensity (i.e., 30–60% of 1RM)
EXERCISE SELECTION	Include major muscle groups targeted through multijoint movements		

Fig. 2 Specific recommendations for resistance training in older adults with different characteristics (RT: resistance training)

groups (e.g., leg press, quadriceps extension, chest press, shoulder press, pulldown, lower-back extension) are recommended; however, multi-joint movements should be performed with caution. In this case, instead of using free weights, these individuals (e.g., beginners or frail older adults) should start with machine-based resistance training or pneumatic resistance equipment. The use of elastic resistance bands has been reported to improve muscle strength in institutionalized older adults (aged 70–75 years) (Motalebi et al. 2018). Rest intervals between sets are suggested to be between 60–180 s; however, longer rest intervals will allow greater neuromuscular recovery in the oldest old.

It is recommended to start with light to moderate loads of 30–60% of 1RM in untrained older adults and increasing loads progressively to 85% of 1RM to ensure safety and adherence (Orssatto et al. 2018). Moreover, including power exercises at high-velocity movements during the concentric phase at moderate intensities (i.e., 40–60% of 1RM) is recommended to promote muscular power and functional tasks in healthy older adults. However, increased movement velocity must be incorporated carefully in the other groups. Contraction velocity is a basic variable to manipulate when designing resistance training programs in older adults. High-speed power training appears to be more efficient in increasing strength and slowing muscular weakness in older populations (Pereira et al. 2012). For velocity training in all age groups, participants should be instructed to perform all the exercises as fast as they are able to; however, the load must be gradually increased.

3.2 New Strategies Used in Combination with Resistance Training in Older Adults

Many older adults are unwilling or unable to perform resistance training exercises for many of the reasons described above. Different strategies to address this issue are available in the literature.

3.2.1 Neuromuscular Electrical Stimulation (NMES)

Different studies have applied high intensity, intermittent electrical stimuli to skeletal muscles to generate involuntary muscle contractions. A recent study by Jandova et al. (2020) investigated the effect of NMES on muscle size and architecture in healthy older individuals. The authors found that after 8 weeks (3 times/week), muscle thickness and CSA of the vastus lateralis increased by 8.6% and 11.4%, respectively, with significant associations with functional tests (i.e., TUG). Evidence suggests that NMES can counteract age-related muscle and functional decline in older people (Kern et al. 2014) with beneficial effects on muscle strength and functional ability. One of the main advantages of this strategy is that it can be applied during disuse situations or even long-term hospitalizations (Maggioni et al. 2010). Consequently, adults at risk of muscle weakness (e.g., prefrail people), very old or hospitalized people, can increase muscle mass and muscle strength with NMES (Jones et al. 2016).

3.2.2 Blood Flow Restriction

Using an inflated cuff around the limb to block venous blood return but not arterial inflow is commonly used to generate metabolic stress and raise anabolic hormones (Grutter Lopes et al. 2019). A recent report showed that strength training performed with blood flow restriction improved strength, muscle mass, IGF-1, endothelial function, and selected inflammatory markers in a 91-year-old sarcopenic sedentary patient. In combination with low-load resistance training, this technique has proven effective for the prevention of muscle loss after one month of unilateral lower limb suspension in young adults (Cook et al. 2010) and was able to improve muscle mass and strength in older subjects (Yasuda et al. 2014). This technique can be very effective in combination with low-intensity resistance training in very old individuals or in clinical populations (Hughes et al. 2017) unable to reach high loads but able to perform exercises using elastic bands or weightlifting at <20% of 1RM. After only 4 weeks of forearm resistance training with blood flow restriction in older adults, increases in muscle strength and size were observed; however, longer training durations or higher volumes were recommended by the authors to evoke vascular adaptations in older adults (Kim et al. 2017).

3.2.3 Whole-Body Vibration (WBV)

WBV elicits involuntary muscle contractions through the activation of stretch reflexes (Pollock et al. 2012), which would make it useful during disuse situations. Numerous systematic reviews have reported promising results with the use of WBV in muscle strength, bone metabolism, walking ability, mobility or body balance in older adults (Lam et al. 2018; Jepsen et al. 2017; Sitjà-Rabert et al. 2012; Rogan et al. 2017). This strategy has been suggested to be safe and feasible in very old nursing home residents (Bautmans et al. 2005). Moreover, recent studies even suggest that the improvements in balance and muscle strength after WBV training in institutionalized older people are equivalent to an exercise program without vibration (Sitjà-Rabert et al. 2015). It appears that WBV, alone or in combination with traditional exercises, provides several benefits to older adults.

3.2.4 Flywheel Resistance Training

Inertial flywheel devices have emerged as an alternative to gravity-dependent weights, allowing participants to produce greater, force, power, and improve muscle mass and neural adaptations, than traditional strength training programs (Norrbrand et al. 2010; Onambélé et al. 2008). It has been suggested that inertial flywheel devices are particularly effective for training the lower-limb muscles in older people (Sañudo et al. 2020). Following a 6-week flywheel resistance exercise training, authors reported significant changes in strength and mean power. Further, these changes were associated with changes in physical function outcomes (i.e.,30-s Chair Sit-Stand Test and walking speed). In another study using this paradigm, improvements in balance and mobility (TUG) were observed in older adults in addition to changes in muscle power, (Sañudo et al. 2019).

3.3 Assessing Strength in Older Adults

When considering the assessment of muscle strength or muscle power assessment, most studies conducted a 1RM test, even in older adults; however, this test is time consuming and might cause injury (Alcazar et al. 2017). According to the European Working Group on Sarcopenia in Older People (EWGSOP), there are different methods available for assessing muscle strength and physical function that also serve to analyze the effectiveness of interventions. In Fig. 3, different tests used to assess these outcomes in older adults are shown.

In clinical practice, tests such as the leg or bench press test have been used for the assessment of muscle mass, muscle strength and physical performance in older adults (Bruyère et al. 2016). In the literature, the most common test is the assessment of handgrip strength. The EWGSOP suggested that values below 27 and 16 kg in grip strength can be considered low for men and women, respectively (Cruz-Jentoft et al. 2019). Another test commonly used for the assessment of muscle strength is the chair stand test (i.e., participants stand up repeatedly from a chair for 30 s) (Cruz-Jentoft et al. 2019). In the geriatric context participants are requested to rise from a seated position five times and this is considered a valid predictor of lower limb limitation and hospitalization events (Cesari et al. 2009). A cutoff score of 17 s for 5 repetitions has been established for older patients (Cruz-Jentoft et al. 2019). Together with the chair stand test, a balance test and the walking speed test over 4 m complete the Short Physical Performance Battery (SPPB) commonly used to determine the level of frailty (Bruyère et al. 2016).

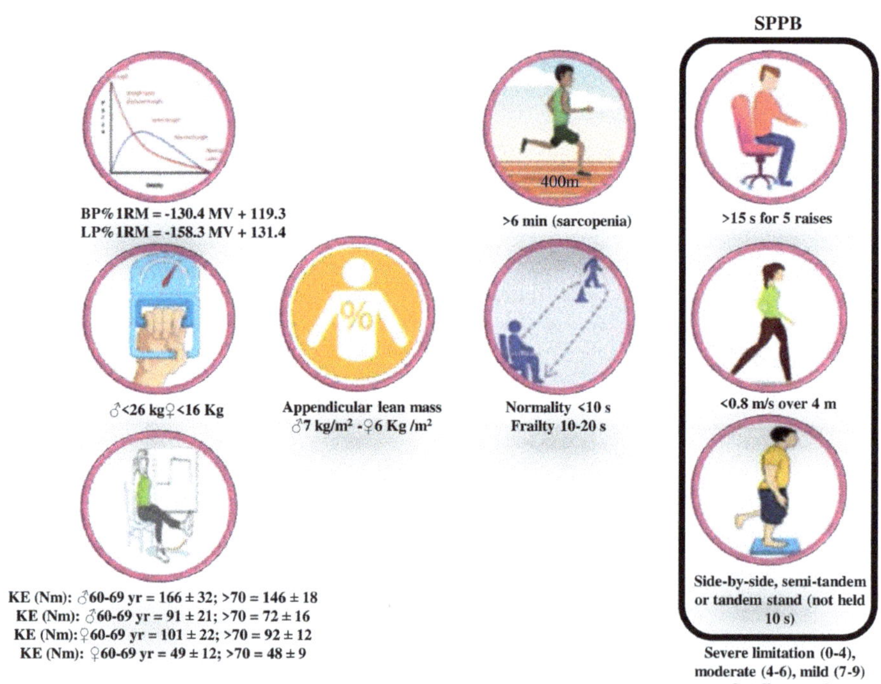

Fig. 3 Examples of test used in the assessment of muscle mass, muscle strength and power, and physical function in older adults

By using this battery, individuals can be classified as persons with disability, with frailty, with pre-frailty or as a robust person (Izquierdo 2019). Gait is the central component of a patient's functional ability to perform basic ADLs. The gait speed test is usually conducted over 4 m and is recommended as a diagnostic tool for sarcopenia with a cutoff point of 0.8 m/s (Cruz-Jentoft et al. 2019). Another functional test available for the assessment of physical performance in older adults is the TUG test which is a good index of mobility and also recommended for assessing sarcopenia. A cutoff of score >20 s has been suggested as an indicator of fall risk (Cruz-Jentoft et al. 2019).

We previously described that impairments in neuromuscular activation affect muscle power and RFD needed for dynamic movements; however, for the measurement of peak forces and rate of force development the number of tests available is low and usually limited to the research context (e.g., isometric torque methods for assessing upper or lower limb strength) (Francis et al. 2017). The assessment of lower limb strength can be measured by the maximal voluntary isometric contraction (MVIC) of both knee extensor and flexor muscle groups, enabling the simultaneous evaluation of the rate of torque development (RTD) or the RFD. These outcomes of knee flexor performance correlate with the number of falls in older people (Bento et al. 2010). Normative data of maximal isokinetic and isometric muscle strength of major muscle groups are available in the literature (Harbo et al. 2012).

For the measurement of muscle mass, the most used and valid test in clinical assessment is dual-energy X-ray absorptiometry (DXA). Thresholds of 7 and 6 kg/m^2 (when adjusted for height square) of appendicular lean mass (i.e., the sum of the muscle mass of the four limbs) have been proposed as cutoff points for sarcopenia in men and women, respectively (Cruz-Jentoft et al. 2019).

4 Filling Gaps

Although much has been written about strength training in older adults, and the recommendation guidelines are updated periodically, there remain some challenges in this area. In particular, and thanks to the development of new technologies, key biological and technical issues can be solved. A recent study highlights the challenges of real-life monitoring of musculoskeletal ageing (Kemp et al. 2018). Further, new strategies in the assessment of muscle strength and power are being defined, allowing a more appropriate training intensity to be prescribed (Alcazar et al. 2017). Finally, and according to the heterogeneity of this population group discussed extensively in this chapter, methodological adaptations in concrete situations could be specified in greater detail.

4.1 In Relation to the Training Intensity

Classical configurations of strength training for older adults include low-velocity (time under tension of 6 s per repetition) and moderate intensity (70–79% of 1RM) (Borde et al. 2015); however, recent advances recommend the prioritization of fast-velocity resistance training for this group (Orssatto et al. 2019). Recent meta-analyses suggest that higher intensities with faster contractions are able to provide greater increases in strength compared with lower intensities (Byrne et al. 2016). It seems that performing muscle actions at higher velocities induce the recruitment of high threshold motor units composed of type II muscle fibres (Aagaard et al. 2010), entailing better neuromuscular and functional adaptations in older adults. Obviously, higher speeds entail a greater risk in this group, so the individualized adjustment of the intensity plays a decisive role.

4.1.1 The Force-Velocity Relationship in Older People

Muscle power has a greater influence on functional mobility in older adults (Byrne et al. 2016). Given that speed of movement is inversely associated with relative intensity, it may be considered a direct indicator of training intensity (Sánchez-Medina and González-Badillo 2011). Thus, the evaluation of the force-velocity (F-V) relationship has been recommended to identify specific neuromuscular deficits in older adults (Alcazar et al. 2017). in view of these parameters, the design of exercise programs can be optimized in order to improve physical performance (Alcazar et al. 2017). The question that arises is how to quantify the F-V relationship in older people. Traditionally, considering that a relatively high number of repetitions against different loads/velocities must be performed for the evaluation of the F-V relationship, this can be a time-consuming and fatiguing task for an older individual (Alcazar et al. 2017). In the aforementioned study, authors compared the reliability and validity of different protocols evaluating the F-V relationship and muscle power in older adults. It was concluded that with only 3 loads the evaluators are able to determine whether the F-V values are sufficiently fitted or an additional attempt must be performed.

Another recent study (Marcos-Pardo et al. 2019) further elucidated ways to estimate the relative load (i.e., %1RM) in older adults. The authors used assessed movement velocity to determine the load during resistance training and proposed regression equations for the bench press and leg press exercises, so that by determining the movement velocity, the load (%1RM) could be determined. Therefore, this can also be considered a useful tool to prescribe the loads during resistance training. In the same context, our lab is trying to do the same in two other functional exercises, the squat with a hexagonal bar and the horizontal bench press. In Fig. 4, an example of the determination of the load that maximized power output in one of these exercises can be observed. In this example, a 71-year-old man performed an incremental test of a squat exercise with a hexagonal bar, starting with 20 kg and progressively increasing by 5 kg. It can be observed how with 35 kg the maximal

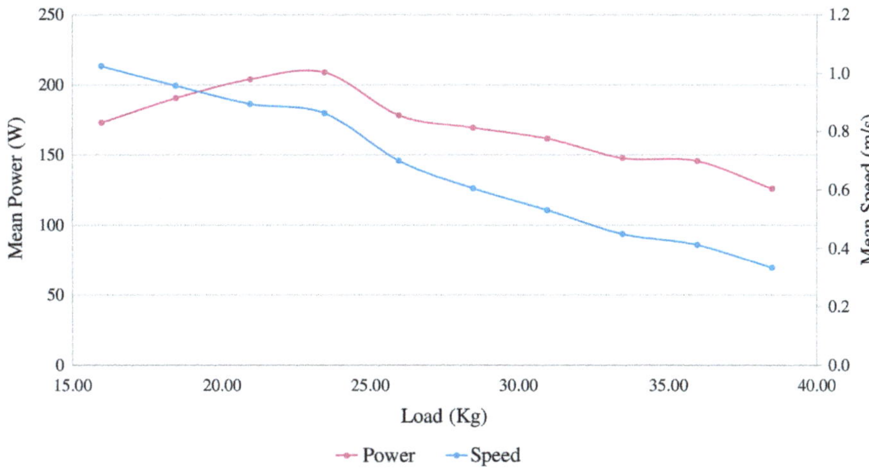

Fig. 4 Example of an incremental test in a squat exercise with a hexagonal bar performed by a 71-year-old man

power output is achieved; therefore, this individual should be advised to use this load in order to improve muscle power and consequently, functional mobility (Byrne et al. 2016).

Technologies are playing an important role for the identification of certain geriatric syndromes, especially for healthcare professionals in the primary care setting. For example, rapid and feasible tools (e.g., mobile applications or wearable devices) will be widely used for this purpose (Sañudo 2017). For example, recent literature uses accelerometer data from the performance of sit-to-stand tests performed by older people that could be reused for testing machine learning for the evaluation of neuromuscular function in older adults (Marques et al. 2020). Electro-Mechanical Systems with acceleration and gyro sensors are also starting to be used to measure muscle strength (Yoshioka et al. 2013).

4.1.2 Resistance Training Programs in Nursing Homes or Clinical Settings

There are numerous alternatives to adapt resistance strength programs to any place or moment. As magnitude of the benefits depends on the type of exercise performed, a recent meta-analysis showed the effectiveness of resistance training for improving not just muscle strength and power, but also functional outcomes in very old (aged 87+) people (Lopez et al. 2018). However, the level of inactivity of some of these older adults preclude them from performing certain activities; despite this, a combination of walking with strengthening exercises has been associated with benefits in functional ability (Martínez-Velilla et al. 2019). Notwithstanding the serious health concerns (e.g., frailty or comorbid conditions), as well as functional disabilities, these programs are being implemented in nursing homes using portable equipment and adapting the exercises (e.g., seated alternatives). A recent study assessed the feasibility of a machine versus free-weight strength training program and its effects on physical performance in nursing home residents and concluded that both methods were similarly effective in improving muscular strength and mobility (Johnen and Schott 2018). However, many individuals become dependent and are reluctant to carry out these programs, consequently, their participation in physical activity decreases dramatically (Valenzuela 2012); therefore, other strategies could be used to prevent the functional decline (Valenzuela et al. 2019b). For example, isometric exercises have been recommended both in institutionalized and hospitalized older adults during bed rest (Kawakami et al. 2001). Other strategies such as blood flow restriction (Clarkson et al. 2017), NMES (O'Connor et al. 2018) or WBV (Orr 2015) have also been suggested to maintain muscle mass and increase functional ability in older patients.

During acute hospitalization, older adults experience a high loss of muscle mass and physical functional capacity, thus, strategies to revert these effects have been recently reported (Sáez de Asteasu et al. 2019). Considering the high prevalence of mobility impairment in the older adults in these facilities, researchers emphasize the importance of muscle power training. However, for older adults with compromised balance and mobility, exercises must be adapted to the seated position. Anthony et al. (Anthony et al. 2013) in their systematic review assessed the effects of chair-based exercises in frail older people who were unable to undertake traditional exercise programs. They suggested a frequency greater than 3 times a week for a duration between 20 and 60 min and incorporating low-intensity exercises. An individualized exercise intervention including low-intensity resistance training (30–60% of 1RM) performed during a short period (mean, 5 days) provided significant improvements in function compared to usual care in acute hospitalized older adults (Martínez-Velilla et al. 2019).

5 Take-Home Messages

1. Resistance training results in hypertrophy of MHC I and II muscle fibres in older adults, although there is a diminished hypertrophic response to resistance training with advancing age. Training increases muscular strength without significant changes of muscle mass in older adults aged ≥ 75 years (Lee et al. 2019).
2. Compared with slow velocity, training at fast velocity using similar loads induces greater improvements in maximal force and greater improvements in power output and explosive force of untrained older adults (Orssattoet al. 2019).
3. Both free weights as well as machine training may be feasible to conduct a strength training program in institutionalized participants.
4. Studies are still needed to determine the optimal dose (e.g., low, moderate, and high intensity) for stimulating skeletal muscle hypertrophy in older muscle.

5. Blood flow restriction has been proposed as an effective option to maximize training-induced adaptations when high-intensity resistance training is not feasible (e.g., in hospitalized or institutionalized individuals).
6. Functional assessments have the advantage of direct relevance to clinical state and quality of life.

References

Aagaard P, Suetta C, Caserotti P, Magnusson SP, Kjaer M (2010) Role of the nervous system in sarcopenia and muscle atrophy with aging: strength training as a countermeasure. Scand J Med Sci Sports 20(1):49–64

Ahlqvist A, Nyfors H, Suhonen R (2016) Factors associated with older people's independent living from the viewpoint of health and functional capacity: a register-based study. Nurs Open 3(2):79–89

Alcazar J, Rodriguez-Lopez C, Ara I, Alfaro-Acha A, Mañas-Bote A, Guadalupe-Grau A et al (2017) The force-velocity relationship in older people: reliability and validity of a systematic procedure. Int J Sports Med 38(14):1097–1104

American College of Sports Medicine (2009) Progression models in resistance training for healthy adults. Med Sci Sport Exerc 41(3):687–708

Andersen LL, Aagaard P (2006) Influence of maximal muscle strength and intrinsic muscle contractile properties on contractile rate of force development. Eur J Appl Physiol 96(1):46–52

Anthony K, Robinson K, Logan P, Gordon AL, Harwood RH, Masud T (2013) Chair-based exercises for frail older people: a systematic review. Biomed Res Int 2013:1–9

Bangsbo J, Blackwell J, Boraxbekk C-J, Caserotti P, Dela F, Evans AB et al (2019) Copenhagen consensus statement 2019: physical activity and ageing. Br J Sports Med 53(14):856–858

Bautmans I, Van Hees E, Lemper J-C, Mets T (2005) The feasibility of Whole Body Vibration in institutionalised elderly persons and its influence on muscle performance, balance and mobility: a randomised controlled trial [ISRCTN62535013]. BMC Geriatr 5:17

Bean JF, Leveille SG, Kiely DK, Bandinelli S, Guralnik JM, Ferrucci L (2003) A comparison of leg power and leg strength within the InCHIANTI study: which influences mobility more? J Gerontol Ser A Biol Sci Med Sci 58(8):M728–M733

Beard JR, Officer A, de Carvalho IA, Sadana R, Pot AM, Michel J-P et al (2016) The world report on ageing and health: a policy framework for healthy ageing. Lancet 387(10033):2145–2154

Bento PCB, Pereira G, Ugrinowitsch C, Rodacki ALF (2010) Peak torque and rate of torque development in elderly with and without fall history. Clin Biomech 25(5):450–454

Bernardo-Filho M, Bemben MG, Taiar R, Sañudo B, Furness T, Clark BC (2020) Editorial: interventional strategies for enhancing quality of life and health Span in older adults. Front Aging Neurosci 11(12):253

Bickel CS, Cross JM, Bamman MM (2011) Exercise dosing to retain resistance training adaptations in young and older adults. Med Sci Sport Exerc. 43(7):1177–1187

Borde R, Hortobágyi T, Granacher U (2015) Dose-response relationships of resistance training in healthy old adults: a systematic review and meta-analysis. Sport Med 45(12):1693–1720

Borzuola R, Giombini A, Torre G, Campi S, Albo E, Bravi M et al (2020) Central and peripheral neuromuscular adaptations to ageing. J Clin Med 9(3):741

Bruyère O, Beaudart C, Reginster J-Y, Buckinx F, Schoene D, Hirani V et al (2016) Assessment of muscle mass, muscle strength and physical performance in clinical practice: an international survey. Eur Geriatr Med 7(3):243–246

Burton E, Farrier K, Lewin G, Pettigrew S, Hill A-M, Airey P et al (2017) Motivators and barriers for older people participating in resistance training: a systematic review. J Aging Phys Act 25 (2):311–324

Byrne C, Faure C, Keene DJ, Lamb SE (2016) Ageing, muscle power and physical function: a systematic review and implications for pragmatic training interventions. Sport Med 46 (9):1311–1332

Campos G, Luecke T, Wendeln H, Toma K, Hagerman F, Murray T et al (2002) Muscular adaptations in response to three different resistance-training regimens: specificity of repetition maximum training zones. Eur J Appl Physiol 88(1–2):50–60

Cesari M, Kritchevsky SB, Newman AB, Simonsick EM, Harris TB, Penninx BW et al (2009) Added value of physical performance measures in predicting adverse health-related events: results from the health, aging and body composition study. J Am Geriatr Soc 57(2):251–259

Chodzko-Zajko WJ, Proctor DN, Fiatarone Singh MA, Minson CT, Nigg CR, Salem GJ et al (2009) Exercise and physical activity for older adults. Med Sci Sports Exerc 41:1510–1530

Clarkson MJ, Conway L, Warmington SA (2017) Blood flow restriction walking and physical function in older adults: a randomized control trial. J Sci Med Sport 20(12):1041–1046

Cook SB, Brown KA, DeRuisseau K, Kanaley JA, Ploutz-Snyder LL (2010) Skeletal muscle adaptations following blood flow-restricted training during 30 days of muscular unloading. J Appl Physiol 109(2):341–349

Cruz-Jentoft AJ, Bahat G, Bauer J, Boirie Y, Bruyère O, Cederholm T et al (2019) Sarcopenia: revised European consensus on definition and diagnosis. Age Ageing 48(1):16–31

Csapo R, Alegre LM (2016) Effects of resistance training with moderate vs heavy loads on muscle mass and strength in the elderly: a meta-analysis. Scand J Med Sci Sports 26(9):995–1006

da Rosa Orssatto LB, de la Rocha Freitas C, Shield AJ, Silveira Pinto R, Trajano GS (2019) Effects of resistance training concentric velocity on older adults' functional capacity: a systematic review and meta-analysis of randomised trials. Exp Gerontol 127:110731

De Souto BP, Rolland Y, Vellas B, Maltais M (2019) Association of long-term exercise training with risk of falls, fractures, hospitalizations, and mortality in older adults: a systematic review and meta-analysis. JAMA Intern Med 179(3):394–405

Feldman HA, Longcope C, Derby CA, Johannes CB, Araujo AB, Coviello AD et al (2002) Age trends in the level of serum testosterone and other hormones in middle-aged men: longitudinal results from the massachusetts male aging study. J Clin Endocrinol Metab 87(2):589–598

Foldvari M, Clark M, Laviolette LC, Bernstein MA, Kaliton D, Castaneda C et al (2000) Association of muscle power with functional status in community-dwelling elderly women. J Gerontol Ser A Biol Sci Med Sci 55(4):M192–M199

Fragala MS, Cadore EL, Dorgo S, Izquierdo M, Kraemer WJ, Peterson MD et al (2019) Resistance training for older adults. J Strength Cond Res 33(8):2019–2052

Francis P, Toomey C, Mc Cormack W, Lyons M, Jakeman P (2017) Measurement of maximal isometric torque and muscle quality of the knee extensors and flexors in healthy 50- to 70-year-old women. Clin Physiol Funct Imaging 37(4):448–455

Frontera WR, Hughes VA, Fielding RA, Fiatarone MA, Evans WJ, Roubenoff R (2000) Aging of skeletal muscle: a 12-yr longitudinal study. J Appl Physiol 88(4):1321–1326

Gerstner GR, Thompsen BJ, Rosenberg JG, Sobolewski EJ, Scharville MJ, Ryan ED (2017) Neural and muscular contributions to the age-related reductions in rapid strength. Med Sci Sport Exerc 49(7):1331–1339

Giné-Garriga M, Roqué-Fíguls M, Coll-Planas L, Sitjà-Rabert M, Salvà A (2014) Physical exercise interventions for improving performance-based measures of physical function in community-dwelling, frail older adults: a systematic review and meta-analysis. Arch Phys Med Rehabil 95(4):753-769.e3

Grutter Lopes K, Bottino DA, Farinatti P, Coelho de Souza M das G, Alves Maranhão P, Soares de Araujo CM et al (2019) Strength training with blood flow restriction—a novel therapeutic approach for older adults with sarcopenia? A case report. Clin Interv Aging 14:1461–9

Guizelini PC, de Aguiar RA, Denadai BS, Caputo F, Greco CC (2018) Effect of resistance training on muscle strength and rate of force development in healthy older adults: a systematic review and meta-analysis. Exp Gerontol 102:51–58
Gujral S, Oberlin L (2021) Is there a dose-response relationship between exercise and cognitive function in older adults? Am J Geriatr Psychiatry 29(2):141–143
Hameed M, Harridge SDR, Goldspink G (2002) Sarcopenia and hypertrophy: a role for insulin-like growth factor-1 in aged muscle? Exerc Sport Sci Rev 30(1):15–19
Harbo T, Brincks J, Andersen H (2012) Maximal isokinetic and isometric muscle strength of major muscle groups related to age, body mass, height, and sex in 178 healthy subjects. Eur J Appl Physiol 112(1):267–275
Hester GM, Ha PL, Dalton BE, VanDusseldorp TA, Olmos AA, Stratton MT, et al (2021) Rate of force development as a predictor of mobility in older adults. J Geriatr Phys Ther 44(2):74–81
Hughes L, Paton B, Rosenblatt B, Gissane C, Patterson SD (2017) Blood flow restriction training in clinical musculoskeletal rehabilitation: a systematic review and meta-analysis. Br J Sports Med 51(13):1003–1011
Ivey FM, Tracy BL, Lemmer JT, NessAiver M, Metter EJ, Fozard JL et al (2000) Effects of strength training and detraining on muscle quality: age and gender comparisons. J Gerontol Ser A Biol Sci Med Sci 55(3):B152–B157
Izquierdo M (2019) Multicomponent physical exercise program: Vivifrail. Nutr Hosp 36(Spec No2):50–56
Izquierdo M, Morley JE, Lucia A (2020) Exercise in people over 85. BMJ, m402
Jandova T, Narici MV, Steffl M, Bondi D, D'amico M, Pavlu D et al (2020) Muscle hypertrophy and architectural changes in response to eight-week neuromuscular electrical stimulation training in healthy older people. Life 10(9):1–15
Jepsen DB, Thomsen K, Hansen S, Jørgensen NR, Masud T, Ryg J (2017) Effect of whole-body vibration exercise in preventing falls and fractures: a systematic review and meta-analysis. BMJ Open 7(12):e018342
Johnen B, Schott N (2018) Feasibility of a machine vs free weight strength training program and its effects on physical performance in nursing home residents: a pilot study. Aging Clin Exp Res 30(7):819–828
Jones S, Man WDC, Gao W, Higginson IJ, Wilcock A, Maddocks M (2016) Neuromuscular electrical stimulation for muscle weakness in adults with advanced disease. Cochrane Database Syst Rev 10(10):CD009419
Katsura Y, Takeda N, Hara T, Takahashi S, Nosaka K (2019) Comparison between eccentric and concentric resistance exercise training without equipment for changes in muscle strength and functional fitness of older adults. Eur J Appl Physiol 119(7):1581–1590
Kawakami Y, Akima H, Kubo K, Muraoka Y, Hasegawa H, Kouzaki M et al (2001) Changes in muscle size, architecture, and neural activation after 20 days of bed rest with and without resistance exercise. Eur J Appl Physiol 84(1–2):7–12
Keadle SK, McKinnon R, Graubard BI, Troiano RP (2016) Prevalence and trends in physical activity among older adults in the United States: a comparison across three national surveys. Prev Med (baltim) 89:37–43
Kemp GJ, Birrell F, Clegg PD, Cuthbertson DJ, De Vito G, van Dieën JH et al (2018) Developing a toolkit for the assessment and monitoring of musculoskeletal ageing. Age Ageing 47 (suppl_4):iv1–iv19
Kern H, Barberi L, Löfler S, Sbardella S, Burggraf S, Fruhmann H et al (2014) Electrical stimulation counteracts muscle decline in seniors. Front Aging Neurosci 6:189
Kim J, Lang JA, Pilania N, Franke WD (2017) Effects of blood flow restricted exercise training on muscular strength and blood flow in older adults. Exp Gerontol 99:127–132
Kryger AI, Andersen JL (2007) Resistance training in the oldest old: consequences for muscle strength, fiber types, fiber size, and MHC isoforms. Scand J Med Sci Sports 17(4):422–430
Kubo K, Kanehisa H, Azuma K, Ishizu M, Kuno S-Y, Okada M et al (2003) Muscle architectural characteristics in young and elderly men and women. Int J Sports Med 24(2):125–130

Lam FM, Chan PF, Liao LR, Woo J, Hui E, Lai CW et al (2018) Effects of whole-body vibration on balance and mobility in institutionalized older adults: a randomized controlled trial. Clin Rehabil 32(4):462–472

Larsson L, Degens H, Li M, Salviati L, Lee Y Il, Thompson W et al (2019) Sarcopenia: aging-related loss of muscle mass and function. Physiol Rev 99(1):427–511

Lavin KM, Roberts BM, Fry CS, Moro T, Rasmussen BB, Bamman MM (2019) The importance of resistance exercise training to combat neuromuscular aging. Physiology 34(2):112–122

Lee H, Kim I-G, Sung C, Jeon T-B, Cho K, Ha Y-C et al (2019) Exercise training increases skeletal muscle strength independent of hypertrophy in older adults aged 75 years and older. Geriatr Gerontol Int 19(3):265–270

Liu C, Chang W-P, Araujo de Carvalho I, Savage KEL, Radford LW, Amuthavalli Thiyagarajan J (2017) Effects of physical exercise in older adults with reduced physical capacity: meta-analysis of resistance exercise and multimodal exercise. Int J Rehabil Res 40(4):303–14

Lopez P, Pinto RS, Radaelli R, Rech A, Grazioli R, Izquierdo M et al (2018) Benefits of resistance training in physically frail elderly: a systematic review. Aging Clin Exp Res 30(8):889–899

Maffiuletti NA, Aagaard P, Blazevich AJ, Folland J, Tillin N, Duchateau J (2016) Rate of force development: physiological and methodological considerations. Eur J Appl Physiol 116 (6):1091–1116

Maggioni MA, Cè E, Rampichini S, Ferrario M, Giordano G, Veicsteinas A et al (2010) Electrical stimulation versus kinesitherapy in improving functional fitness in older women: a randomized controlled trial. Arch Gerontol Geriatr 50(3):e19-25

Manini T, Marko M, VanArnam T, Cook S, Fernhall B, Burke J et al (2007) Efficacy of resistance and task-specific exercise in older adults who modify tasks of everyday life. J Gerontol Ser A Biol Sci Med Sci 62(6):616–623

Marcos-Pardo PJ, González-Hernández JM, García-Ramos A, López-Vivancos A, Jiménez-Reyes P (2019) Movement velocity can be used to estimate the relative load during the bench press and leg press exercises in older women. Peer J 7:e7533

Marques MC, Izquierdo M, Pereira A (2013) High-speed resistance training in elderly people. Strength Cond J 35(2):23–29

Marques DL, Pires IM, Neiva HP, Marinho DA, Marques MC (2020) Accelerometer data from the performance of sit-to-stand test by elderly people. Data Br 106328

Martínez-Velilla N, Casas-Herrero A, Zambom-Ferraresi F, Sáez de Asteasu ML, Lucia A, Galbete A et al (2019) Effect of exercise intervention on functional decline in very elderly patients during acute hospitalization. JAMA Intern Med 179(1):28

McNeil CJ, Vandervoort AA, Rice CL (2007) Peripheral impairments cause a progressive age-related loss of strength and velocity-dependent power in the dorsiflexors. J Appl Physiol 102(5):1962–1968

Motalebi SA, Cheong LS, Iranagh JA, Mohammadi F (2018) Effect of low-cost resistance training on lower-limb strength and balance in institutionalized seniors. Exp Aging Res 44(1):48–61

Nakamura Y, Tanaka K, Yabushita N, Sakai T, Shigematsu R (2007) Effects of exercise frequency on functional fitness in older adult women. Arch Gerontol Geriatr 44(2):163–173

Negaresh R, Ranjbar R, Habibi A, Mokhtarzade M, Fokin A, Gharibvand MM (2017) The effect of resistance training on quadriceps muscle volume and some growth factors in elderly and young men. Adv Gerontol Uspekhi Gerontol 30(6):880–887

Nilwik R, Snijders T, Leenders M, Groen BBL, van Kranenburg J, Verdijk LB et al (2013) The decline in skeletal muscle mass with aging is mainly attributed to a reduction in type II muscle fiber size. Exp Gerontol 48(5):492–498

Norrbrand L, Pozzo M, Tesch PA (2010) Flywheel resistance training calls for greater eccentric muscle activation than weight training. Eur J Appl Physiol 110(5):997–1005

O'Connor D, Brennan L, Caulfield B (2018) The use of neuromuscular electrical stimulation (NMES) for managing the complications of ageing related to reduced exercise participation. Maturitas 113:13–20

Onambélé GL, Maganaris CN, Mian OS, Tam E, Rejc E, McEwan IM et al (2008) Neuromuscular and balance responses to flywheel inertial versus weight training in older persons. J Biomech 41(15):3133–3138
Orr R (2015) The effect of whole body vibration exposure on balance and functional mobility in older adults: a systematic review and meta-analysis. Maturitas 80(4):342–358
Orssatto LBR, Moura BM, Bezerra ES, Andersen LL, Oliveira SN, Diefenthaeler F (2018) Influence of strength training intensity on subsequent recovery in elderly. Exp Gerontol 106:232–239
Orssatto da LB R, Cadore EL, Andersen LL, Diefenthaeler F (2019) Why fast velocity resistance training should be prioritized for elderly people. Strength Cond J 41(1):105–14
Palmer TB, Thiele RM, Williams KB, Adams BM, Akehi K, Smith DB et al (2015) The identification of fall history using maximal and rapid isometric torque characteristics of the hip extensors in healthy, recreationally active elderly females: a preliminary investigation. Aging Clin Exp Res 27(4):431–438
Palmer TB, Thiele RM, Conchola EC, Smith DB, Thompson BJ (2016) A preliminary study of the utilization of maximal and rapid strength characteristics to identify chair-rise performance abilities in very old adults. J Geriatr Phys Ther 39(3):102–109
Papa E V, Dong X, Hassan M (2017a) Skeletal muscle function deficits in the elderly: current perspectives on resistance training. J Nat Sci 3(1)
Papa EV, Dong X, Hassan M (2017b) Resistance training for activity limitations in older adults with skeletal muscle function deficits: a systematic review. Clin Interv Aging 12:955–961
Pereira A, Izquierdo M, Silva AJ, Costa AM, Bastos E, González-Badillo JJ et al (2012) Effects of high-speed power training on functional capacity and muscle performance in older women. Exp Gerontol 47(3):250–255
Peterson MD, Sen A, Gordon PM (2011) Influence of resistance exercise on lean body mass in aging adults. Med Sci Sport Exerc 43(2):249–258
Pollock RD, Woledge RC, Martin FC, Newham DJ (2012) Effects of whole body vibration on motor unit recruitment and threshold. J Appl Physiol 112(3):388–395
Ramírez-Campillo R, Castillo A, de la Fuente CI, Campos-Jara C, Andrade DC, Álvarez C et al (2014) High-speed resistance training is more effective than low-speed resistance training to increase functional capacity and muscle performance in older women. Exp Gerontol 58:51–57
Reeves ND, Narici MV, Maganaris CN (2004a) Effect of resistance training on skeletal muscle-specific force in elderly humans. J Appl Physiol 96(3):885–892
Reeves ND, Narici MV, Maganaris CN (2004b) In vivo human muscle structure and function: adaptations to resistance training in old age. Exp Physiol 89(6):675–689
Reeves ND, Narici MV, Maganaris CN (2006) Musculoskeletal adaptations to resistance training in old age. Man Ther 11(3):192–196
Reid KF, Fielding RA (2012) Skeletal muscle power. Exerc Sport Sci Rev 40(1):4–12
Ribeiro A, Carneiro N, Nascimento M, Gobbo L, Schoenfeld BJ, Achour Júnior A et al (2015) Effects of different resistance training frequencies on flexibility in older women. Clin Interv Aging 10:531–538
Roberts BM, Lavin KM, Many GM, Thalacker-Mercer A, Merritt EK, Bickel CS et al (2018) Human neuromuscular aging: sex differences revealed at the myocellular level. Exp Gerontol 106:116–124
Rogan S, Taeymans J, Radlinger L, Naepflin S, Ruppen S, Bruelhart Y et al (2017) Effects of whole-body vibration on postural control in elderly: an update of a systematic review and meta-analysis. Arch Gerontol Geriatr 73:95–112
Sáez de Asteasu ML, Martínez-Velilla N, Zambom-Ferraresi F, Casas-Herrero Á, Lucía A, Galbete A et al (2019) Physical exercise improves function in acutely hospitalized older patients: secondary analysis of a randomized clinical trial. J Am Med Dir Assoc 20(7):866–73
Sánchez-Medina L, González-Badillo JJ (2011) Velocity loss as an indicator of neuromuscular fatigue during resistance training. Med Sci Sport Exerc 43(9):1725–1734

Sañudo B (2017) Nuevas tecnologías aplicadas a la actividad física y el deporte, p 162. Editorial Aranzadi

Sañudo B, González-Navarrete Á, Álvarez-Barbosa F, de Hoyo M, Del Pozo J, Rogers ME (2019) Effect of flywheel resistance training on balance performance in older adults. A randomized controlled trial. J Sports Sci Med 18(2):344–350

Sañudo B, de Hoyo M, McVeigh JG (2020) Improved muscle strength, muscle power, and physical function after flywheel resistance training in healthy older adults. J Strength Cond Res 1

Schaap LA, Pluijm SMF, Deeg DJH, Visser M (2006) Inflammatory markers and loss of muscle mass (sarcopenia) and strength. Am J Med 119(6):526.e9-526.e17

Serra-Rexach JA, Bustamante-Ara N, Hierro Villarán M, González Gil P, Sanz Ibáñez MJ, Blanco Sanz N et al (2011) Short-term, light- to moderate-intensity exercise training improves leg muscle strength in the oldest old: a randomized controlled trial. J Am Geriatr Soc 59(4):594–602

Shafiee G, Keshtkar A, Soltani A, Ahadi Z, Larijani B, Heshmat R (2017) Prevalence of sarcopenia in the world: a systematic review and meta-analysis of general population studies. J Diabetes Metab Disord 16(1):21

Silva RB, Aldoradin-Cabeza H, Eslick GD, Phu S, Duque G (2017) The effect of physical exercise on frail older persons: a systematic review. J Frailty Aging 6(2):91–96

Sitjà-Rabert M, Rigau D, Fort Vanmeerghaeghe A, Romero-Rodríguez D, Bonastre Subirana M, Bonfill X (2012) Efficacy of whole body vibration exercise in older people: a systematic review. Disabil Rehabil 34(11):883–893

Sitjà-Rabert M, Martínez-Zapata MJ, Fort Vanmeerhaeghe A, Rey Abella F, Romero-Rodríguez D, Bonfill X (2015) Effects of a whole body vibration (WBV) exercise intervention for institutionalized older people: a randomized, multicentre, parallel, clinical trial. J Am Med Dir Assoc 16(2):125–131

Snijders T, Nederveen JP, Joanisse S, Leenders M, Verdijk LB, van Loon LJC et al (2017) Muscle fibre capillarization is a critical factor in muscle fibre hypertrophy during resistance exercise training in older men. J Cachexia Sarcopenia Muscle 8(2):267–276

Steib S, Schoene D, Pfeifer K (2010) Dose-response relationship of resistance training in older adults. Med Sci Sport Exerc 42(5):902–914

Straight CR, Lindheimer JB, Brady AO, Dishman RK, Evans EM (2016) Effects of resistance training on lower-extremity muscle power in middle-aged and older adults: a systematic review and meta-analysis of randomized controlled trials. Sport Med 46(3):353–364

Straight CR, Fedewa MV, Toth MJ, Miller MS (2020) Improvements in skeletal muscle fiber size with resistance training are age-dependent in older adults: a systematic review and meta-analysis. J Appl Physiol 129(2):392–403

Thompson BJ, Ryan ED, Herda TJ, Costa PB, Herda AA, Cramer JT (2014) Age-related changes in the rate of muscle activation and rapid force characteristics. Age (omaha) 36(2):839–849

Toth MJ, Matthews DE, Tracy RP, Previs MJ (2005) Age-related differences in skeletal muscle protein synthesis: relation to markers of immune activation. Am J Physiol Metab 288(5):E883–E891

Valenzuela T (2012) Efficacy of progressive resistance training interventions in older adults in nursing homes: a systematic review. J Am Med Dir Assoc 13(5):418–428

Valenzuela PL, Castillo-García A, Morales JS, Izquierdo M, Serra-Rexach JA, Santos-Lozano A et al (2019a) Physical exercise in the oldest old. Compr Physiol 9(4):1281–1304

Valenzuela PL, Castillo-García A, Morales JS, Izquierdo M, Serra-Rexach JA, Santos-Lozano A et al (2019b) Physical exercise in the oldest old. Compr Physiol 9(4):1281–1304

Valenzuela PL, Castillo-García A, Morales JS, Izquierdo M, Serra-Rexach JA, Santos-Lozano A et al (2019c) Physical exercise in the oldest old. In: Comprehensive physiology, pp 1281–304. Wiley

von Haehling S, Morley JE, Anker SD (2010) An overview of sarcopenia: facts and numbers on prevalence and clinical impact. J Cachexia Sarcopenia Muscle 1(2):129–133

Yasuda T, Fukumura K, Fukuda T, Uchida Y, Iida H, Meguro M et al (2014) Muscle size and arterial stiffness after blood flow-restricted low-intensity resistance training in older adults. Scand J Med Sci Sports 24(5):799–806

Yoshioka S, Nagano A, Hay DC, Tabata I, Isaka T, Iemitsu M et al (2013) New method of evaluating muscular strength of lower limb using MEMS acceleration and gyro sensors. J Robot Mechatronics 25(1):153–161

Borja Sañudo received the B.S. and M.S. degrees in physical activity and health from the International University of Andalusia in 2007 and the Ph.D. degree in sport science from University of Seville, Spain, in 2009. From 2010, Borja is full professor at the Department of Physical Education and Sport (University of Seville, Spain). His major research interests are centered on the role of exercise and other lifestyle factors for promoting improvements in physiological function, quality of life and disease-free survival in clinical populations and improvements in athletes' performance. Since 2005, he has been working on a number of studies with clinical populations (mainly fibromyalgia, Type 2 diabetes and obesity) and athletes (whole body vibration and exercise and the use of technologies in sports). He is the author of five books, more than 100 articles in peer-reviewed journals, and more than 80 international presentations.

Michael E. Rogers, Ph.D. earned his Ph.D. in Exercise Physiology from Kent State University in Ohio. He is a Professor, as well as the Department Chair for the Department of Human Performance Studies at Wichita State University (WSU). He teaches graduate courses such as Exercise Physiology, Applied Physiology, and Advanced Exercise Testing. He has extensive experience and expertise with older populations. His primary research interests concern the effects of balance and resistance training on muscle and bone health, fall prevention, and functional capacity in older adults.

Currently, he is conducting studies on the efficacy of community-based exercise programs in Wichita and Nagoya, Japan. He has received funding for his research from the National Research Council, American Federation for Aging Research, Thera-Band, and the State of Kansas. He has an impressive record of presentations at national and international conferences.

Dr. Rogers was one of six people appointed nationally to the Steering Committee for the American College of Sports Medicine Special Interest Group on Aging. Recently he was named to the Executive Committee of the International Society for Aging and Physical Activity. In addition he serves as the web site master for this international organization. He is also responsible for the formation of a sistership between Wichita State University and Nagoya University in Japan.

Resistance Training for Children and Adolescents

Emilio Villa-González and Avery D. Faigenbaum

Abstract Regular participation in youth resistance training (RT) programs has been shown to elicit favorable short-term influences on diverse health dimensions. However, traditional concerns and misconceptions persist, although a compelling body of research indicated that supervised resistance exercises can be a safe and effective protocol of training for children and adolescents. The World Health Organization supports participation in a variety of physical activities including those that strengthen muscle and bone. Thus, the current chapter will address the benefits of RT programs in young people, mainly focused on youth sports practice. Besides, it will be addressed specific aspects to consider within the sports environment, where there is a considerable variation in growth, maturity, and performance status among individuals of the same chronologic age, especially during the pubertal years. Finally, practical strategies to develop safe and effective RT programs for athletes of different levels will be presented.

Keywords Strength training · Maturation · Performance · Youth

1 Introduction

1.1 Operational Definitions

Prior to discussing the literature surrounding youth resistance training, it is pertinent to define key terminologies used throughout this chapter:

- **Resistance training** is a collective term that refers to method of conditioning that involve the progressive use of a wide range of resistive loads different

E. Villa-González (✉)
Department of Physical and Sports Education, University of Granada, Granada, Spain
e-mail: evilla@ugr.es

A. D. Faigenbaum
Department of Health and Exercise Science, The College of New Jersey, Ewing, USA

A. Muñoz-López et al. (eds.), *Resistance Training Methods*, Lecture Notes in Bioengineering, https://doi.org/10.1007/978-3-030-81989-7_17

movements velocities, and a variety of training modalities. Methods of resistance training include the use of body weight, free weights, weigh machines, elastics bands, medicine balls and manual resistance. The term resistance training should be distinguished from the sport of weightlifting, which involve performing exercises such as snatch and clean and jerk lifts in competition (Faigenbaum et al. 2019a).

- **Childhood** represents the developmental period of life from the end of infancy to the beginning of adolescence. The term children refers to girls and boys (generally up to the age of 11 and 13 years, respectively) who have not developed secondary sex characteristics (Lloyd et al. 2012).
- **Adolescence** refers to a period of life between childhood and adulthood. Although adolescence is a more difficult period to define in terms of chronological age due to differential maturation rates (Malina et al. 2004) girls 12–18 years and boys 14–18 years are generally considered adolescents.
- **Youth** represent global terms which include both children and adolescents (Lloyd et al. 2012).
- **Growth** is typically viewed as a quantifiable change in body composition, the size of the body as a whole or the size of specific regions of the body (Beunen and Malina 2008).
- **Maturation** refers to the highly variable timing and tempo of progressive change within the human body from childhood to adulthood, and which, in addition to growth, influences overall physical performance capabilities.
- **Physical fitness** is a state of health and well-being and, more specifically, the ability to perform aspects of sports, occupations and daily activities.
- **Physical Literacy** is the motivation, confidence, physical competence, knowledge, and understanding to value and take responsibility for engagement in physical activities for life (Zwolski et al. 2017).

2 From Theory

2.1 Secular Trends in Muscular Fitness in Modern day Youth

Monitoring of physical fitness (PF) in youth is important, because PF is known as one of the most relevant resources for health and is regarded as one of the foundations of an active lifestyle (Eberhardt et al. 2020). A recent systematic review analyzing 24 secular trend studies from 16 countries, concluding that a general decline of PF (including endurance, strength, flexibility) over time was observed in most of the studies (Eberhardt et al. 2020). Systematic monitoring of PF is thus needed to assess and design interventions and programs aiming to maintain or increase PF. However, only few studies exist that were designed to investigate secular trends in PF of children and adolescents with a sufficiently large sample size, an investigation period with several measuring points, and a uniform methodology over years. For instance, low levels of muscular fitness (i.e., muscular strength, muscular power and local muscular endurance) in children and adolescents are associated with poor motor competence, functional limitations and adverse health outcomes (Smith et al. 2014; García-Hermoso et al. 2019) and recent findings indicate that measures of muscular strength and power in modern-day youths are lower than in previous generations (Chulvi-Medrano et al. 2020; Moliner-Urdiales et al. 2010; Sandercock and Cohen 2019; Faigenbaum et al. 2019b). Likewise, Sandercock and Cohen reported a decline in muscular fitness (bent-arm hang, sit-ups and handgrip) using allometric equations in 10-year-old English children from 1998 to 2014, and noted this trend was independent of secular changes in body size (Sandercock and Cohen 2019). A similar trend in muscular fitness was observed in Spanish adolescents between 2001–2002 and 2006–2007 (Moliner-Urdiales et al. 2010), and in an international sample of children and adolescents between 1964 and 2017 (Kaster et al. 2020). Lower levels of muscular strength and power in modern day youth appear to be consequent to lifestyles characterized by reduced physical activity (PA) and increased sedentary behaviors (Sandercock and Cohen 2019; Faigenbaum et al. 2019b). Since muscular strength and fundamental movement skill (FMS) proficiency are considered foundational for ongoing participation in PA across the lifespan (Hulteen et al. 2018; Smith et al. 2019), ongoing efforts are needed to target neuromuscular deficits in modern day youth.

2.2 *Benefits of Youth Resistance Training*

2.2.1 Health Benefits

The World Health Organization (WHO) supports participation in a variety of physical activities including those that strengthen muscle and bone (WHO). Since contemporary youth are not as active as they should be (Ekelund et al. 2011; Guthold et al. 2010; Nyberg et al. 2009), children and adolescents should be encouraged to participate regularly in play, games, sports and planned exercise in the context of school and community activities. Not only is PA essential for normal growth and development, but also youth programmes that enhance muscular strength and FMS performance early in life appear to build the foundation for an active lifestyle later in life (Barnett et al. 2013; Barnett et al. 2009; Lopes et al. 2011). Regular participation in youth RT programme has been shown to elicit favorable short-term influences on diverse health dimensions (Lloyd et al. 2014). Childhood may be the opportune time for the bone remodeling process to respond to the tensile and compressive forces correlated with weight-bearing exercise (Ishikawa et al. 2013; Behringer et al. 2014), including RT which can increase bone mineral density during the growing years.

Regular participation in youth fitness programmes designed to enhance muscular fitness may also improve cardiovascular and metabolic health (Ortega et al. 2008; Lau et al. 2010). A longitudinal study found that males able to maintain higher levels of muscular power from childhood to adulthood tend to have a healthier glucose homeostasis profile (fasting insulin, HOMA2-β and HOMA2-IR) during adulthood (Fraser et al. 2019). Moreover, higher childhood levels of muscular strength, muscular power and a combined muscular fitness score have been found to be associated with a reduced risk of metabolic syndrome, independent of cardiorespiratory fitness (Fraser et al. 2016). Of potential relevance, findings from a recent meta-analysis indicate that higher levels of upper- and lower-body muscular strength are associated with a lower risk of mortality in adults (García-Hermoso et al. 2018a). Participation in strength-building programs can have favorable effects on other variables including body composition, quality of life, and psychological well-being (Lloyd et al. 2014; Collins et al. 2019). However, when youth do not have regular opportunities to engage in structured exercise programs that include some type of RT, they may be more likely to experience negative health outcomes and less likely to engage regularly in moderate to vigorous PA (Smith et al. 2019).

Reviews and longitudinal studies have noted evidence for the positive association between motor competence and PA (Lopes et al. 2011; Logan et al. 2015; Holfelder and Schott 2014). Motor competence is the degree to which an individual can perform goal-directed human movement (Robinson et al. 2015). Although national and international youth PA guidelines typically focus on the promotion of aerobic activities (e.g. running, cycling, etc.) (Martinez-Gomez et al. 2010; Ottevaere et al. 2011; Myer et al. 2015), more recent data have highlighted the unique and complementary benefits of muscle-strengthening physical activities for children and adolescents (Smith et al. 2014; Lloyd et al. 2014). Early exposure to engaging strength and skill-based activities might be particularly important for ongoing participation in moderate to vigorous PA (Myer et al. 2015; Brian et al. 2020). Such experiences may encouragement perceptions of competence and confidence, which can, in turn, help to maintain participation in exercise and sport activities throughout the lifecourse. In addition to facilitating safe and effective participation in RT, developing competence and confidence in RT skills might support self-efficacy beliefs, promote positive physical self-perceptions, and enhance physical literacy (Zwolski et al. 2017; Smith et al. 2018). Thus, a growing body of evidence indicates observable health and fitness benefits when increasing muscular strength and motor competence among young people.

In addition, given the growing prevalence of youth who are overweight and obese and associated health-related problems, the effect of RT on the metabolic health, body composition and injury risk profile of children and adolescents with excess body fat has received increased attention (Collins et al. 2018; Suh et al. 2011; Heijden et al. 2010; McHugh 2010). The study of Collins et al., which included data from 18 controlled trials, which evaluated 8 outcomes related to weight status, suggested that an isolated RT intervention may have an effect on weight status in youth (Collins et al. 2018). Consistent with these findings, a recent systematic review and meta-analysis investigated the effects of concurrent exercise (i.e., resistance plus aerobic training) versus aerobic training alone on anthropometric and metabolic outcomes in children and adolescents who were obese (García-Hermoso et al. 2018b) Notably, concurrent exercise (>24 weeks) improved the body composition and cardiometabolic profiles more so than isolated aerobic training (García-Hermoso et al. 2018b). Thus, practitioners and clinicians working with youth who are obese should consider exercise interventions that integrate aerobic and RT training, rather than isolated aerobic exercise when designing interventions for this population.

2.2.2 Sport Performance Benefits

Athletic training should begin in childhood, so that future athletes can progressively and systematically enhance their physical and psychological readiness to play (Bompa and Buzzichelli 2015). The optimal time for sports could be defined as one in which a child achieves the necessary maturity to learn and understand a given task. In the context of sports, this also includes physical, neurological, cognitive, psychological and social factors. Physical performance is commonly measured as the outcome (product) of standardized motor tasks requiring speed, agility, balance, flexibility, explosive strength, local muscular endurance, and static muscular strength (Bompa and Buzzichelli 2015).

2.3 Growth/Maturation and Athletic Performance

Growth refers to measurable changes in size, physique and body composition, and various systems of the body, whereas maturation denotes to progress toward the mature state. Maturation is variable among bodily systems and also in timing and tempo of progress. The processes of growth and maturation are related, and both influence physical performance. Chronologic age is the common reference in studies of growth and performance. However, there is considerable variation in growth, maturity, and performance status among individuals of the same chronologic age, especially during the pubertal years (Malina et al. 2004; Beunen and Malina 2008). For instance, isometric strength increases linearly with age during childhood and the transition into adolescence in both sexes. At approximately 13 years, strength development accelerates considerably in boys (adolescent spurt), but continues to increase linearly in girls through about 15 years with less evidence for a clear adolescent spurt, although data vary among specific strength tests (Beunen and Malina 2008). Sex differences in strength are consistent, although small, through childhood and the transition into adolescence. Thereafter, the differences become increasingly larger so that at the age of 16 years and later only a few girls perform at the same level as the average boy. Strength is related to body size and muscle mass, so that sex differences might relate to a size advantage in boys. During childhood and adolescence boys tend to have greater strength per unit body size, especially in the upper body and trunk than girls, but corresponding sex differences in lower extremity strength are negligible. The disproportionate strength increase is most apparent during male adolescence, and is greater in the upper extremities than in the trunk or lower extremities (Beunen and Malina 2008).

In addition, individual differences in maturity status at a given age and in the timing of the adolescent spurt influence growth status and performance. Furthermore, youth who are successful in sport tend to differ, on average, in maturity status and rate compared with the general population. Variation in maturity status influences body dimensions, composition and proportions, and also performance. The effect of variation in maturity status can operate through associated variation in body size and/or composition and through a direct influence on performance.

During and after puberty, testosterone secretion in males is associated with observable gains in fat-free mass and linear growth (Malina et al. 2004). During this period, RT-induced gains in muscular strength and power may be associate with changes in hypertrophic factors due to hormonal production (Kraemer et al. 1989). However, the internal hormonal can differ greatly between adolescents of the same chronological age; therefore, youth fitness specialist should except interindividual differences in muscle size before and after the RT in young lifters. In the case of girls, the smaller amount of testosterone limits the magnitude of gains in muscle mass following RT (Malina et al. 2004).

Training-induced gains in strength and power during childhood appear to derive primarily from neuromuscular mechanisms rather than hypertrophic factors (i.e., increased muscle glycogen storage and myofibrillar hypertrophy). In order to fully distingue the effects of RT on muscle mass from expected gains due to growth and maturation, we may need to design more advanced training protocols, observe longer training periods, and develop sensitive measuring techniques that are ethically appropriate for youth (Faigenbaum et al. 2019a). Previous data indicated that gains induced in children by RT related with increases in motor unit activation and changes in motor unit coordination, recruitment, and firing frequency or decrease in electromechanical delay, improvements in stiffness properties (Granacher et al. 2011; Ozmun et al. 1994).

The relationship between growth/maturation variables and physical performance is complicated, since at times the expected effects of detraining (for example loss of muscle strength) could be masked by the gains in muscle strength due to growth and maturation (Faigenbaum et al. 1996). Available data indicates that training-induced gains in muscular strength and power during childhood and adolescents are impermanent and that gains tend to regress toward untrained control-group values during a detraining period (Faigenbaum et al. 1996; Faigenbaum et al. 2013a). Another study found that an integrative neuromuscular training program (INT), a training model that incorporates general and specific strength and conditioning activities, improved the countermovement vertical jump height in young volleyball players, and that these gains were maintained during the 12 week detraining period (Nunes et al. 2019). However, the precise physiological mechanism responsible for these observations during detraining in childhood and adolescents remain uncertain, but possibilities include changes in neuromuscular functioning and motor coordination.

2.4 *Fundamental Movement Skills and Sport Performance*

Developing appropriate levels of motor competence may be a critical antecedent for promoting PA and other positive health trajectories across the lifespan (Robinson et al. 2015; Stodden et al. 2008). Various measures of gross motor skills have been used for decades. In children and adolescents, this research has focused primarily on 'traditional' FMS, which has been commonly divided into categories including locomotor (e.g., running, hopping), object control (e.g., catching, kicking), and stability (e.g., balance, body roll) skills. New insights into the design of long-term physical development plans highlighted the importance of improving muscular strength and motor skills performance early in life in order to give children to best possible chance of participating persistently and successfully in exercise and sport activities (Behm et al. 2017; Lloyd et al. 2016b; Granacher et al. 2016). Meta-analytical and recent findings indicate that RT is an effective method for enhancing general motor skills, such as jumping, running, throwing, etc., during childhood and adolescents (Menezes et al. 2020; Behringer et al. 2011). In this line, a new recent model (Hulteen et al. 2018) extends previous motor development work by (Biddle et al. 2011) broadening the classification of movement skills important for PA participation, referred to as 'foundational movement skills' rather than fundamental movement skills. 'Foundational' refers to 'an underlying base or support', maximizing opportunities for participation in PA. The new model involved activities such as RT, which require competency in specific coordinative movement patterns (e.g., bodyweight squat, push-up or lunge) that do not easily fit into the traditional classifications of FMS. Thus, skill development across time should be viewed in the context of how it leads to skillful performance and in terms of how various movement forms support and maintain a lifetime of physical activity.

Additionally, previous evidence indicated that regular participation in periodized program of RT improved several performance variables in a variety of sports, such as soccer swimming, rhythmic gymnastics, runners, badminton, basketball, etc. (Faigenbaum et al. 2019a). Thus, RT has been proven to be an effective strategy for enhancing motor skills performance in youth.

2.5 *Reduction and Risk of Injuries*

Although the total elimination of sport-related and physical activity-related injuries is an unrealistic goal, multifaceted training programmes that include general and specific strength and conditioning activities may help to reduce the likelihood of injuries in youth (Lloyd et al. 2014). A compelling body of evidence, including position statements from the International Olympic Committee (Bergeron et al. 2015) and the National Strength and Conditioning Association (Lloyd et al. 2016a), support regular participation in RT as a means of reducing sports-related injury risk among youth.

Young athletes and nonathletes alike require a basal level of and power to build a strong foundation for a sustainable active lifestyle (Faigenbaum et al. 2016). In the meantime, the number of youth reporting overuse-type injuries has become a growing concern for sports health professionals, a consequence of overexposure to high volume and intensity of sport-specific training without adequate rest (Jayanthi et al. 2015). Stronger youth become more efficient "movers" and will likely move with more confidence and competence in their physical abilities, resulting in enhanced physical literacy (Zwolski et al. 2017; Faigenbaum and Rebullido 2018). RT among youth aged 6–18 years elicits improvements in muscular strength, power, running, speed, kicking velocity, endurance, dynamic balance, flexibility, and general motor performance and these gains make young athletes more resistant to sports-related injuries (Zwolski et al. 2017). A previous meta-analysis (Lauersen et al. 2014) of children and adolescent athletes indicated that RT reduces sports-related injuries, both overuse and acute, by up to 66%. Moreover, and beyond the RT benefits on motor skill development, the enhancement of strength can address ensuing muscular imbalances that may heighten risk of overuse or overexposure injury in the future (Lloyd et al. 2015a). Another meta-analysis on the effects of preventative neuromuscular training intervention on ACL injury risk in young female athletes named "strengthening," in addition to proximal control exercises and multi exercise genres, as one of the most efficacious injury prevention interventions for this at-risk population (Sugimoto et al. 2015).

In discussion of injuries from RT itself, it is critical to put this into the context of the sports these young individuals are participating in (Myers et al. 2017). With effective supervision and training, as well as a properly designed lifting program, the rates of injury are quite low. A recent review article (Myers et al. 2017) indicated that in a fully examining the risks of RT it is quite apparent that, though some risk of injury does exist, this is comparable to that of sports these children are already participating in. Furthermore, all injury risk can be minimized with effective training program development, qualified supervision, and technique driven progression (Faigenbaum and Myer 2010).

Finally, practitioners working with young athletes should consider known risk factors, such as youths' movement ability, low levels of PA and poor muscular strength, in order to implement effective preventive strategies, including the selection of appropriate assessment tools to aid in the identification and training of players at a heightened risk of injury (Read et al. 2018).

3 From Practice

3.1 General Recommendations Before Start Resistance Training

Although factors such as heredity, training experience, and health habits (e.g., nutrition and sleep) will influence the rate and magnitude of adaptation that occurs, seven fundamental principles that determine the effectiveness of youth RT programs are the principles of (a) the Progression, (b) Regularity, (c) Overload, (d) Creativity, (e) Enjoyment, (f) Socialization, and (g) Supervision. These basic principles can be remembered as the PROCESS of youth resistance training (Faigenbaum and McFarland 2016). When working with children and adolescents it is important to remember that the goal of the program should not be limited to increasing muscular strength. Improving motor skills, fostering new social networks, and promoting healthy behaviors in a supportive environment are equally important (Faigenbaum and McFarland 2016).

Also, when designing and implementing youth fitness programs, we must consider the relationship between participant cognitive, physical development and training experience. Regarding cognitive age, a child must have the intellectual capacity to comprehend instructions and listen to constructive feedback about the performance of a skill or adherence to program guidelines. For instance, there is no evidence that indicates a minimum age for participation in a RT program as INT. However, it is generally agreed that participants must be able to follow coaching instructions and be able to handle the attention demands of a training program. Several data indicated that a child that is deemed ready for structured sports participation (about age 7 or 8 years) would typically be ready for some type of RT (Myer et al. 2013). However, a recent clinical report noted that most children begin participating in sports activities about 5–7 years of age (Stricker et al. 2020), and it is reasonable that they can also begin some type of RT with, for example, body weight movements.

Children and adolescents can improve their PF and enhance their athletic abilities while working toward a common goal in a supportive environment. Practitioners who model appropriate behaviors and develop a teaching philosophy are better prepared to designs exercises programs and teach the technical aspects of complex exercises to the young people. The most successful youth fitness specialist are effective teachers who understand the fundamentals principles of pediatric exercise science and developmental physiology. Table 1 represent pedagogy strategies to successful RT programs, following an adaptations of the Youth Coach Dozen proposed by (Faigenbaum et al. 2019a), where authors identified several rules to carried out a successful fitness program.

3.2 *Resistance Training Program Variables*

Youth fitness specialist need to determine individual goals and establish realistic performance standards that are appropriate and meaningful. Proper prescription is particularly important in for untrained youth, who may overestimate their physical abilities. The program variables that should be considered when designing resistance RT programs include choice and order of exercises, training intensity and volume load, rest intervals between sets and exercises, repetition velocity, and training frequency (Lloyd et al. 2014). Table 2 presents a general example of these training variables prescribed to beginner, intermediate and advances youth lifters.

Table 1 Pedagogy strategies to successful RT programs

	Pedagogy strategy	Explanation	RT application
1	Safe exercise environment	Spacious, uncluttered, well ventilated and lit area. Order and locations of exercise equipment's, appropriately dress	Materials ordered ensuring safety (bars, weight plates, accessory materials, etc.), properly padded floors or strength machines adapted to young people
2	Stay connected	Leadership of instructors who stay connected to the participants during the program	To establish intrasession strategies to maintain participants' attention (games, auditory signals, attentional strategies such as timers)
3	Be enthusiastic	Positive and passionate about exercise, fitness and sports, positive learning environment, motivations for engaging in the activities	Role model for youth It is beneficial for young people that the coach demonstrate some exercises
4	Foster creativity	Imagination, collaborate with peers, creative energy, stimulating and fun exercises, etc.	Warm-up in pairs, small groups or large groups based on fun and effective games to develop and reinforce motor skills and muscle strength
5	Understand the process	Engage participants in activities that are both physically and mentally challenging	To set challenges within the exercises themselves or mini weekly or group team challenges
6	Deliver clear instructions	Appropriate communication (tone, pronunciation., etc.) by fitness professionals, use analogies, demonstrations and coaching cues	It is advisable using several feedbacks (verbal, visual…), including "scaffolding technique", finally limiting the number of feedbacks to improve the individual motor development process
7	Diversify the portfolio	Variety of fun and challenging skills and activities	It is recommended to include activities that incorporate various training modes, body movements, as well as different strategies that develop muscular fitness
8	Learn for mistakes	Recognize mistakes as valuable parts of the learning process that provide interesting opportunities to the participants	An interesting strategy is to perform a movement self-assessment, or perform the movement assessment in pairs limiting the number of feedbacks to improve the individual motor development process
9	Be patient	Specialist need to be patient in the training practice and progressions of the youth programs	It is recommended to use an observation sheet of adherence to RT training in order to know the level and the youth's progression process

(continued)

Table 1 (continued)

	Pedagogy strategy	Explanation	RT application
10	Maximize recovery	Recovery schedule well planned and well balanced	It is recommended to establish at least 24 h of rest between 2 RT sessions. Also, it is interesting to establish complementary recommendations to improve the quality of sleep and diet
11	Think long-term	Long term approach to optimize training adaptations and enhance the development of the youth	A supplemental youth athlete injury prevention plan can be incorporated into general strength training
12	Enjoy the game	Fun and learning skills during an exercise session to facilities sustainment of desired behaviors	It is important to include new exercises, which also encourage their creativity and create a challenge for young people

Adaptation of the "Youth Coach Dozen" proposed by (Faigenbaum et al. 2019a)

Table 2 General example of the training variables prescribed to beginner, intermediate and advances youth lifters

Training status	Exercise recommendation	Sets	Repetitions	Load	Repetition velocity	Rest intervals	Frequency
Beginner	Introductory period with initial exposure to RT guidelines, safety and basic movements	1–2	Varied	<60%	Controlled	1–2 min	1–2
Basic	Basic movements throughout the range of motion, body weight training and exercises with light loads	1–2	8–12	60–70% 1RM	Controlled	1–2 min	2–3
Intermediate	RT with free weight and weigh machines	2–3	4–8	70–80% 1RM	Medium	<1 min	3–4
Athlete	Advanced RT movements and olympic lifts	3–5	2–8	80–95% 1RM	Fast	30 sec–1 min	4–5

3.2.1 Exercise Choice

Some of the resistance modes available to those prescribing youth RT programmes include bodyweight, weight machines, free weights (i.e., barbells and dumbbells), elastic resistance bands, medicine balls or even manual RT; all of which have been proven to elicit physiological adaptation and/or performance enhancement when used in youth RT programmes (Lloyd et al. 2014). The selection of the resistance modality will largely depend on the technical ability and baseline fitness levels of the individual, the level of coaching expertise, the overall goal of the training programme and the availability of equipment. Once the child can demonstrate appropriate technical competency, they can be introduced to more advanced exercises that challenge the child in terms of coordination and require greater levels and rates of force production (Faigenbaum et al. 2019a). Multijoint exercises with free weighs (e.g., lunges) and dynamic whole-body movements (e.g., plyometrics) are particularly beneficial for enhancing motor skill performance and muscle power in youth (Ramírez-Campillo et al. 2014; Chaouachi et al. 2014).

In the case of weightlifting exercises, which by their nature are more complex movements, researchers have previously suggested that early exposure should focus on technical development using modified equipment and light external loads to aid in motor control development (Faigenbaum et al. 2009b).

3.2.2 Training Volume and Intensity

Volume refers to the total number of times an exercise is performed within a training session multiplied by the resistance used (kg), whereas intensity is commonly referring to the resistance that is required to overcome during a repetition (Lloyd et al. 2014). Both training intensity and volume load significantly influence the outcome of RT programs in youth. In order to reduce unexpected events, participants must first learn how to perform each exercise correctly with a light load and then gradually progress the training intensity and volume load (Lloyd et al. 2014; Lesinski et al. 2016).

With the aim of reducing the risk of injury while training at an appropriate RT intensity, coaches should prescribe the appropriate training intensity, i.e., a percentage of an individuals' one repetition-maximum (1RM) (Lloyd et al. 2014). In this sense, previous research indicates that maximal strength and power testing of children (Faigenbaum et al. 2003) and adolescents (Faigenbaum et al. 2012) is safe and reliable when standardized protocols are implemented and monitored by qualified professionals, although predictive equations that estimate 1RM values (Horvat et al. 2007; Kravitz et al. 2003) or RPE scales (Robertson et al. 2005) could also be used in young population.

Based on the international consensus statement of 2014 (Lloyd et al. 2014), youth should begin RT with one or two sets of light intensity (<60% 1RM) on 6–10 exercises, prescribing basic exercises and light training loads for beginners, performing a between 6 and 12 repetitions per set. In order to optimize strength gains, youth RT programs should progress to two to three sets of 8–15 repetitions with loads up to 80% of 1RM (Behringer et al. 2010). Finally, if the goal is to improve the athletic performance, then the dose-response relationship to RT could include five sets of 6 to 8 repetitions at a training intensity of 80–89% of 1RM (Lesinski et al. 2016). In addition, the ability to perform controlled movements at a high velocity is crucial during childhood, when neural plasticity and motor coordination are most sensitive to change, although a moderate training velocity may be recommended for untrained youth.

3.2.3 Rest Interval Between Sets and Exercises

Research indicates that children can recover more quickly from fatigue-inducing RT (Faigenbaum et al. 2008; Zafeiridis et al. 2005), and are less likely to suffer muscle damage following this form of exercise, owing to the increased pliability of their muscle tissue (Eston et al. 2003). Rest periods of approximately 1 min should suffice for most children when performing a moderate to-intensity RT program. However, this may need to be increased to 2–3 min as the intensity of training increases, especially if the exercises require high levels of skill, force or power production (e.g., weightlifting or plyometric exercises). Within session RT performance should be monitored by youth fitness specialists who provide real-time feedback to ensure the proper exercise technique.

3.2.4 Frequency

Training frequency typically refers to the number of sessions performed within a week. Previous research has indicated that 2–3 sessions per week on non-consecutive days is most appropriate in order to develop muscular strength levels in children and adolescents (Faigenbaum et al. 2009b). Behringer and colleagues (2010) substantiated these recommendations, indicating that across 42 studies (where mean training frequency was 2.7 ± 0.8 sessions/week), training frequency was significantly correlated with RT effects. However, training frequency may increase as children go through adolescence and approach adulthood, especially for youth in competitive sport. In that case, youth should be monitored closely for early signs of stress intolerance, including sore muscles and changes in motivation (Faigenbaum et al. 2019a).

4 Filling Gaps

4.1 Concerns and Misconceptions About Youth Resistance Training

Although compelling body of research indicated that supervised resistance exercises can be a safe and effective protocol of training for children and adolescents

(Granacher et al. 2016; Faigenbaum and Myer 2010; Stricker et al. 2020; Behringer et al. 2010), traditional concerns and misconceptions persist (McGladrey et al. 2014; Fröberg et al. 2014). One of the most usual concerns associated with youth resistance training (hereafter referred to as RT) relates to the potential for injury to the physis (area of cartilage between metaphysis and epiphysis) or growth plate located near the ends of long bones, due to bone in youth has not yet ossified and is susceptible of injury. While case studies of injuries were reported in the literature in the 1970s and 1980s, most of the reported injuries were due to inappropriate loads or unsupervised RT (Jenkins and Mintowt-Czyz 1986; Gumbs et al. 1982). If participants are taught how to properly perform RT and if the progression is based on RT skill competency, then the risk of injury to the growth plate is minimal. Notably, the sport-specific forces placed on the developing skeleton of young athletes, who regularly participates in sports are likely to be greater both in magnitude and duration than those resulting from RT (McNitt-Gray et al. 2001; Malina et al. 2013).

Another common misconception is that gains cannot be induced by RT during childhood because children have insufficient levels of circulating androgens. However, untrained youth who participate in well-designed programs can generally be expected to experience gains in muscular strength of about 30–40% following the first two or three months of RT (Lloyd et al. 2014; Faigenbaum et al. 2009a). Moreover, these training-induced adaptations will likely set the stage for even greater gains in strength and power later in life (Faigenbaum et al. 2013b). For instance, previous results from tracking analysis showed that muscular fitness phenotypes tracked moderately between youth and young adulthood. So, it seems that youth with low muscular fitness are at increased risk of maintaining a low muscular fitness level into adulthood (Fraser et al. 2017).

Additionally, there is a traditional belief among some youth (especially adolescent females) that RT could dramatically increase muscle size (hypertrophy). However, children and females lack of adequate levels of circulating androgens to stimulate large increases in muscle size consequent to RT, and there is scientific evidence describing the neuromuscular mechanisms (e.g., changes in motor unit coordination, recruitment and firing frequency) that are primarily responsible for RT induced gains in muscle strength in youth (Granacher et al. 2011; Ozmun et al. 1994).

4.2 Resistance Training in the Long-Term Athlete's Development Model

A popular athlete development model in which sensitive periods were proposed is the long-term athlete development (LTAD) model by Balyi and Way (2005) which was updated in 2013 (Balyi et al. 2013). Basically, the authors simplified the physical attributes of sports into 5 general motor abilities of suppleness (flexibility), speed, skills, stamina (endurance), and strength, and proposed sensitive periods based on biological and chronological age for boys and girls. For instance, in the original model the ages 15–16 years in boys and 14–15 years in girls respectively, was proposed as sensitive period to train strength, in both cases after peak height velocity (PHV, the phase in which growth is fastest). Thus, it is assumed that training of speed or aerobic capacity outside these sensitive periods could result in adaptations that are smaller in magnitude and therefore has a reduced effect on performance. The LTAD model states that strength is always trainable but recommends the optimal "window of trainability" for boys is 12–18 months following peak height velocity, while for girls it is immediately after peak height velocity or at the onset of the menarche (Balyi and Hamilton 2004). However, the original model supporting the LTAD model's optimal "window of trainability" for strength is speculative and there is a lack of empirical evidence upon which the model is based (Ford et al. 2011).

A recent study (Van Hooren and Croix 2020) has critically evaluated the rationale of the widely adopted generic sensitive periods for youth training. This study indicated that the LTAD model by Balyi et al. (2013) may have positive influences on sports practice such as creating awareness on the risks of early specialization, awareness on biological rather than chronological age, and a focus on long-term success rather than short-term success, although the lack of validation of the sensitive periods has been reported as a barrier to implement the model by coaches. Thus, athlete development models and practitioners should therefore not rely on generic sensitive periods to train youth athletes. Despite the general acceptance of the LTAD model by sporting associations, recent criticisms from the academic fields have questioned its rigid view of athletic development and the fact that the model lacks any real empirical evidence (Lloyd et al. 2015b).

More recently, researchers created the Youth Physical Development (YPD) model, which used existing empirical research from the development of individual components of fitness to establish an overall long-term strategy for physical development across childhood and adolescence (Lloyd et al. 2016a). The introduction of the YPD model moved away from "athlete-centered" terminology to place emphasis on the long-term development of physical abilities, mainly focus on muscular strength and movement competency for both children and adolescents. According to the approach presented by this model, although the development of athleticism has traditionally been viewed as a goal for aspiring "young athletes," it is crucial that strength and conditioning coaches, personal trainers, teachers, parents, and medical professionals adopt a systematic approach to long-term athletic development for youth of all ages, abilities, and aspirations. Figure 1 presents the 10 pillars of successful long-term athletic, which summarize the key recommendations detailed within the position statement proposed by Lloyd et al. (2016a). The two keys of this model are qualified instruction and training early in life with regular opportunities for structured and unstructured/free play activities that are fun and engaging, both represent with different colors in the figure. Finally, this study concluded indicating that a better understanding of the training process in youth, the manner in which training interacts with growth and maturation, and how long-term approaches to athletic development influence physical performance, health and well-being, and injury risk are key areas that require further study.

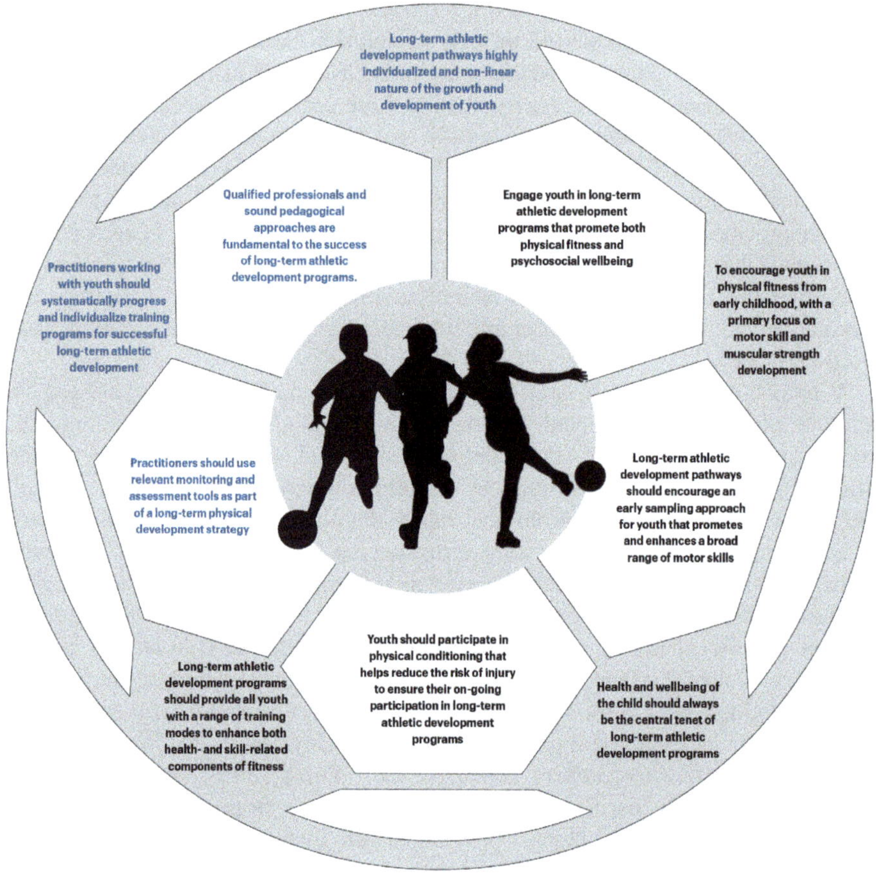

Fig. 1 10 pillars of successful long-term athletic development. Blue color letter represents "qualified instruction", whereas black color represents "training early in life with regular opportunities for structured and unstructured/free play activities that are fun and engaging", both as global pillars of LTAD (Lloyd et al. 2015b)

5 Take-Home Messages and Practical Resources

1. Supervised resistance exercises can be a safe and effective protocol of training for children and adolescents.
2. There is a considerable variation in growth, maturity, and performance status among individuals that should be considered in youth RT programs.
3. Training early in life with regular opportunities for structured and unstructured/free play activities that are fun. Further, it is important to include new exercises, which also encourage their creativity and create a challenge for young people.
4. It is recommended to include activities that incorporate various training modes, body movements, as well as different strategies that develop muscular fitness.

5. Engage youth in long-term athletic development programs that promote both physical fitness and phychological wellbeing.
6. Practioners should use relevant monitoring and assessment tools as part of a long-term physical development strategy.
7. A supplemental youth athlete injury prevention plan can be incorporated into general strength training.

References

Balyi I, Hamilton A (2004) Long-term athlete development: trainability in childhood and adolescence. Olympic Coach 16(1):4–9

Balyi I, Way R (2005) The role of monitoring growth in long-term athlete development. Can Sport Life 8–10

Balyi I, Way R, Higgs C (2013) Long-term athlete development. Human Kinetics

Barnett LM, Van Beurden E, Morgan PJ, Brooks LO, Beard JR (2009) Childhood motor skill proficiency as a predictor of adolescent physical activity. J Adolesc Heal 44(3):252–259

Barnett L, Cliff K, Morgan P, van Beurden E (2013) Adolescents' perception of the relationship between movement skills, physical activity and sport. Eur Phys Educ Rev 19(2):271–285

Behm DG, Young JD, Whitten JHD, Reid JC, Quigley PJ, Low J et al (2017) Effectiveness of traditional strength vs. power training on muscle strength, power and speed with youth: a systematic review and meta-analysis. Front Physiol 8:423

Behringer M, vom Heede A, Yue Z, Mester J (2010) Effects of resistance training in children and adolescents: a meta-analysis. Pediatrics 126(5):e1199–e1210

Behringer M, vom Heede A, Matthews M, Mester J (2011) Effects of strength training on motor performance skills in children and adolescents: a meta-analysis. Pediatr Exerc Sci 23(2):186–206

Behringer M, Gruetzner S, McCourt M, Mester J (2014) Effects of weight-bearing activities on bone mineral content and density in children and adolescents: a meta-analysis. J Bone Miner Res 29(2):467–478

Bergeron MF, Mountjoy M, Armstrong N, Chia M, Côté J, Emery CA et al (2015) International Olympic Committee consensus statement on youth athletic development. Br J Sports Med 49 (13):843–851

Beunen G, Malina RM (2008) Growth and biologic maturation: relevance to athletic performance. young athlete Malden, pp 3–17. MA Blackwell Publ

Biddle SJH, Gorely T, Pearson N, Bull FC (2011) An assessment of self-reported physical activity instruments in young people for population surveillance: project ALPHA. Int J Behav Nutr Phys Act 8(1):1

Bompa T, Buzzichelli C (2015) Periodization training for sports, 3e. Human kinetics

Brian A, Getchell N, True L, De Meester A, Stodden DF (2020) Reconceptualizing and operationalizing seefeldt's proficiency barrier: applications and future directions. Sport Med 1–12

Chaouachi A, Hammami R, Kaabi S, Chamari K, Drinkwater EJ, Behm DG (2014) Olympic weightlifting and plyometric training with children provides similar or greater performance improvements than traditional resistance training. J Strength Cond Res 28(6):1483–1496

Chulvi-Medrano I, Pombo M, Saavedra-García MÁ, Rebullido TR, Faigenbaum AD (2020) A 47-year comparison of lower body muscular power in spanish boys: a short report. J Funct Morphol Kinesiol. 5(3):64

Collins H, Fawkner S, Booth JN, Duncan A (2018) The effect of resistance training interventions on weight status in youth: a meta-analysis. Sport Med 4(1):41

Collins H, Booth JN, Duncan A, Fawkner S, Niven A (2019) The effect of resistance training interventions on 'the self'in youth: a systematic review and meta-analysis. Sport Med 5(1):29

Eberhardt T, Niessner C, Oriwol D, Buchal L, Worth A, Bös K (2020) Secular trends in physical fitness of children and adolescents: a review of large-scale epidemiological studies published after 2006. Int J Environ Res Public Health 17(16):5671

Ekelund U, Tomkinson G, Armstrong N (2011) What proportion of youth are physically active? Measurement issues, levels and recent time trends. Br J Sports Med 45(11):859–865

Eston R, Byrne C, Twist C (2003) Muscle function after exercise-induced muscle damage: considerations for athletic performance in children and adults. J Exerc Sci Fit 1(2):85–96

Faigenbaum AD, Myer GD (2010) Resistance training among young athletes: safety, efficacy and injury prevention effects. Br J Sports Med 44(1):56–63

Faigenbaum AD, McFarland JE (2016) Resistance training for kids: right from the start. Acsms Health Fit J 20(5):16–22

Faigenbaum AD, Rebullido TR (2018) Understanding physical literacy in youth. Strength Cond J 40(6):90–94

Faigenbaum AD, Westcott WL, Micheli LJ, Outerbridge AR, Long CJ, LaRosa-Loud R et al (1996) The effects of strength training and detraining on children. J Strength Cond Res 10:109–114

Faigenbaum AD, Milliken LA, Westcott WL (2003) Maximal strength testing in healthy children. J Strength Cond Res 17(1):162–166

Faigenbaum AD, Ratamess NA, McFarland J, Kaczmarek J, Coraggio MJ, Kang J et al (2008) Effect of rest interval length on bench press performance in boys, teens, and men. Pediatr Exerc Sci 20(4):457–469

Faigenbaum AD, Farrell AC, Radler T, Zbojovsky D, Chu DA, Ratamess NA et al (2009a) "Plyo Play": a novel program of short bouts of moderate and high intensity exercise improves physical fitness in elementary school children. Phys Educ 66(1)

Faigenbaum AD, Kraemer WJ, Blimkie CJR, Jeffreys I, Micheli LJ, Nitka M et al (2009b) Youth resistance training: updated position statement paper from the national strength and conditioning association. J Strength Cond Res 23:S60-79

Faigenbaum AD, McFarland JE, Herman R, Naclerio F, Ratamess NA, Kang J et al (2012) Reliability of the one repetition-maximum power clean test in adolescent athletes. J Strength Cond Res Strength Cond Assoc 26(2):432

Faigenbaum AD, Farrell AC, Fabiano M, Radler TA, Naclerio F, Ratamess NA et al (2013a) Effects of detraining on fitness performance in 7-year-old children. J Strength Cond Res 27(2):323–330

Faigenbaum AD, Lloyd RS, Myer GD (2013b) Youth resistance training: past practices, new perspectives, and future directions. Pediatr Exerc Sci 25(4):591–604

Faigenbaum AD, Lloyd RS, MacDonald J, Myer GD (2016) Citius, Altius, Fortius: beneficial effects of resistance training for young athletes: narrative review. Br J Sports Med 50(1):3–7

Faigenbaum AD, Lloyd RS, Oliver JL (2019a) Essentials of youth fitness. Human Kinetics Publishers

Faigenbaum AD, Rebullido TR, Peña J, Chulvi-Medrano I (2019b) Resistance exercise for the prevention and treatment of pediatric dynapenia. J Sci Sport Exerc 1–9

Ford P, De Ste CM, Lloyd R, Meyers R, Moosavi M, Oliver J et al (2011) The long-term athlete development model: physiological evidence and application. J Sports Sci 29(4):389–402

Fraser BJ, Huynh QL, Schmidt MD, Dwyer T, Venn AJ, Magnussen CG (2016) Childhood muscular fitness phenotypes and adult metabolic syndrome. Med Sci Sports Exerc 48(9)

Fraser BJ, Schmidt MD, Huynh QL, Dwyer T, Venn AJ, Magnussen CG (2017) Tracking of muscular strength and power from youth to young adulthood: longitudinal findings from the childhood determinants of adult health study. J Sci Med Sport 20(10):927–931

Fraser BJ, Blizzard L, Schmidt MD, Dwyer T, Venn AJ, Magnussen CG (2019) The association between muscular power from childhood to adulthood and adult measures of glucose homeostasis. Scand J Med Sci Sports 29(12):1909–1916

Fröberg A, Alricsson M, Ahnesjö J (2014) Awareness of current recommendations and guidelines regarding strength training for youth. Int J Adolesc Med Health 26(4):517–523

García-Hermoso A, Cavero-Redondo I, Ramírez-Vélez R, Ruiz JR, Ortega FB, Lee D-C et al (2018a) Muscular strength as a predictor of all-cause mortality in an apparently healthy population: a systematic review and meta-analysis of data from approximately 2 million men and women. Arch Phys Med Rehabil 99(10):2100–2113

García-Hermoso A, Ramírez-Vélez R, Ramírez-Campillo R, Peterson MD, Martínez-Vizcaíno V (2018b) Concurrent aerobic plus resistance exercise versus aerobic exercise alone to improve health outcomes in paediatric obesity: a systematic review and meta-analysis. Br J Sports Med 52(3):161–166

García-Hermoso A, Ramírez-Campillo R, Izquierdo M (2019) Is muscular fitness associated with future health benefits in children and adolescents? A systematic review and meta-analysis of longitudinal studies. Sport Med 49(7):1079–1094

Granacher U, Goesele A, Roggo K, Wischer T, Fischer S, Zuerny C et al (2011) Effects and mechanisms of strength training in children. Int J Sports Med 32(05):357–364

Granacher U, Lesinski M, Büsch D, Muehlbauer T, Prieske O, Puta C et al (2016) Effects of resistance training in youth athletes on muscular fitness and athletic performance: a conceptual model for long-term athlete development. Front Physiol 7:164

Gumbs VL, Segal D, Halligan JB, Lower G (1982) Bilateral distal radius and ulnar fractures in adolescent weight lifters. Am J Sports Med 10(6):375–379

Guthold R, Cowan MJ, Autenrieth CS, Kann L, Riley LM (2010) Physical activity and sedentary behavior among schoolchildren: a 34-country comparison. J Pediatr 157(1):43–49

Holfelder B, Schott N (2014) Relationship of fundamental movement skills and physical activity in children and adolescents: a systematic review. Psychol Sport Exerc 15(4):382–391

Horvat M, Franklin C, Born D (2007) Predicting strength in high school women athletes. J Strength Cond Res 21(4):1018

Hulteen RM, Morgan PJ, Barnett LM, Stodden DF, Lubans DR (2018) Development of foundational movement skills: a conceptual model for physical activity across the lifespan. Sport Med 48(7):1533–1540

Ishikawa S, Kim Y, Kang M, Morgan DW (2013) Effects of weight-bearing exercise on bone health in girls: a meta-analysis. Sport Med 43(9):875–892

Jayanthi NA, LaBella CR, Fischer D, Pasulka J, Dugas LR (2015) Sports-specialized intensive training and the risk of injury in young athletes: a clinical case-control study. Am J Sports Med 43(4):794–801

Jenkins NH, Mintowt-Czyz WJ (1986) Bilateral fracture-separations of the distal radial epiphyses during weight-lifting. Br J Sports Med 20(2):72

Kaster T, Dooley FL, Fitzgerald JS, Walch TJ, Annandale M, Ferrar K et al (2020) Temporal trends in the sit-ups performance of 9,939,289 children and adolescents between 1964 and 2017. J Sports Sci 1–11

Kraemer WJ, Fry AC, Frykman PN, Conroy B, Hoffman J (1989) Resistance training and youth. Pediatr Exerc Sci 1(4):336–350

Kravitz LEN, Akalan C, Nowicki K, Kinzey SJ (2003) Prediction of 1 repetition maximum in high-school power lifters. J Strength Cond Res 17(1):167–172

Lau PWC, Kong Z, Choi C, Clare CW, Chan DFY, Sung RYT et al (2010) Effects of short-term resistance training on serum leptin levels in obese adolescents. J Exerc Sci Fit 8(1):54–60

Lauersen JB, Bertelsen DM, Andersen LB (2014) The effectiveness of exercise interventions to prevent sports injuries: a systematic review and meta-analysis of randomised controlled trials. Br J Sports Med 48(11):871–877

Lesinski M, Prieske O, Granacher U (2016) Effects and dose-response relationships of resistance training on physical performance in youth athletes: a systematic review and meta-analysis. Br J Sports Med 50(13):781–795

Lloyd RS, Faigenbaum AD, Myer GD, Stone M, Oliver J, Jeffreys I et al (2012) UKSCA position statement: youth resistance training. Prof Strength Cond 26:26–39

Lloyd RS, Faigenbaum AD, Stone MH, Oliver JL, Jeffreys I, Moody JA et al (2014) Position statement on youth resistance training: the 2014 international consensus. Br J Sports Med 48 (7):498–505

Lloyd RS, Oliver JL, Faigenbaum AD, Howard R, Croix MBADS, Williams CA, et al (2015a) Long-term athletic development, part 2: barriers to success and potential solutions. J Strength Cond Res 29(5):1451–64

Lloyd RS, Oliver JL, Faigenbaum AD, Howard R, Croix MBADS, Williams CA et al (2015b) Long-term athletic development-part 1: a pathway for all youth. J Strength Cond Res 29 (5):1439–1450

Lloyd RS, Cronin JB, Faigenbaum AD, Haff GG, Howard R, Kraemer WJ et al (2016a) National Strength and Conditioning Association position statement on long-term athletic development. J Strength Cond Res 30(6):1491–1509

Lloyd RS, Radnor JM, Croix MBADS, Cronin JB, Oliver JL (2016b) Changes in sprint and jump performances after traditional, plyometric, and combined resistance training in male youth pre-and post-peak height velocity. J Strength Cond Res 30(5):1239–1247

Logan SW, Webster EK, Getchell N, Pfeiffer KA, Robinson LE (2015) Relationship between fundamental motor skill competence and physical activity during childhood and adolescence: a systematic review. Kinesiol Rev 4(4):416–426

Lopes VP, Rodrigues LP, Maia JAR, Malina RM (2011) Motor coordination as predictor of physical activity in childhood. Scand J Med Sci Sports 21(5):663–669

Malina RM, Bouchard C, Bar-Or O (2004) Growth, maturation, and physical activity. Human kinetics

Malina RM, Baxter-Jones ADG, Armstrong N, Beunen GP, Caine D, Daly RM et al (2013) Role of intensive training in the growth and maturation of artistic gymnasts. Sport Med 43(9):783–802

Martinez-Gomez D, Ruiz JR, Ortega FB, Veiga OL, Moliner-Urdiales D, Mauro B et al (2010) Recommended levels of physical activity to avoid an excess of body fat in European adolescents: the HELENA Study. Am J Prev Med 39(3):203–211

McGladrey BW, Hannon JC, Faigenbaum AD, Shultz BB, Shaw JM (2014) High school physical educators' and sport coaches' knowledge of resistance training principles and methods. J Strength Cond Res 28(5):1433–1442

McHugh MP (2010) Oversized young athletes: a weighty concern. Br J Sports Med 44(1):45–49

McNitt-Gray JL, Hester DME, Mathiyakom W, Munkasy BA (2001) Mechanical demand and multijoint control during landing depend on orientation of the body segments relative to the reaction force. J Biomech 34(11):1471–1482

Menezes GB, Alexandre DRO, Pinto JCBL, Assis TVL, Faigenbaum AD, Mortatti AL (2020) Effects of integrative neuromuscular training on motor performance in prepubertal soccer players. J Strength Cond Res

Moliner-Urdiales D, Ruiz JR, Ortega FB, Jiménez-Pavón D, Vicente-Rodriguez G, Rey-López JP et al (2010) Secular trends in health-related physical fitness in Spanish adolescents: the AVENA and HELENA studies. J Sci Med Sport 13(6):584–588

Myer GD, Lloyd RS, Brent JL, Faigenbaum AD (2013) How young is "too young" to start training? Acsms Health Fit J 17(5):14

Myer GD, Faigenbaum AD, Edwards NM, Clark JF, Best TM, Sallis RE (2015) Sixty minutes of what? A developing brain perspective for activating children with an integrative exercise approach. Br J Sports Med 49(23):1510–1516

Myers AM, Beam NW, Fakhoury JD (2017) Resistance training for children and adolescents. Transl Pediatr 6(3):137

Nunes A, Cattuzzo MT, Faigenbaum AD, Mortatti AL (2019) Effects of integrative neuromuscular training and detraining on countermovement jump performance in youth volleyball players. J Strength Cond Res 10:1–6

Nyberg GA, Nordenfelt AM, Ekelund U, Marcus C (2009) Physical activity patterns measured by accelerometry in 6-to 10-yr-old children. Med Sci Sports Exerc 41(10):1842–1848

Ortega FB, Ruiz JR, Castillo MJ, Sjöström M (2008) Physical fitness in childhood and adolescence: a powerful marker of health. Int J Obes 32(1):1–11

Ottevaere C, Huybrechts I, Benser J, De Bourdeaudhuij I, Cuenca-Garcia M, Dallongeville J et al (2011) Clustering patterns of physical activity, sedentary and dietary behavior among European adolescents: the HELENA study. BMC Public Health 11(1):328

Ozmun J, Mikesky A, Surburg P (1994) Neuromuscular adaptations following prepubescent strength training. Med Sci Sport Exerc. 26(4):510–514

Ramírez-Campillo R, Meylan C, Álvarez C, Henríquez-Olguín C, Martínez C, Cañas-Jamett R et al (2014) Effects of in-season low-volume high-intensity plyometric training on explosive actions and endurance of young soccer players. J Strength Cond Res 28(5):1335–1342

Read PJ, Jimenez P, Oliver JL, Lloyd RS (2018) Injury prevention in male youth soccer: current practices and perceptions of practitioners working at elite English academies. J Sports Sci 36 (12):1423–1431

Robertson RJ, Goss FL, Andreacci JL, Dube JJ, Rutkowski JJ, Frazee KM et al (2005) Validation of the children's OMNI-resistance exercise scale of perceived exertion. Med Sci Sport Exerc 37 (5):819–826

Robinson LE, Stodden DF, Barnett LM, Lopes VP, Logan SW, Rodrigues LP et al (2015) Motor competence and its effect on positive developmental trajectories of health. Sport Med 45 (9):1273–1284

Sandercock GRH, Cohen DD (2019) Temporal trends in muscular fitness of english 10-year-olds 1998–2014: an allometric approach. J Sci Med Sport 22(2):201–205

Smith JJ, Eather N, Morgan PJ, Plotnikoff RC, Faigenbaum AD, Lubans DR (2014) The health benefits of muscular fitness for children and adolescents: a systematic review and meta-analysis. Sport Med 44(9):1209–1223

Smith JJ, DeMarco M, Kennedy SG, Kelson M, Barnett LM, Faigenbaum AD et al (2018) Prevalence and correlates of resistance training skill competence in adolescents. J Sports Sci 36 (11):1241–1249

Smith JJ, Eather N, Weaver RG, Riley N, Beets MW, Lubans DR (2019) Behavioral correlates of muscular fitness in children and adolescents: a systematic review. Sport Med 49(6):887–904

Stodden DF, Goodway JD, Langendorfer SJ, Roberton MA, Rudisill ME, Garcia C et al (2008) A developmental perspective on the role of motor skill competence in physical activity: an emergent relationship. Quest 60(2):290–306

Stricker PR, Faigenbaum AD, McCambridge TM (2020) Resistance training for children and adolescents. Pediatrics 145(6)

Sugimoto D, Myer GD, Foss KDB, Hewett TE (2015) Specific exercise effects of preventive neuromuscular training intervention on anterior cruciate ligament injury risk reduction in young females: meta-analysis and subgroup analysis. Br J Sports Med 49(5):282–289

Suh S, Jeong I-K, Kim MY, Kim YS, Shin S, Kim SS et al (2011) Effects of resistance training and aerobic exercise on insulin sensitivity in overweight Korean adolescents: a controlled randomized trial. Diabetes Metab J 35(4):418–426

Van Der Heijden G, Wang ZJ, Chu Z, Toffolo G, Manesso E, Sauer PJJ et al (2010) Strength exercise improves muscle mass and hepatic insulin sensitivity in obese youth. Med Sci Sports Exerc 42(11):1973

Van Hooren B, Croix MDS (2020) Sensitive periods to train general motor abilities in children and adolescents: do they exist? A critical appraisal. Strength Cond J

WHO (2010) Global recommendations on physical activity for health, p 60. Geneva World Health Organanisation

Zafeiridis A, Dalamitros A, Dipla K, Manou V, Galanis N, Kellis S (2005) Recovery during high-intensity intermittent anaerobic exercise in boys, teens, and men. Med Sci Sport Exerc 37 (3):505–12

Zwolski C, Quatman-Yates C, Paterno MV (2017) Resistance training in youth: laying the foundation for injury prevention and physical literacy. Sports Health. 9(5):436–443

Emilio Villa is an associate professor in the Department of Physical and Sports Education of the University of Granada (Spain) and has Ph.D. in the Biomedicine Program (University of Granada). He is a specialist in the area of physical exercise in the pediatric population, especially from a public health and preventive medicine approach. He has co-authored over 100 scholarly articles, 17 book chapters, and 3 books. He is a member of the Editorial Board of scientific journal such as BMC Public Health (UK) and Sustainability (Switzerland).

Avery D. Faigenbaum is a full professor in the Department of Health and Exercise Science at the College of New Jersey. He is a recognized leader in the field of pediatric exercise science, with a special emphasis on youth resistance training and preventive medicine. He has co-authored over 250 scholarly articles, 50 book chapters, and 10 books including *Essentials of Youth Fitness*. He is a fellow of the National Strength and Conditioning Association and of the American College of Sports Medicine and was awarded the 2017 Boyd Epley Award for Lifetime Achievement from the National Strength and Conditioning Association.

Resistance Training in Women

Beatriz Bachero-Mena and Susana Moral-González

Abstract Resistance training (RT) is one of the most popular methods of exercise for improving physical fitness. The current interest in RT by women is evidenced by the great number of women who now train and the growth of female training contests. It is well documented that long-term systematic RT causes increased muscle strength and cardiovascular function. Regarding women, evidence has shown other related benefits such as an increased bone mineral density, improvements in maternal health and perinatal outcomes during pregnancy, changes in body composition, improvements in health-related outcomes in old age, and the treatment and risk reduction for multiple chronic diseases. In this chapter, we will cover the importance of RT training in women and its associated increase in general physical fitness and so in quality of life. We will describe the physiological mechanisms related to resistance training in women, and some gender differences. We will also describe the main effects and characteristics of RT programs in women, and focus on the potential benefits of resistance exercise during pregnancy and post-partum. Finally, we will try to provide some recommendations specific to women RT based on current research.

Keywords Resistance exercise · Muscle strength · Physical fitness · Health · Women

1 Introduction

Resistance training (RT) is one of the most popular methods of exercise for improving physical fitness (American College of Sports Medicine 2009). The current interest in RT by women is evidenced by the great number of women who

B. Bachero-Mena (✉)
University of Seville, Sevilla, Spain
e-mail: bbachero@us.es

S. Moral-González
European University of Madrid, Madrid, Spain

A. Muñoz-López et al. (eds.), *Resistance Training Methods*, Lecture Notes in Bioengineering, https://doi.org/10.1007/978-3-030-81989-7_18

now train and the growth of female training contests. It is well documented that long-term systematic RT causes increased skeletal muscle size and strength in both men and women of different ages. Regarding women, RT has shown an evidence bringing many benefits, not only related to an increased muscle strength (Gentil et al. 2016) or cardiovascular function (Kelly et al. 2008) but also regarding to bone mineral density (Xu et al. 2016), improvements in maternal health and perinatal outcomes during pregnancy (Perales et al. 2016), changes in body composition (Paoli et al. 2015), improvements in health-related outcomes in old age (Barbalho et al. 2017) and in breast cancer survivors (Dos Santos et al. 2017), and the treatment and risk reduction for multiple chronic diseases including metabolic syndrome, fibromyalgia, and rheumatoid arthritis (Hurley et al. 2011).

Muscular adaptations to RT are important to manage conditions where muscle weakness may limit function, such as muscle wasting disorders, prolonged bed rest, aging, and rehabilitation. In that sense, low muscle strength has been related to lower functional capacity, and muscle strength has been shown to be inversely associated with risk of mortality and cardiovascular diseases (Timpka et al. 2014).

Human muscle strength reaches its peak between the ages of 20–30 years. However, with increasing age, especially at the onset of the sixth decade, a steeper decline in maximal strength begins in both genders (Häkkinen and Häkkinen 1991). This decrease in maximal strength is related to a great extent to the reduction in muscle mass in both men and women, since ageing is associated with alterations in hormone balance, especially with decreased androgen levels (Häkkinen and Pakarinen 1993), and often also with a decline in the volume of normal physical activities and/or with a decrease in the loads (loading intensity) of these activities. Aging in women is of particular interest due to menopause. In this regard, menopause has shown to bring a physiological decline, mostly due to a reduction in bone mineral density (BMD), which can be attributed to estrogen deficiency (Douchi et al. 2002). These reductions in BMD suppose a big risk for older women to suffer from osteoporosis, and this illness could lead to balance and gait issues, a higher risk of injury, subsequent financial costs, and even a higher risk of mortality (Quirino et al. 2012). More so, a decrease in muscle strength in combination with reduced BMD can further impair balance and mobility, leading to a decline in functional capacity (Karinkanta et al. 2009). Thus, it becomes apparent of the need for RT to attenuate the decline in lean mass, muscle mass, and BMD that accompany aging and inactivity in both men and women, especially in women due to menopause.

In this chapter, we will cover the importance of RT training in women, not only because of its associated increase in general physical fitness and so in quality of life, but also regarding the treatment and risk reduction of mortality and other multiple chronic diseases. We will also describe the main effects and characteristics of RT programs in women, and will focus on the potential benefits of resistance exercise during pregnancy and post-partum. Finally, we will try to provide some recommendations specific to women RT based on current research.

2 From Theory

It is well-known that RT improves muscle size and strength in both males and females (Hubal et al. 2005a). When comparing training-induced changes in muscular strength between sexes, it is important to distinguish between absolute and relative measures. Although it has been shown that absolute hypertrophy and gains in strength are larger in males after RT, it seems that relative increases in both muscular and lower-body strength are similar between the sexes, and relative gains in upper-body strength may be larger in females.

2.1 Physiological Mechanisms Related to Resistance Training in Women. Gender Differences

Sex differences in skeletal muscle mass and distribution are well known (Janssen et al. 2000). Females often present less total and lean body mass, a higher body fat percentage, and a smaller muscle fiber cross-sectional area (Roberts et al. 2018). Also, they have a greater proportion of type I fibers in the vastus lateralis and the biceps brachii (Roberts et al. 2018; Miller et al. 1993). These physiological differences could explain some of the differences observed in strength or hypertrophy gains in men and women, and may potentially influence programs design, and the subsequent adaptations experienced.

In this line, significant differences with respect to maximal strength values can be observed between men and women. Cross-sectional studies have indicated that, in general, women present lower absolute strength values than men, being these differences higher in upper body (strength ratio women/men = 0.56) than in lower body (ratio = 0.72) (Holloway et al. 1990). On the other hand, with respect to relative strength, either such related to body mass or such related to muscle fiber cross-sectional area, the differences between men and women decrease or even disappear in the lower body, while in the upper body these differences remain quite high (Holloway et al. 1990). However, the reasons why men present higher maximal strength values are not clear. It has been suggested that the higher size, weight, percentage of muscle mass, muscle fiber size, differences on fiber distribution (Holloway et al. 1990; Staron et al. 1994) and testosterone concentration (Weiss et al. 1983) of men could explain partially or to a large extent those differences. In fact, it is well established that females have lower levels of testosterone, free-testosterone, and insulin-like growth factor-binding protein 1 (IGF-1) compared with males (Roberts et al. 2018). Testosterone concentrations in women may vary with training and women with relatively high levels of testosterone may have more chance for an increase in muscle size and strength (Weiss et al. 1983).

Concerning the long-term responses to RT, many studies have compared the adaptations of males and females using the same training protocol. Most of the studies have found a greater increase in absolute strength in males than in females

(Dias et al. 2005). However, regarding relative strength, many found that the relative increase in muscle strength and hypertrophy were similar between sexes (Staron et al. 1994; Abe et al. 2000; Hunter 1985; Hurlbut et al. 2002; Kosek et al. 2006; Roth et al. 2001). While others have found a greater relative strength increase in females (Cureton et al. 1988; Hubal et al. 2005b; Stock and Thompson 2014). The first data to indicate there may be differences in strength changes between sexes comes from the research of Wilmore (1974). This author found that strength was similar when normalized to body weight after 10 weeks of intensive RT, showing that relative upper-body strength increased 29% in females compared with 17% in males, whereas relative increases in lower body strength were similar. Similarly, in one of the largest studies to date, with 342 females and 243 males, Hubal et al. (2005b) found higher relative upper strength increases in females than in males.

In a recent review, Roberts et al. (2020) analyzed studies that compared the long-term responses to RT for strength or hypertrophy in young to middle-aged males and females using the same RT protocol. They found that males and females adapted to RT with similar effect sizes for hypertrophy and lower-body strength, but females had a larger effect for relative upper-body strength, suggesting that it is possible that untrained females display a higher capacity to increase upper-body strength than males. A possible explanation for this result may be that the initial muscular strength recorded for the women was lower than that for the men, since it has been already well established in the literature that initial strength levels might influence adaptations induced by RT (Cormie et al. 2010). Similarly, Ahtiainen et al. (2016), showed that heterogeneity of muscle size and strength responses in lower body were similar between men and women and subjects at different ages after a 24-week periodized RT program. However, when comparing the responses between males and females it is important to consider some well-known set of differences such as hormonal fluctuations or pre-training levels of muscle size and strength, which are generally greater in males, independent of training status (Roth et al. 2001).

2.2 *Effects of Resistance Training in Women*

In the past, it was considered that specific RT adaptations in women were fewer than in men, and so, that the benefits obtained from this activity were also less for them (Wells 1978). Nevertheless, and despite sex-related differences, it has been reported that women and men respond to strength training in very similar ways from their individual pretraining baselines. Although women on the average have smaller bodies than men, have less absolute muscle mass and smaller individual muscle fibers, and display approximately two-thirds of the absolute overall strength and power of men, when considering unit for unit, female muscle tissue has shown to be similar in force output to male muscle tissue, and there is quite evidence to support similar, proportional increases for both sexes in strength performance and hypertrophy of muscle fiber relative to pretraining status (Holloway et al. 1990; Cureton et al. 1988).

Many studies have reported that women can perform almost identical RT programs to men, and that men and women showed similar results after a RT program (Gentil et al. 2016). However, their acute responses have been shown to differ, especially regarding fatigability (Hill et al. 2018) and muscle recovery (Flores et al. 2011). Flores et al. (2011), showed that men and women had similar relative strength loss after a traditional concentric and eccentric isoinertial resistance exercise protocol; however, three days after exercise, strength was still depressed in women. They also found that recovery from muscle thickness was longer in women when compared with that in men. On the other hand, men and women developed and dissipated muscle soreness in a similar manner (Flores et al. 2011). Taking into account the above-mentioned, it might be important to program resting days between RT sessions in a gender-specific manner, since for the same muscle group trained, women might need longer rest periods between training sessions than men do (Flores et al. 2011).

On the other hand, despite the increasing relevance of women in sport in recent decades and the improvement of practicing RT by women, few studies have analyzed the effects of different training intensities on strength and physical performance gains in physically active women. In a recent review, Hagstrom et al. (2020) examined the effects of RT on adaptations in muscular strength and hypertrophy in healthy women adults (18–50 years). The main findings of this review were that RT had a significant effect on muscular strength and hypertrophy in untrained healthy adult females. Gains in muscular strength were approximately 25% in the upper body and 27% in the lower body. Other adaptations to RT included average gains of 3.3% in lean mass, which was equivalent to approximately 1.45 kg (range 0.4–3.3 kg). These adaptations occurred following participation in programs with an average duration of 15 weeks. Another interesting result of this review was that prescription variables related to both training frequency and volume, but not load, showed to be significant contributors to the magnitude of upper and lower body strength gains in females. Although significant muscular hypertrophy occurred following RT, the authors did not find any difference within the different moderators (i.e., light vs heavy load; low vs high volume) for the magnitude of gains.

In another recent study (Barbalho et al. 2019), the effects of different volumes of RT on muscle performance and hypertrophy were compared in trained women. Although all groups significantly increased all muscle thickness measures and 10RM tests after 24 weeks of RT, authors concluded that 5–10 sets per week might be sufficient for attaining gains in muscle size and strength in trained women, at least when all sets are closely supervised and performed to muscle failure. The authors concluded that there appears no further benefit by performing higher exercise volumes.

The benefits of RT in an older population are unquestionable, however, there are some disparities regarding the ideal training volume (i.e., number of sets, repetitions, and load) (Hass et al. 2001). Hunter et al. (2001) compared the effects of high-resistance training (whole-body RT, including elbow flexion and extension, seated row, overhead press, leg extension and curl, bench press, and sit ups), 3 times per week at 80% maximum strength (1RM) with 3 times per week variable-resistance training (once-weekly training at 80%, 65%, and 50% 1RM) in older adults. They observed similar increases in strength and fat free mass (FFM), however, the variable resistance group decreased difficulty of performing daily tasks over those who trained intensely three times per week, suggesting that training too intensely or too frequently may result in increased fatigue, and consequently a reduced training adaptation in older women due to insufficient time to recover.

A number of studies in the older adult literature have demonstrated that high-speed power training is more effective to prevent functional decline than traditional low-speed RT (Sayers and Gibson 2010; Cuoco et al. 2004). Also, different training volumes have been compared in the literature. In this sense, low volume training (1 set per exercise) compared to high volume training (3 sets per exercise) performed twice a week for 13 weeks induced similar improvements in maximal dynamic strength for knee extensors and elbow flexion, muscular activation of the vastus medialis and the biceps brachii, and muscle thickness for the knee extensors and elbow flexors in elderly women (Radaelli et al. 2013). The authors suggested that during the initial months of training, elderly women can significantly increase upper- and lower-body strength by utilizing low volume training. In addition, low-intensity resistance exercise protocols have shown to be more effective for older adults by increasing adherence rates (Williams 2008). However, after longer periods of training, larger muscle groups may require greater training volume to provide further strength gains (American College of Sports Medicine 2009).

2.3 The Influence of the Menstrual Cycle on Muscle Strength Performance

A key consideration when analyzing muscle strength performance or specific adaptations to RT in women is that, unlike men, women are exposed to continuous variations in serum concentrations of several female sex steroid hormones during the menstrual cycle. The main four female sex hormones are: estrogen, progesterone, follicle stimulating hormone, and luteinizing hormone, and the fluctuations of these hormones are a key factor to regulate the patterns of the ovulatory cycle. In that sense, basal serum levels of those hormones will be altered depending on the moment of the menstrual cycle. A normal menstrual cycle is characterized by two main phases: the follicular phase and the luteal phase, separated by a short period of ovulation. The early follicular phase is characterized by low serum concentration of both estrogen and progesterone; whereas maximal estrogen concentration, low progesterone levels and increased luteinizing hormone occur during the late follicular phase (Kraemer et al. 1991; Janse de Jonge 2003). The luteal phase begins once the luteinizing hormone has returned to basal levels and is characterized by high concentrations of both estrogen and progesterone. It has been suggested that skeletal muscle performance might vary with alterations in hormone production during the different phases of the menstrual cycle. However, research investigating muscle contractile strength throughout the menstrual cycle has shown conflicting results. Sarwar et al. (1996) found an 11% increase in quadriceps and handgrip maximum voluntary isometric force during the late follicular phase. Similarly, Phillips et al. (1996) reported an increase in adductor pollicis strength during the follicular phase. On the contrary, Greeves et al. (1985) reported the greatest strength during the mid-luteal phase and suggested that progesterone may be also implicated in the regulation of strength production. In a recent study (Romero-Moraleda et al. 2019), authors have suggested that eumenorrheic females have similar muscle strength and power performance in the smith machine half-squat exercise during the different phases of the menstrual cycle.

So, while some authors have suggested that estrogen may have a strengthening effect on skeletal muscle (Sarwar et al. 1996; Phillips et al. 1996), others have found negative relationships between estrogen concentration and strength (Bassey et al. 1995). Nevertheless, most of them have found no change in the muscle contractile characteristic strength across the different phases of the menstrual cycle (Janse de Jonge 2003; Romero-Moraleda et al. 2019; Lebrun et al. 1995). These conflicting findings can be explained by some methodological shortcomings such as the use of different methods to estimate the phases of the menstrual cycle and the use of different methods for the measurement of muscle strength performance.

On the other hand, adaptations to long-term RT have also been analyzed, finding greater changes in maximum isometric muscle force and muscle diameters when the frequency of the RT was high in the follicular phase and low in the luteal phase as compared to training with a low frequency in the follicular phase and high in the luteal phase, and suggesting that the improved adaptations were the result of the anabolic effect of estrogen rather than an improved muscle capacity for training in the follicular phase (Sung et al. 2014).

Obviously, it is also important to consider the negative symptoms, both physical (bloating, tiredness, stomach and lower back pain) and emotional (mood swings, depressed), related to the menstrual cycle (Oxfeldt et al. 2020), that affect physical training and even competitions in women.

3 From Practice

3.1 General Resistance Training Recommendations in Women

Resistance exercise has proven to be a safe and effective method for improving general physical conditioning, health or performance, in individuals with different needs, goals, or abilities. However, most of the research aiming to examine the acute and chronic responses to different training protocols and stimulus has been carried in men. So, general recommendations given about RT should be interpreted cautiously when involving women populations. In a recent review (Hagstrom et al. 2020), authors drew evidence-based conclusions regarding training expectations in female populations. For most of the studies following average RT programs in untrained females, typical prescriptive parameters included a frequency of three sessions per week, and the performance of three sets of each exercise for approximately ten repetitions. For the upper body, authors indicated that women should perform 3-4 sets per exercise, on 2–4 training days per week for the best strength gains. Moreover, this volume can be accrued across the range of training loads (i.e., light and heavy weights), and prescription methods (i.e., failure or non-failure sets), because neither of these variables moderated the magnitude of upper body strength gains. Similarly, for the lower body, that review found that women should perform lower body exercise on 2–4 training days per week, with a goal of high-volume accrual across the week for the best strength gains. Within-session prescription variables such as sets per exercise, load, and prescription method (failure vs non-failure) did not influence strength gains. Thus, the available evidence would suggest that lower body strength gains in women can be achieved with a variety of prescription combinations, although frequency and total weekly volume must be emphasized. A continuing dose response above four sessions per week may be present; however, due to the lack of high-frequency studies, these authors could not draw conclusions as to the upper limit of this relationship. These findings showed that hypertrophy can be achieved through a variety of exercise prescriptions and are in line with the previous research in females.

Although traditional strength training, including the use of weight machines, has been shown to induce positive changes in strength and FFM in older women adults (Elliott et al. 2002), other alternative methods of RT to the traditional use of weight machines, such as elastic bands or aquatic devices, have shown also their effectiveness at improving body composition and physical capacity (Colado et al. 2012).

In another review (Kendall and Fairman 2014), authors examined current physical activity guidelines, and provided some recommendations specific to strength training in women based on current research. They considered a key factor for improving women's adherence to RT the participation in a supervised or group-based resistance exercise program, which may allow to attain higher intensities, and so greater health benefits. ACSM recommends that older adults perform RT at least 2 non-consecutive days per week, including 8–10 exercises involving all the major muscle groups at moderate intensity (selecting a weight that allows 10–15 repetitions of each exercise), with 2–3 min of rest between each set. Additionally, beginners and those with low levels of strength could start RT with lighter intensity (40–50% 1-RM) to improve strength, power, and balance (2011). For optimal results from a resistance program, the focus should be on full-body, compound movements (bench press, squat, pull-ups, etc.).

In recent years, coaches and sport scientists have started using velocity data for prediction and monitoring changes in strength. However, very few studies have analyzed the effects of different training programs on muscle strength and power performance in women as assessed using barbell velocity measures across different intensities and loading zones. Mora-Custudio et al. (2016) analyzed the effects of two RT protocols differing on the load (light vs. moderate) on neuromuscular performance in physically active women. They found that a low-load (40–60%

1RM) training program produced similar or more beneficial effects on neuromuscular performance than a moderate load training (65–80% 1RM), suggesting that faster training velocities compared to slower training velocities may provide a greater degree of transfer of strength to sport related performance, and highlighting the importance of lifting the load at maximal voluntary velocity also in women, in order to maximize the neuromuscular performance.

3.2 Resistance Training Recommendations During Pregnancy and Post-Partum

Pregnant, healthy women are recommended to do 30 min or more of light to moderate exercise a day on most, if not all, days of the week (Wolfe and Davies 2003). Some of the benefits from exercise during pregnancy are related with reduced back pain (Garshasbi and Faghih 2005), improved health perception (Barakat et al. 2011), and weight gain control (Lebrun et al. 2013). Other benefits reported (Mottola et al. 2018) have been associated with: (1) fewer newborn complications (i.e., large for gestational age) and; (2) maternal health benefits (i.e., decreased risk of pre-eclampsia, gestational hypertension, gestational diabetes, caesarean section, instrumental delivery, urinary incontinence, excessive gestational weight gain and depression; improved blood glucose; decreased total gestational weight gain; and decreased severity of depressive symptoms and lumbopelvic pain). Also, it has been demonstrated that physical activity is not associated with miscarriage, stillbirth, neonatal death, preterm birth, preterm/prelabour rupture of membranes, neonatal hypoglycaemia, low birth weight, birth defects, induction of labour or birth complications. Therefore, prenatal physical activity should be considered a front-line therapy for reducing the risk of pregnancy complications and enhancing maternal physical and mental health.

Most of the investigations of the effects of exercise during pregnancy have focused mostly on the effect of aerobic exercise, however, fewer studies have examined the efficacy and safety of resistance exercise during pregnancy (Barakat et al. 2009). Petrov et al. (2015) evaluated the health effect and safety of moderate-to-vigorous intensity resistance exercise using free weights in healthy women during pregnancy, with regard to health-related quality of life, pain location, physical strength, body weight gain, blood pressure, functional status, activity level, and childbirth outcomes. These authors concluded that supervised, moderate-to-vigorous resistance exercise does not compromise the health status of healthy pregnant women or the fetus during pregnancy, but instead appears to be an appropriate form of exercise in healthy pregnancy. The supervised training program consisted of warm-up, heel raises, non-deep squats, and pelvic-lift and static-abdominal training. Authors also suggested that good trunk- and pelvic-floor muscle control is very important during moderate-to-vigorous exercise in pregnancy, especially during exercises when the hip is in flexion and a heavier load is placed on pelvic floor muscles and cervix (for example squats or cycling in a forward leaning position). In addition, resistance exercise performed with elastic bands has shown to be effective in improving glycaemic control in women with gestation diabetes mellitus (Barros et al. 2010), with no adverse impact on the delivery, or the newborn (Barakat et al. 2009). Although the exercise recommendations in pregnant women do not provide specific guidance for vigorous intensity exercise (Wolfe and Davies 2003), in their review, Nascimento et al. (2012) supported the promotion of moderate-to-vigorous prenatal resistance exercise to the exercise recommendations.

Furthermore, it has been shown an association between newborn's birth weight and mother's pre-pregnancy body weight. Barakat et al. (2009) suggested that light intensity resistance exercise training performed during the second and third trimester of pregnancy might attenuate the adverse effect of high maternal body weight before pregnancy on the newborn's birth size, as well as prevent the gravida from excessive gestational weight gain. On the other hand, exercise during pregnancy may reduce the risk of low birthweight outcome (Leiferman and Evenson 2003) leading to a hypothesis that exercise enhances growth in the feto–placental unit and reduces the risk of preterm birth (Jukic et al. 2012). Price et al. (2012) showed that in comparison with pregnant women who remain sedentary during the gestation period, active pregnant women improve muscular strength and fitness. This increased muscle strength may attenuate the cardiovascular response to a given load during physical activities of daily living because the load would represent a lower percentage of the maximal muscle contraction.

The 2019 Canadian Guideline for Physical Activity throughout Pregnancy (Mottola et al. 2018) provided specific recommendations regarding physical activity in pregnant women, indicating the quality of the evidence informing the recommendations and the strength of the recommendations:

1. All women without contraindication should be physically active throughout pregnancy.
2. Pregnant women should accumulate at least 150 min of moderate-intensity physical activity each week to achieve clinically meaningful health benefits and reductions in pregnancy complications.
3. Physical activity should be accumulated over a minimum of 3 days per week; however, being active every day is encouraged.
4. Pregnant women should incorporate a variety of aerobic and RT activities to achieve greater benefits. Adding yoga and/or gentle stretching may also be beneficial.
5. Pelvic floor muscle training (eg, Kegel exercises) may be performed on a daily basis to reduce the risk of urinary incontinence. Instruction on the proper technique is recommended to obtain optimal benefits.
6. Pregnant women who experience light-headedness, nausea or feel unwell when they exercise flat on their back should modify their exercise position to avoid the supine position.

In general, more physical activity (frequency, duration and/or volume) has been associated with greater benefits; however, physical activity below the recommendations also incurred some benefits. The findings of two systematic reviews also demonstrated that combining aerobic exercise and RT during pregnancy is more effective at improving health outcomes than interventions focused on aerobic exercise alone (Perales et al. 2016; Davenport et al. 2018). Although evidence has not been identified regarding the safety or additional benefit of exercising at levels significantly above the recommendations, it has been recommended high-intensity physical activity only in a monitored environment. Moderate-intensity physical activity is recommended throughout pregnancy. In addition, concerning elite athletes, the International Olympic Committee (IOC) has released a series of recommendations to guide elite athletes during and following pregnancy (Bø et al. 2017). Also, it has been suggested that a warm-up and cool-down period should be included in any physical activity regimen, especially for pregnant women since ligaments become more relaxed during pregnancy due to increasing hormone levels and may impact on the range of movement, thereby increasing the risk of injury (Wolfe et al. 2000).

Finally, studies have demonstrated reduced postpartum urinary incontinence in women involved in specific postpartum pelvic floor exercises performed with a vaginal device providing resistance or feedback, likely by increasing strength (Harvey 2003). There is consistent evidence for improved mood with both acute and chronic physical activity. Improvements include increased vigor, reduced fatigue, reduced stress and anxiety, decreased symptoms of negative mood and depression, and improved self-concept (Dishman 1998). Evidence also suggests that a postpartum exercise intervention may be effective in reducing symptoms of depression (Armstrong and Edwards 2003).

4 Filling Gaps

Most of the studies analyzing the effects of RT in women on strength and general physical performance and health, have focused on unique groups intervention, with the same training protocols (volume and intensities), or compared to control groups or men groups performing the same protocols. However, very few investigations have focused on the effects of different training variables configurations, in order to study adaptations to different training programs in women. The optimal configuration of RT requires the combination of different training variables, including volume, load or relative intensity, frequency, type and order of exercise, rest periods between sets and exercise, contraction type and movement velocity (Kraemer and Ratamess 2004). Of all these training variables, training load has shown to be the most important factor to consider when designing an RT program aimed at improving neuromuscular performance (Kraemer and Ratamess 2004). Although the question of the load magnitudes in RT has been widely analyzed in the scientific literature, the vast majority of the studies have centered on men populations. And still, there is a lack of consensus about the load or load range that produces the greatest improvements in strength and muscle power. So, considering the lack of studies involving women, it seems still necessary to study the effects of different RT programs in women, in order to get more information about the responses, and so, to promote the best exercise training programs in women.

Recently, velocity-based training (VBT) has been proposed as a valid method to objectively quantify and adjust training intensity with high precision on a daily basis. This approach is based on the extremely close relationship, between %1-RM and barbell velocity, which has been initially studied in bench press exercise (González-Badillo and Sánchez-Medina 2010), and then in squat exercise (Sánchez-Medina et al. 2017). This relationship has been also studied in other exercises such as the prone bench pull (Sánchez-Medina et al. 2014), pull-up (Sánchez-Moreno et al. 2017), leg-press (Conceição et al. 2016), and hip thrust (de Hoyo et al. 2019). However, the load-velocity relationship in resistance exercises has been analyzed almost exclusively in men. Very few studies have analyzed the load-velocity relationship in women (Balsalobre-Fernández et al. 2018; García-Ramos et al. 2019; Torrejón et al. 2019). These studies were conducted on upper-body exercises (bench press, inclined bench press, and military press) and showed that the load-velocity relationship was strong and linear in both sexes but the velocity associated with each %1-RM was higher in men compared to women. A recent study (Pareja-Blanco et al. 2020) examined the validity of using bar velocity to estimate relative load in squat and bench-press exercises for both young men and women. They observed close relationships between bar velocity and relative load in both sexes for both squat and bench press exercises. Although men's equation applied to women showed a high level of agreement, lower bias and higher level of agreement was observed when a sex-specific equation was applied in women. Therefore, the general equations previously published for the main exercises, such as bench press or squat may not be suitable for women. This result entails important concerns when designing and prescribing velocity based resistance training, since if the training load is prescribed with the same velocity value for both men and women, the training stimulus for men and women would be different. Hence, the expected outcomes may not be the same, and ultimately it would not be possible to extrapolate findings from training studies conducted in men to women (Pareja-Blanco et al. 2020). Thus, it is reinforced the necessity of conducting more studies regarding the effects of different VBT programs in female populations.

5 Take-Home Messages

1. While men usually have higher levels of absolute muscle mass and strength, relative strength gains actually tend to be larger in women, at least in the short term. This is especially true for younger women and upper body strength gains.

2. Long-term, relative rates of muscle growth and strength gains are probably roughly equal for men and women, though women may make slightly larger gains, relative to their initial levels, across their entire training career.
3. There are key differences between men and women that impact training and recovery, and that should be taken into account when prescribing RT programs.
4. High-speed RT has shown to be more effective to prevent functional decline than traditional low-speed RT in older women.
5. The influence of the menstrual cycle on muscle strength performance has shown conflicting findings. However, it is important taking into account the negative symptoms, both physical and emotional, related to the menstrual cycle.
6. Moderate prenatal RT should be considered a front-line therapy for reducing the risk of pregnancy complications and enhancing maternal physical and mental health. RT + aerobic training seems to be the best exercise combination.
7. Considering the low number of studies involving women when compared to men, it seems necessary to study the effects of different RT programs configurations (differing on intensities, volumes, exercises, etc.) in women, in order to get more information about the responses, and so, to promote the best exercise training programs for women.
8. VBT has been proposed as a valid method to objectively quantify and adjust training intensity with high precision on a daily basis. However, specific adjusted equations should be found for women populations in the different exercises.

References

Abe T, DeHoyos DV, Pollock ML, Garzarella L (2000) Time course for strength and muscle thickness changes following upper and lower body resistance training in men and women. Eur J Appl Physiol 81(3):174–180. https://doi.org/10.1007/s004210050027 (PMID: 10638374)

Ahtiainen JP, Walker S, Peltonen H, Holviala J, Sillanpää E, Karavirta L et al (2016) Heterogeneity in resistance training-induced muscle strength and mass responses in men and women of different ages. Age (Dordr) 38(1):10. https://doi.org/10.1007/s11357-015-9870-1 (Epub 2016 Jan 15. PMID: 26767377; PMCID: PMC5005877)

American College of Sports Medicine (2009) American College of Sports Medicine position stand. Progression models in resistance training for healthy adults. Med Sci Sports Exerc 41(3):687–708. https://doi.org/10.1249/MSS.0b013e3181915670 (PMID: 19204579)

Armstrong K, Edwards H (2003) The effects of exercise and social support on mothers reporting depressive symptoms: a pilot randomized controlled trial. Int J Ment Health Nurs 12(2):130–138. https://doi.org/10.1046/j.1440-0979.2003.00229.x (PMID: 12956024)

Balsalobre-Fernández C, García-Ramos A, Jiménez-Reyes P (2018) Load–velocity profiling in the military press exercise: effects of gender and training. Int J Sports Sci Coach 13:743–750

Barakat R, Lucia A, Ruiz JR (2009) Resistance exercise training during pregnancy and newborn's birth size: a randomised controlled trial. Int J Obes (Lond) 33(9):1048–1057. https://doi.org/10.1038/ijo.2009.150 (Epub 2009 Jul 28 PMID: 19636320)

Barakat R, Pelaez M, Montejo R, Luaces M, Zakynthinaki M (2011) Exercise during pregnancy improves maternal health perception: a randomized controlled trial. Am J Obstet Gynecol 204(5):402.e1-402.e7. https://doi.org/10.1016/j.ajog.2011.01.043 (Epub 2011 Feb 26 PMID: 21354547)

Barbalho MSM, Gentil P, Izquierdo M, Fisher J, Steele J, Raiol RA (2017) There are no no-responders to low or high resistance training volumes among older women. Exp Gerontol 1 (99):18–26. https://doi.org/10.1016/j.exger.2017.09.003 (Epub 2017 Sep 14 PMID: 28918362)

Barbalho M, Coswig VS, Steele J, Fisher JP, Paoli A, Gentil P (2019) Evidence for an upper threshold for resistance training volume in trained women. Med Sci Sports Exerc 51(3):515–522. https://doi.org/10.1249/MSS.0000000000001818 (PMID: 30779716)

Bassey EJ, Coates L, Culpan J et al (1995) Natural variations in oestrogen and FSH levels in eumenorrheic women in negative association with voluntary muscle strength [abstract]. J Physiol 489P:28P

Bø K, Artal R, Barakat R, Brown WJ, Davies GAL, Dooley M et al (2017) Exercise and pregnancy in recreational and elite athletes: 2016/17 evidence summary from the IOC expert group meeting, Lausanne. Part 4-Recommendations for future research. Br J Sports Med 51 (24):1724–1726. https://doi.org/10.1136/bjsports-2017-098387 (Epub 2017 Sep 25. PMID: 28947674)

Colado JC, Garcia-Masso X, Rogers ME, Tella V, Benavent J, Dantas EH (2012) Effects of aquatic and dry land resistance training devices on body composition and physical capacity in postmenopausal women. J Hum Kinet 32:185–195. https://doi.org/10.2478/v10078-012-0035-3 (Epub 2012 May 30. PMID: 23487349; PMCID: PMC3590866)

Conceição F, Fernandes J, Lewis M, Gonzaléz-Badillo JJ, Jimenéz-Reyes P (2016) Movement velocity as a measure of exercise intensity in three lower limb exercises. J Sports Sci 34 (12):1099–1106. https://doi.org/10.1080/02640414.2015.1090010 (Epub 2015 Sep 22 PMID: 26395837)

Cormie P, McGuigan MR, Newton RU (2010) Influence of strength on magnitude and mechanisms of adaptation to power training. Med Sci Sports Exerc 42(8):1566–1581. https://doi.org/10.1249/MSS.0b013e3181cf818d (PMID: 20639724)

Cuoco A, Callahan DM, Sayers S, Frontera WR, Bean J, Fielding RA (2004) Impact of muscle power and force on gait speed in disabled older men and women. J Gerontol A Biol Sci Med Sci 59(11):1200–1206. https://doi.org/10.1093/gerona/59.11.1200 (PMID: 15602076)

Cureton KJ, Collins MA, Hill DW, McElhannon FM Jr (1988) Muscle hypertrophy in men and women. Med Sci Sports Exerc 20(4):338–344. https://doi.org/10.1249/00005768-198808000-00003 (PMID: 3173042)

Davenport MH, Meah VL, Ruchat SM, Davies GA, Skow RJ, Barrowman N et al (2018) Impact of prenatal exercise on neonatal and childhood outcomes: a systematic review and meta-analysis. Br J Sports Med 52(21):1386–1396. https://doi.org/10.1136/bjsports-2018-099836 (PMID: 30337465)

de Barros MC, Lopes MA, Francisco RP, Sapienza AD, Zugaib M (2010) Resistance exercise and glycemic control in women with gestational diabetes mellitus. Am J Obstet Gynecol 203 (6):556.e1-556.e6. https://doi.org/10.1016/j.ajog.2010.07.015 (Epub 2010 Sep 22 PMID: 20864072)

de Hoyo M, Núñez FJ, Sañudo B, Gonzalo-Skok O, Muñoz-López A, Romero-Boza S et al (2019) Predicting loading intensity measuring velocity in barbell hip thrust exercise. J Strength Cond Res https://doi.org/10.1519/JSC.0000000000003159 (Epub ahead of print. PMID: 31009439)

Dias RM, Cyrino ES, Salvador ES, Nakamura FY, Pina FLC, Oliveira AR (2005) Impact of an eight-week weight training program on the muscular strength of men and women. Rev Bras Med Esporte 11(4):224–228

Dishman RK (1998) Physical activity and mental health. In: Encyclopedia of Mental Health, vol 3, pp 171–1883. Academic Press, Friedman HS. San Diego, CA

Dos Santos WDN, Gentil P, de Moraes RF, Ferreira Júnior JB, Campos MH, de Lira CAB et al (2017) Chronic effects of resistance training in breast cancer survivors. Biomed Res Int 2017:8367803. https://doi.org/10.1155/2017/8367803. Epub 2017 Aug 1. PMID: 28835898; PMCID: PMC5557266

Douchi T, Yamamoto S, Yoshimitsu N, Andoh T, Matsuo T, Nagata Y (2002) Relative contribution of aging and menopause to changes in lean and fat mass in segmental regions. Maturitas 42(4):301–306. https://doi.org/10.1016/s0378-5122(02)00161-5 (PMID: 12191853)

Elliott KJ, Sale C, Cable NT (2002) Effects of resistance training and detraining on muscle strength and blood lipid profiles in postmenopausal women. Br J Sports Med. 36(5):340–344. https://doi.org/10.1136/bjsm.36.5.340 (PMID:12351331;PMCID:PMC1724556)

Flores DF, Gentil P, Brown LE, Pinto RS, Carregaro RL, Bottaro M (2011) Dissociated time course of recovery between genders after resistance exercise. J Strength Cond Res 25 (11):3039–3044. https://doi.org/10.1519/JSC.0b013e318212dea4 (PMID: 21804429)

Garber CE, Blissmer B, Deschenes MR, Franklin BA, Lamonte MJ, Lee IM et al (2011) American College of Sports Medicine. American College of Sports Medicine position stand. Quantity and quality of exercise for developing and maintaining cardiorespiratory, musculoskeletal, and neuromotor fitness in apparently healthy adults: guidance for prescribing exercise. Med Sci Sports Exerc 43(7):1334–1359. https://doi.org/10.1249/MSS.0b013e318213fefb (PMID: 21694556)

García-Ramos A, Suzovic D, Pérez-Castilla A (2019) The load-velocity profiles of three upper-body pushing exercises in men and women. Sports Biomech 1–13. https://doi.org/10.1080/14763141.2019.1597155 (Epub ahead of print. PMID: 31296136)

Garshasbi A, Faghih Zadeh S (2005) The effect of exercise on the intensity of low back pain in pregnant women. Int J Gynaecol Obstet 88(3):271–275. https://doi.org/10.1016/j.ijgo.2004.12.001 (Epub 2005 Jan 16 PMID: 15733880)

Gentil P, Steele J, Pereira MC, Castanheira RP, Paoli A, Bottaro M (2016) Comparison of upper body strength gains between men and women after 10 weeks of resistance training. PeerJ. 11 (4):e1627. https://doi.org/10.7717/peerj.1627 (PMID:26893958;PMCID:PMC4756754)

González-Badillo JJ, Sánchez-Medina L (2010) Movement velocity as a measure of loading intensity in resistance training. Int J Sports Med 31(5):347–352. https://doi.org/10.1055/s-0030-1248333 (Epub 2010 Feb 23 PMID: 20180176)

Greeves JP, Cable NT, Reilly T (1985) The relationship between maximal muscle strength and reproductive hormones during the night. J Appl Physiol 59(6):1911–1917

Hagstrom AD, Marshall PW, Halaki M, Hackett DA (2020) The effect of resistance training in women on dynamic strength and muscular hypertrophy: a systematic review with meta-analysis. Sports Med 50(6):1075–1093. https://doi.org/10.1007/s40279-019-01247-x (PMID: 31820374)

Häkkinen K, Häkkinen A (1991) Muscle cross-sectional area, force production and relaxation characteristics in women at different ages. Eur J Appl Physiol Occup Physiol 62(6):410–414. https://doi.org/10.1007/BF00626612 (PMID: 1893904)

Häkkinen K, Pakarinen A (1993) Muscle strength and serum testosterone, cortisol and SHBG concentrations in middle-aged and elderly men and women. Acta Physiol Scand 148(2):199–207. https://doi.org/10.1111/j.1748-1716.1993.tb09549.x (PMID: 8352031)

Harvey MA (2003) Pelvic floor exercises during and after pregnancy: a systematic review of their role in preventing pelvic floor dysfunction. J Obstet Gynaecol Can 25(6):487–498. https://doi.org/10.1016/s1701-2163(16)30310-3 (PMID: 12806450)

Hass CJ, Feigenbaum MS, Franklin BA (2001) Prescription of resistance training for healthy populations. Sports Med 31(14):953–964. https://doi.org/10.2165/00007256-200131140-00001 (PMID: 11735680)

Hill EC, Housh TJ, Smith CM, Schmidt RJ, Johnson GO (2018) Gender- and muscle-specific responses during fatiguing exercise. J Strength Cond Res 32(5):1471–1478. https://doi.org/10.1519/JSC.0000000000001996 (PMID: 29334581)

Holloway JB, Baechle TR (1990) Strength training for female athletes. A review of selected aspects. Sports Med 9(4):216–228. https://doi.org/10.2165/00007256-199009040-00003 (PMID: 2183330)

Hubal MJ, Gordish-Dressman H, Thompson PD, Price TB, Hoffman EP, Angelopoulos TJ et al (2005a) Variability in muscle size and strength gain after unilateral resistance training. Med Sci Sports Exerc 37(6):964–972 (PMID: 15947721)

Hubal MJ, Gordish-Dressman H, Thompson PD, Price TB, Hoffman EP, Angelopoulos TJ et al (2005b) Variability in muscle size and strength gain after unilateral resistance training. Med Sci Sports Exerc 37(6):964–972 (PMID: 15947721)

Hunter GR (1985) Research: changes in body composition, body build and performance associated with differentweight training frequencies inmales and females. Natl Strength Cond J 7:26–28
Hunter GR, Wetzstein CJ, McLafferty CL Jr, Zuckerman PA, Landers KA, Bamman MM (2001) High-resistance versus variable-resistance training in older adults. Med Sci Sports Exerc 33 (10):1759–1764. https://doi.org/10.1097/00005768-200110000-00022 (PMID: 11581563)
Hurlbut DE, Lott ME, Ryan AS, Ferrell RE, Roth SM, Ivey FM et al (2002) Does age, sex, or ACE genotype affect glucose and insulin responses to strength training? J Appl Physiol (1985) 92(2):643–650. https://doi.org/10.1152/japplphysiol.00499.2001 (PMID: 11796676)
Hurley BF, Hanson ED, Sheaff AK (2011) Strength training as a countermeasure to aging muscle and chronic disease. Sports Med 41(4):289–306. https://doi.org/10.2165/11585920-000000000-00000 (PMID: 21425888)
Janse de Jonge XA (2003) Effects of the menstrual cycle on exercise performance. Sports Med 33 (11):833–851. https://doi.org/10.2165/00007256-200333110-00004 (PMID: 12959622)
Janssen I, Heymsfield SB, Wang ZM, Ross R (2000) Skeletal muscle mass and distribution in 468 men and women aged 18–88 yr. J Appl Physiol (1985) 89(1):81–88. https://doi.org/10.1152/jappl.2000.89.1.81. Erratum in: J Appl Physiol (1985) 116(10):1342 (PMID: 10904038)
Jukic AM, Evenson KR, Daniels JL, Herring AH, Wilcox AJ, Hartmann KE (2012) A prospective study of the association between vigorous physical activity during pregnancy and length of gestation and birthweight. Matern Child Health J 16(5):1031–1044. https://doi.org/10.1007/s10995-011-0831-8 (PMID:21674218;PMCID:PMC3386423)
Karinkanta S, Heinonen A, Sievänen H, Uusi-Rasi K, Fogelholm M, Kannus P (2009) Maintenance of exercise-induced benefits in physical functioning and bone among elderly women. Osteoporos Int 20(4):665–674. https://doi.org/10.1007/s00198-008-0703-2 (Epub 2008 Aug 12 PMID: 18696173)
Kelly CM, Burnett AF, Newton MJ (2008) The effect of strength training on three-kilometer performance in recreational women endurance runners. J Strength Cond Res 22(2):396–403. https://doi.org/10.1519/JSC.0b013e318163534a (PMID: 18550953)
Kendall KL, Fairman CM (2014) Women and exercise in aging. J Sport Health Sci 3(3):170–178. https://doi.org/10.1016/j.jshs.2014.02.001
Kosek DJ, Kim JS, Petrella JK, Cross JM, Bamman MM (2006) Efficacy of 3 days/wk resistance training on myofiber hypertrophy and myogenic mechanisms in young vs. older adults. J Appl Physiol (1985) 101(2):531–544. https://doi.org/10.1152/japplphysiol.01474.2005 (Epub 2006 Apr 13. PMID: 16614355)
Kraemer WJ, Ratamess NA (2004) Fundamentals of resistance training: progression and exercise prescription. Med Sci Sports Exerc 36(4):674–688. https://doi.org/10.1249/01.mss.0000121945.36635.61 (PMID: 15064596)
Kraemer WJ, Gordon SE, Fleck SJ, Marchitelli LJ, Mello R, Dziados JE et al (1991) Endogenous anabolic hormonal and growth factor responses to heavy resistance exercise in males and females. Int J Sports Med 12(2):228–235. https://doi.org/10.1055/s-2007-1024673 (PMID: 1860749)
Lamina S, Agbanusi E (2013) Effect of aerobic exercise training on maternal weight gain in pregnancy: a meta-analysis of randomized controlled trials. Ethiop J Health Sci 23(1):59–64 (PMID: 23559839; PMCID: PMC3613816)
Lebrun CM, McKenzie DC, Prior JC, Taunton JE (1995) Effects of menstrual cycle phase on athletic performance. Med Sci Sports Exerc 27(3):437–444 (PMID: 7752873)
Leiferman JA, Evenson KR (2003) The effect of regular leisure physical activity on birth outcomes. Matern Child Health J 7(1):59–64. https://doi.org/10.1023/a:1022545718786 (PMID: 12710801)
Miller AE, MacDougall JD, Tarnopolsky MA, Sale DG (1993) Gender differences in strength and muscle fiber characteristics. Eur J Appl Physiol Occup Physiol 66(3):254–262. https://doi.org/10.1007/BF00235103 (PMID: 8477683)
Mora-Custodio R, Rodríguez-Rosell D, Pareja-Blanco F, Yañez-García JM, González-Badillo JJ (2016) Effect of low- vs. moderate-load squat training on strength, jump and sprint performance in physically active women. Int J Sports Med 37(6):476–482. https://doi.org/10.1055/s-0042-100471 (Epub 2016 Mar 18. PMID: 26990723)

Mottola MF, Davenport MH, Ruchat SM, Davies GA, Poitras VJ, Gray CE et al (2018) 2019 Canadian guideline for physical activity throughout pregnancy. Br J Sports Med 52(21):1339–1346. https://doi.org/10.1136/bjsports-2018-100056 (PMID: 30337460)

Nascimento SL, Surita FG, Cecatti JG (2012) Physical exercise during pregnancy: a systematic review. Curr Opin Obstet Gynecol 24(6):387–394. https://doi.org/10.1097/GCO.0b013e328359f131 (PMID: 23014142)

Oxfeldt M, Dalgaard LB, Jørgensen AA, Hansen M (2020) Hormonal contraceptive use, menstrual dysfunctions, and self-reported side effects in elite athletes in Denmark. Int J Sports Physiol Perform 1–8. https://doi.org/10.1123/ijspp.2019-0636 (Epub ahead of print. PMID: 32957078)

Paoli A, Moro T, Bianco A (2015) Lift weights to fight overweight. Clin Physiol Funct Imaging 35 (1):1–6. https://doi.org/10.1111/cpf.12136 (Epub 2014 Feb 24 PMID: 24612071)

Pareja-Blanco F, Walker S, Häkkinen K (2020) Validity of using velocity to estimate intensity in resistance exercises in men and women. Int J Sports Med 41(14):1047–1055. https://doi.org/10.1055/a-1171-2287 (Epub 2020 Jul 20 PMID: 32688415)

Perales M, Santos-Lozano A, Ruiz JR, Lucia A, Barakat R (2016) Benefits of aerobic or resistance training during pregnancy on maternal health and perinatal outcomes: a systematic review. Early Hum Dev 94:43–48. https://doi.org/10.1016/j.earlhumdev.2016.01.004 (Epub 2016 Feb 3 PMID: 26850782)

Petrov Fieril K, Glantz A, Fagevik Olsen M (2015) The efficacy of moderate-to-vigorous resistance exercise during pregnancy: a randomized controlled trial. Acta Obstet Gynecol Scand 94(1):35–42. https://doi.org/10.1111/aogs.12525 (Epub 2014 Nov 13 PMID: 25287282)

Phillips SK, Sanderson AG, Birch K, Bruce SA, Woledge RC (1996) Changes in maximal voluntary force of human adductor pollicis muscle during the menstrual cycle. J Physiol 496(Pt 2)(Pt 2):551–557. https://doi.org/10.1113/jphysiol.1996.sp021706 (PMID: 8910237; PMCID: PMC1160898)

Price BB, Amini SB, Kappeler K (2012) Exercise in pregnancy: effect on fitness and obstetric outcomes-a randomized trial. Med Sci Sports Exerc 44(12):2263–2269. https://doi.org/10.1249/MSS.0b013e318267ad67 (PMID: 22843114)

Quirino MA, Modesto-Filho J, de Lima Vale SH, Alves CX, Leite LD, Brandão-Neto J (2012) Influence of basal energy expenditure and body composition on bone mineral density in postmenopausal women. Int J Gen Med 5:909–915. https://doi.org/10.2147/IJGM.S36823 (Epub 2012 Nov 5. PMID: 23204854; PMCID: PMC3508566)

Radaelli R, Botton CE, Wilhelm EN, Bottaro M, Lacerda F, Gaya A et al (2013) Low- and high-volume strength training induces similar neuromuscular improvements in muscle quality in elderly women. Exp Gerontol 48(8):710–716. https://doi.org/10.1016/j.exger.2013.04.003 (Epub 2013 Apr 19 PMID: 23603619)

Roberts BM, Lavin KM, Many GM, Thalacker-Mercer A, Merritt EK, Bickel CS et al (2018) Human neuromuscular aging: sex differences revealed at the myocellular level. Exp Gerontol 106:116–124. https://doi.org/10.1016/j.exger.2018.02.023 (Epub 2018 Feb 24. PMID: 29481967; PMCID: PMC6031257)

Roberts BM, Nuckols G, Krieger JW (2020) Sex differences in resistance training: a systematic review and meta-analysis. J Strength Cond Res 34(5):1448–1460. https://doi.org/10.1519/JSC.0000000000003521 (PMID: 32218059)

Romero-Moraleda B, Coso JD, Gutiérrez-Hellín J, Ruiz-Moreno C, Grgic J, Lara B (2019) The influence of the menstrual cycle on muscle strength and power performance. J Hum Kinet 21(68):123–133. https://doi.org/10.2478/hukin-2019-0061 (PMID:31531138;PMCID:PMC6724592)

Roth SM, Ivey FM, Martel GF, Lemmer JT, Hurlbut DE, Siegel EL et al (2001) Muscle size responses to strength training in young and older men and women. J Am Geriatr Soc 49 (11):1428–1433. https://doi.org/10.1046/j.1532-5415.2001.4911233.x (PMID: 11890579)

Sánchez-Medina L, González-Badillo JJ, Pérez CE, Pallarés JG (2014) Velocity- and power-load relationships of the bench pull vs. bench press exercises. Int J Sports Med 35(3):209–216. https://doi.org/10.1055/s-0033-1351252 (Epub 2013 Jul 30. PMID: 23900903)

Sánchez-Medina L, Pallarés JG, Pérez CE, Morán-Navarro R, González-Badillo JJ (2017) Estimation of relative load from bar velocity in the full back squat exercise. Sports Med Int

Open 1(2):E80–E88. https://doi.org/10.1055/s-0043-102933 (PMID: 30539090; PMCID: PMC6226068)

Sánchez-Moreno M, Rodríguez-Rosell D, Pareja-Blanco F, Mora-Custodio R, González-Badillo JJ (2017) Movement velocity as indicator of relative intensity and level of effort attained during the set in pull-up exercise. Int J Sports Physiol Perform 12(10):1378–1384. https://doi.org/10.1123/ijspp.2016-0791 (Epub 2017 Dec 21 PMID: 28338365)

Sarwar R, Niclos BB, Rutherford OM (1996) Changes in muscle strength, relaxation rate and fatiguability during the human menstrual cycle. J Physiol 493(Pt 1)(Pt 1):267–272. https://doi.org/10.1113/jphysiol.1996.sp021381 (PMID: 8735711; PMCID: PMC1158967)

Sayers SP, Gibson K (2010) A comparison of high-speed power training and traditional slow-speed resistance training in older men and women. J Strength Cond Res 24(12):3369–3380. https://doi.org/10.1519/JSC.0b013e3181f00c7c (PMID: 21068681)

Staron RS, Karapondo DL, Kraemer WJ, Fry AC, Gordon SE, Falkel JE et al (1994) Skeletal muscle adaptations during early phase of heavy-resistance training in men and women. J Appl Physiol (1985) 76(3):1247–1255. https://doi.org/10.1152/jappl.1994.76.3.1247 (PMID: 8005869)

Stock MS, Thompson BJ (2014) Sex comparisons of strength and coactivation following ten weeks of deadlift training. J Musculoskelet Neuronal Interact 14(3):387–397 (PMID: 25198235)

Sung E, Han A, Hinrichs T, Vorgerd M, Manchado C, Platen P (2014) Effects of follicular versus luteal phase-based strength training in young women. Springerplus 11(3):668. https://doi.org/10.1186/2193-1801-3-668 (PMID:25485203;PMCID:PMC4236309)

Timpka S, Petersson IF, Zhou C, Englund M (2014) Muscle strength in adolescent men and risk of cardiovascular disease events and mortality in middle age: a prospective cohort study. BMC Med 14(12):62. https://doi.org/10.1186/1741-7015-12-62 (PMID:24731728;PMCID:PMC4006633)

Torrejón A, Balsalobre-Fernández C, Haff GG, García-Ramos A (2019) The load-velocity profile differs more between men and women than between individuals with different strength levels. Sports Biomech 18(3):245–255. https://doi.org/10.1080/14763141.2018.1433872 (Epub 2018 Mar 21 PMID: 29558855)

Weiss LW, Cureton KJ, Thompson FN (1983) Comparison of serum testosterone and androstenedione responses to weight lifting in men and women. Eur J Appl Physiol Occup Physiol 50(3):413–419. https://doi.org/10.1007/BF00423247 (PMID: 6683165)

Wells CL (1978) The female athlete: myths and superstitions put to rest. In: Burke EJ (ed) Toward an understanding of human performance: readings in exercise physiology for the coach and athlete, pp 37–40. Movement, Ithaca, NY

Williams DM (2008) Exercise, affect, and adherence: an integrated model and a case for self-paced exercise. J Sport Exerc Psychol 30(5):471–496. https://doi.org/10.1123/jsep.30.5.471 (PMID: 18971508;PMCID:PMC4222174)

Wilmore JH (1974) Alterations in strength, body composition and anthropometric measurements consequent to a 10-week weight training program. Med Sci Sports 6(2):133–138 (PMID: 4461973)

Wolfe LA (2000) Pregnant women and endurance exercise. In: Shephard RJ, Astrand PO (eds) Endurance in sport, 2nd edn. Blackwell Science, London, pp 531–546

Wolfe LA, Davies GA (2003) Canadian guidelines for exercise in pregnancy. Clin Obstet Gynecol. 46 (2):488–495. https://doi.org/10.1097/00003081-200306000-00027 (PMID: 12808398)

Xu J, Lombardi G, Jiao W, Banfi G (2016) Effects of exercise on bone status in female subjects, from young girls to postmenopausal women: an overview of systematic reviews and meta-analyses. Sports Med 46(8):1165–1182. https://doi.org/10.1007/s40279-016-0494-0 (PMID: 26856338)

Beatriz Bachero-Mena has M.Sc. and Ph.D. in Sports Sciences from Pablo de Olavide University (Seville). She is a professor at the Department of Physical Education and Sport at University of Seville (Spain) from 2016. Some other research topics are related to the determinant factors of the sports performance in track and field athletes, importance of strength levels on physical and athletic performance in middle-distance athletes, monitoring training load in athletes, resisted sprint training, neuromuscular adaptations to acute and chronic exercise, and physiological responses to acute and chronic exercise. She works also as a collaborator of several research projects at the Physical Performance and Sports Research Center of Pablo de Olavide University, and she is an author of several papers published in international scientific, peer-reviewed journals. From 2015, she is a Track and Field National Coach and works with athletes from national level in sprint and middle-distance disciplines.

Susana Moral-González is M.Sc. and Ph.D. in Physical Activity and Sports Sciences. She is professor of Exercise Techniques in the European University of Madrid (Spain). She is a member of the research team in Exercise and Nutritional Science of the SDSU University-San Diego (USA), a member of the "EHFA Technical Experts Group (TEG)" of the European Health and Fitness Association, and a member of the Cycling Health and Performance Research Team. She has authored various publications (books and articles) related to physical activity and health and has presented at national and international conventions.

Supplementation and Ergogenic Aids for Enhancing Muscular Strength Production

Raúl Domínguez, Alireza Naderi, and Antonio Jesús Sánchez-Oliver

Abstract This chapter presents two essential premises to abolish the classic bilateral relationship between nutrition and strength: the relationship of strength production with hypertrophy, focusing in the mechanical and physiological mechanisms involved on strength production (i); and, the relationship between sport, strength and nutrition, avoiding using the term strength sport modalities that present an old wrong conceptualization, in as much as strength training is not only specific for the strength sport modalities (ii). This chapter will focus on those type of sport supplements that provide ergogenic effects for enhancing muscular strength production, following the two premises previously described. The ergogenic effects of the sport supplements is dependent on the mechanical and metabolic demands of each modality of exercise. Therefore, in the present chapter, it has been reviewed the sport supplements with possible ergogenic aids to increase muscle strength, including the mechanism of action related to increased muscle force production, including a correct guidelines with the optimal dosage and timing to administer these supplements. We selected caffeine, creatine, β-alanine, Sodium bicarbonate and Nitrate as the five top sports supplements with the most evidence-based use in the sports nutrition, improving neuromuscular function through various metabolic pathways that directly/indirectly lead to muscle strength gain.

Keywords Sport performance · Sport nutrition · Neuromuscular adaptation · Caffeine · Creatine · β-alanine · Sodium bicarbonate · Nitrate

R. Domínguez (✉) · A. J. Sánchez-Oliver
Departamento de Motricidad Humana y Rendimiento Deportivo, Universidad de Sevilla, C/Pirotecnia s/n. 41013, Sevilla, Spain
e-mail: rdherrera@us.es

A. Naderi
Department of Sport Physiology, Islamic Azad University, N 35, Abshar avenue, Valiasr street, Tehran, Iran

A. Muñoz-López et al. (eds.), *Resistance Training Methods*, Lecture Notes in Bioengineering, https://doi.org/10.1007/978-3-030-81989-7_19

1 Introduction

Strength is the most important physical fitness component in the training phases of athletic populations which exists different demonstrations of strength according to its objective and context. This chapter presents two essential premises to abolish the classic bilateral relationship between nutrition and strength. The first premise is to abolish the only relationship of strength production with hypertrophy that it is confused, incoherent or inopportune. Thereby, the relationship between nutrition and strength must be focused since the mechanical and physiological mechanisms involved on strength production (Gonzalez-Badillo and Ribas-Serna 2019). Thus, both mechanical and physiological mechanisms is considered as protagonist in neuromuscular system. Consequently, The main aim of this chapter is to examine the ergogenic effects of sport supplements on the neuromuscular adaptation based on three mechanisms of action involving on the strength production: (i) increase the magnitude of the force generated by the neuromuscular system; (ii) increase the rhythm or cadence on the ratio of force development; (iii) increase the time/force relationship (Gonzalez-Badillo and Ribas-Serna 2019). Therefore, the relationship between strength and nutrition can not be reflected by the muscle mass that is frequently more related with esthetic than sport performance.

The second premise is based on the evolution of the relationship between nutrition and strength that require a reflection and refound the way in which sport, strength and nutrition coming together. Firstly, it is necessary to avoid using the term *strength sport modalities* that present an old wrong conceptualization. In this way, in sport performance area, sports training have evolved for training the main and secondaries capacities in all kind of sports aimed by the principles of generality, variety or transference, or seeking the prevention of injuries, for example. Actually the strength training is not only specific for the strength sport modalities. In this sense, the main objective is to optimize the applied force in every sport action, namely the strength produced in the muscle or muscle group in a period of time or velocity in sport context or during a sport action as jumps, sprints, throwing (Gonzalez-Badillo and Ribas-Serna 2019). Thus, nutrition as one of the critical factors in muscle strength enhancement must be considered for increasing the applied force in every strength demanding sports action.

Many athletes consume sport supplements for enhancing their sport performance (Muñoz et al. 2020), due to the little distance between the success and sport defeats (Mujika et al. 2002). The International Olympic Committee (IOC) in a position stand published in 2018 defined a sport supplement as a food, food component, nutrition or non-food compound that is purposefully ingested in addition to the habitually consumed diet with the aim of achieving a specific health and/or performance benefit (Maughan et al. 2018a). This chapter will focus on those type of sport supplements that provide ergogenic effects during exercise based on the above classification (Garthe and Maughan 2018).

Before election of consuming a sport supplement, three important criteria such as, safety, efficacy and legality must be considered. Thus, this chapter only will focus on those sport supplements under category of ergogenic aids that their safety, efficacy and legality have been well-documented in the literature (AIS 2019; Kerksick et al. 2018). Nevertheless, there are usually an unsuitable use of sport supplements based on a low security, efficacy and legality is observed as the non-favourable contexts. Therefore, attending to the premises of safety, efficacy and legality of the sport supplements, it is necessary to define the context that justify the consume (Maughan et al. 2018b; Mata-Ordoñez et al. 2018; Thomas et al. 2016), attending to the individualized and contextualized situations (Mata-Ordoñez et al. 2018; Caraballo et al. 2020). Also, it is important to consider that the use of sport supplements are not sufficient for counteracting an unsuitable food selection or an incorrect eating habit with the only exception when the changes on the diet is needed (Maughan et al. 2018b). While following a balanced diet based on the sports nutrition guidelines can support the benefits of sport supplement based on the scientific evidence as optimizing sport performance, delaying fatigue, modifying body composition or increasing health status (Rawson et al. 2018). Collectively, we selected caffeine, creatine, β-alanine, Sodium bicarbonate and Nitrate as the five top sports supplements with the most evidence-based use in the sports nutrition (Maughan et al. 2018a; Garthe and Maughan 2018; Maughan et al. 2018b) improving neuromuscular adaptation through various metabolic pathways that directly/indirectly lead to muscle strength gain.

2 From Theory

2.1 Caffeine

In the 1970s, the firsts studies reported an improvement in the time to exhaustion on cycle ergometer explained the ergogenicity of caffeine (CAF) related to a delayed muscle glycogen depletion and lipolysis enhancement, However, actually, CAF's mechanisms of action is attributed to the antagonistic effect of adenosine binding to adenosine receptors A_1 and A_{2a} which will reduce the parasympathetic activity, increases the synthesis of neurotransmitters in the brain (mainly dopamine) and catecholamines. Therefore, the tension, vigour and the perception of vitality have been shown to be increased after CAF ingestion and modify the relation between training load and rate of perceived effort (RPE) that could influence the ergogenic effect of CAF on both, moderately and high-trained athletes (Jodra et al. 2020). However, the recruitment of motor units at the neuromuscular level during physical exercise increase by CAF ingestion through mediation in the muscle force production potentiating the muscular strength to the same load (López-González et al. 2018). In addition, CAF ingestion improves the Na^+–K^+ pump activity and increase the rate of calcium (Ca^{2+}) release from the sarcoplasmic reticulum (Weber and Herz 1968). Also, higher bioactivity of Ca^{2+} in the myoplasm increases the number of cross-bridges and, consequently the force of muscle production. These peripheral effects of CAF could mediate in the increase of power output produced at the same electromyographic activity reported in highly trained athletes (San Juan et al. 2019).

Thus, the peripheral effect of CAF at the neuromuscular level plays the main role in the force production and its effects on central nervous system (CNS) less is known in sports actions with high force demands. In the literature, two recent meta-analysis have been reported that CAF increases the maximal force measured as one maximum repetition (1-RM) (Grgic et al. 2018) and the movement velocity with the submaximal load as lower as upper limb during the execution of resistance exercise (Raya-González et al. 2020).

2.2 Creatine Supplementation

Creatine (CR) is synthesized from three amino acids, glycine, arginine and methionine in the kidney, pancreas and liver. Approximately 95% of CR is stored in skeletal muscle with the remainder found in other tissues such as testes and the brain (~5%). Once CR is taken up by the muscle, ~70% is converted to phosphocreatine (PCr) while the remainder is found as free CR (Hultman et al. 1996). Approximately 1–2 g of CR is broken down each day and converted to creatinine, thus the daily requirements of CR is ~2 g d^{-1}, which can be replaced exogenously through food sources (e.g., red meat, seafood, and poultry), dietary supplementation, and/or endogenously (Cooper et al. 2012). Since dietary sources primarily involve animal-based products, vegans and vegetarians typically have lower CR content within the muscle and tend to respond more favourably to supplementation (Kaviani et al. 2020). Muscle CR plays several key roles within the muscle including an increased capacity to resynthesize ATP (Volek and Kraemer 1996), aid in shuttling of energy between the mitochondria and cytosol (Wallimann et al. 1984), buffers intracellular H^+ (Quinn et al. 1992), and alters muscle protein balance (primarily via an attenuation of muscle protein breakdown) (Persky and Brazeau 2001).

Creatine supplementation, combined with resistance exercise, has been shown to be an effective supplement for increasing muscle mass and strength (Candow and Chilibeck 2008; Alves et al. 2013), especially in athletes with a higher proportion of type II muscle fibres, due to their higher capacity for creatine storage (Syrotuik and Bell 2004). Augmented intramuscular CR and PCr stores enhance energy regeneration and myofibrillar cross-bridge cycling potentially due to alterations in calcium handling that may improve force production (Lanhers et al. 2017). Lanhers and colleagues systematically reviewed the literature and found that creatine enhanced upper and lower body strength (Lanhers et al. 2015, 2017). Chilibeck et al. (2017), observed that those who resistance trained and ingested creatine gained ~1.37 kg more muscle compared to resistance training and placebo (Chilibeck et al. 2017). Overall, creatine is pleiotropic and appears to have several effects within the muscle, and it is clear that creatine can increase both muscular strength and muscle mass when combined with exercise.

2.3 β-alanine Supplementation

β-alanine (BA) is a non-proteogenic amino acid synthesized in the liver from the degradation of thymine, cytosine and uracil, but BA can be ingested through different dietary sources of animal origin or supplements. The ergogenic effects of BA on physical performance are mediated by its capacity to increase muscle carnosine concentrations. Carnosine is a di-peptide formed by BA and L-histidine which BA is known as a rate-limiting precursor for carnosine synthesis (Stellingwerff et al. 2012). In this sense, it has been reported a significant increase in muscle carnosine of the vastus lateralis, gastrocnemius and soleus muscles followed by 4 weeks of BA, but not L-histidine supplementation alone (Blancquaert et al. 2017). Carnosine is the most abundant protein at intramuscular level acting as an antioxidant and neurotransmitter, but its main function is related to the intramuscular acid-base balance and the muscle contractility (Domínguez et al. 2015).

Furthermore, at intramuscular level, carnosine plays a role as a shuttle, transporting Ca^{2+} from the sarcoplasmic reticulum to the myoplasm (Dutka and Lamb 2004) where it is captured and transported a hydrogen ion (H^+) to the extracellular space (Swietach et al. 2013). The increases of Ca^{2+} bioavailability on myoplasma facilitates the cross-bridges formation and muscle force production. Nevertheless, in acidic environment caused by H^+ production during muscle contractions, an increase in muscle carnosine levels favour the flux of H^+ outside the sarcolemma preserving the intramuscular pH. Therefore, considering that the accumulation of intracellular H^+ inhibit the activity of the phosphofructokinase affecting the glycolytic and phosphocreatine contribution in muscle metabolism (Sahlin and Harris 2011). Thus, improving muscular strength during resistance training could be potentiated by BA supplementation through optimization of muscle carnosine concentration. Regarding the effects of BA supplementation on muscle force production, in a study, an increase in the number of repetitions with a load corresponding to the 65% 1-RM in leg press was observed (Outlaw et al. 2016) while the other studies have shown an improvement in the 1-RM on back squat (Hoffman et al. 2006) and the maximal power output during a back squat incremental load test (Maté-Muñoz et al. 2018).

2.4 Sodium Bicarbonate Supplementation

Sodium bicarbonate ($NaHCO_3$) is a supplement that is supported from increasing blood levels of $NaHCO_3$, raising the extracellular pH and the plasma alkalosis. After $NaHCO_3$, supplementation the gradient concentration between muscle fibre and blood is elevated, facilitating the H^+ flow due to the lack of permeability of $NaHCO_3$ to the sarcolemma (Lancha Junior et al. 2015). In line, $NaHCO_3$ reduces H^+ produced during high-intensity effort by 62% (Medbo and Tabata 1993). Thus, the main goal of $NaHCO_3$ supplementation is increasing the alkalosis blood levels before and during exercise with a high anaerobic glycolysis contribution (McNaughton et al. 2008). In this sense, the flux of H^+ is increased after an elevation of $NaHCO_3$ blood levels, regulating intramuscular pH, favouring the contribution to the glycolytic pathway and delaying the onset of fatigue (Granier et al. 1996). This is a very useful nutritional strategy in exercises characterized by the glycolytic metabolism like resistance exercise to enhance muscular strength based on the lactate threshold which is localized above 27–36% 1-RM (Domínguez et al. 2018a). In theory, force production can be decreased by intracellular H^+ accumulation leading to an impairment in cross-bridge actomyosin attachment. Thus, resistance-trained athletes can benefit from this supplement's ergogenic effects via increasing extracellular buffering capacity (Grgic et al. 2020).

Regarding the specific studies that have analysed the effect of $NaHCO_3$ supplementation on resistance exercise, it has been reported a higher training volume on a training session that consisted of 3 sets of four leg exercises with a load of 10–12-RM after 300 mg kg^{-1} of $NaHCO_3$ supplementation (Carr et al. 2013) in contrast, in another study an improvement in back squat and bench sets until failure with a load corresponding at 80% 1RM was reported (Duncan et al. 2014). The recent meta-analysis found that $NaHCO_3$ supplementation can improve the number of repetitions performed with a given load during resistance training, the time to maintain isometric force production at a maximum load and isometric total work done (Grgic et al. 2020). Therefore, although no study have analysed the effect of $NaHCO_3$ supplementation on muscular strength production, but this supplement could increase the training volume during resistance training and, consequently, potentiate training adaptation that favours muscular strength.

2.5 *Nitrate Supplementation*

The ergogenic effect of nitrate (NO_3^-) supplementation, typically administered in the form of beetroot juice (BJ) that is a dietary source riches in NO_3^-, is attributed to its stepwise reduction to nitrite (NO_2^-) and, subsequently, to nitric oxide (NO). therefore, initially in the oral cavity the dietary NO_3^- is reduced by commensal anaerobic bacteria to NO_2^-, followed by one-electron reduction to NO by various NO_2^- reductases in the tissue and blood, especially under conditions of acidosis, hypoxia and occur intramuscularly during exercise (San Juan et al. 2020). The effects of NO_3^- supplementation on physical exercise are mediated by its capacity of increasing NO levels. In this sense, the ergogenic effects of NO_3^- are dependent on the magnitude of the NO_2^- blood levels after NO_3^- supplementation (Coggan et al. 2018).

NO is a potent vasodilator and increase blood flow to muscles, promoting an enhancement in the nutrients and gas exchange in the muscle as an optimization in the mitochondrial respiration and mitochondrial biogenesis that could justify the ergogenic effect of NO_3^- supplementation on the running economy and time to exhaustion during endurance sport modalities (Domínguez et al. 2017b). NO_3^-

supplementation also has been shown to reduce VO_2 during submaximal intensity exercise (Domínguez et al. 2017b), and diminish the reduction of phosphocreatine (PCr) levels for a determined work (Bailey et al. 2010). This higher efficiency on PCr metabolism has been proposed as a potential mechanism for considering ergogenic effect of NO_3^- supplements during the strength training session (Domínguez et al. 2018b). In theory, NO_3^- supplementation may increase the content of the calcium-handling proteins dihydropyridine receptor (DHPR) and calsequestrin (CASQ) in mouse fast-twitch muscle (Hernández et al. 2012). An increase in calcium-handling proteins activity promotes Ca^{2+} bioavailability in myoplasm and, subsequently the number of actin-myosin cross-bridges and muscle power production is augmented (Domínguez et al. 2017a). There are a few studies have examined NO_3^- supplementation during resistance training that only those studies using the doses higher than 6.4 mmol of NO_3^- reported an ergogenic effect on power output production and the number of repetitions during resistance exercise (San Juan et al. 2020).

3 From Practice

3.1 Caffeine

CAF ingestion can elevate blood caffeine levels and reach the peak value when ingested 15–45 min prior to exercise as the optimal timing for enhancing the muscle force production. Regarding the dosage, the most studies used 6 mg kg^{-1}, but the ergogenic effect of CAF on muscle productions is initiated with a dosage of 3 mg kg^{-1} and present a relationship between the magnitude of the improvements and the amount of CAF (Grgic et al. 2019a). Related to the possible side effects, CAF could induce tachycardia, hypertension, attention deficits, restlessness, distress, and sleep disturbances. Based on the relationship between the dose and the secondary effects, it is recommended to supplement with a dose of 3–9 mg kg^{-1}, 30–60 min prior to exercise (Goldstein et al. 2010) and the dosages can be increased in those athletes with habitual caffeine intake (Grgic et al. 2019b) in order to gain muscle force during strength training.

3.2 Creatine Supplementation

Creatine is a well-known ergogenic aid to augment acute and chronic exercise (Kreider et al. 2017), however individual responses appear to vary (Syrotuik and Bell 2004). Syrotuik and Bell (2004) found that responders had lower initial intramuscular total CR stores (possibly due to lower creatine dietary intake) and had a greater proportion of fast-twitch muscle fibres (Syrotuik and Bell 2004). Despite these individual responses, there are two supplementation protocols that are often used in the literature, which are both known to saturate muscle CR stores over time. The first protocol includes a loading phase (20 g day^{-1} [0.3 g kg^{-1} day^{-1}] of creatine for 5–7 days, provided in 4 equal doses) followed by a maintenance dose thereafter (3–5 g day^{-1} [0.03 g kg^{-1} day^{-1}]) (Harris et al. 1992). The second protocol known to be effective at augmenting performance is a dose of 0.1 g kg^{-1} day^{-1} of creatine without a loading phase (Candow et al. 2015). Both protocols increase CR intramuscular stores by ~20% (dependent on initial baseline levels) (Kreider et al. 2017). This increase in intramuscular CR content will bring water into the muscle cell due to the osmotic shift and typically increases body mass 1–2%. This increase in water content stimulates a cascade of events within the muscle which appears to be important in stimulating muscle growth (particularly when combined with exercise) (Kreider et al. 2017). When CR is co-ingested with carbohydrates or protein (Green et al. 1996; Steenge et al. 2000), known to stimulate insulin, there is an increased creatine uptake. Recently, there has been an interest in optimal timing of creatine to augment resistance training adaptations. There is a small number of studies comparing creatine ingested before and after training. No individual study found a significant difference for gains in strength or muscle mass when comparing creatine ingested before or after training, which was further supported by a recent within-subject 8-week training study (Mills et al. 2020). In addition, creatine ingested during training (i.e., 0.0055 g·kg^{-1} immediately after each set for a total of 0.1 g·kg^{-1}·day^{-1}) is another alternative and effective strategy to augment resistance training adaptations. Therefore, it appears that timing is not a major factor when athletes are attempting to augment adaptations with creatine over time, and thus athletes can select to ingest creatine either before, during, or after training (Mills et al. 2020).

3.3 β-alanine Supplementation

It is well-documented that BA supplementation as the limiting factor can increase intramuscular carnosine content (Stellingwerff et al. 2012) through the dose and duration-dependent effects (Harris et al. 2006). In line, 6–10 weeks BA supplementation with a dose of 4–6.4 g day^{-1}increased muscle carnosine by 60–80% in comparison with basal values (Hill et al. 2007). Two other studies also have reported that muscle carnosine can be augmented followed by a moderate dose of 3 g·day-1 up to 12 g day^{-1} from BA supplementation during 4–2 weeks, respectively (Stellingwerff et al. 2012; Church et al. 2017). Its seems that skeletal muscle has a large capacity for muscle carnosine optimization in a non-linear manner, irrespective of gender, and low baseline value of carnosine levels (Rezende et al. 2020), with a slow washout period, earlier studies in this regards reported a time range from 6 to 20 weeks (Stellingwerff et al. 2012; Baguet et al. 2009). In addition, most recent study also found a complete washout from 12 to 16 weeks, followed by 8 weeks of 6.4 g day^{-1} BA supplementation (Yamaguchi et al. 2020).

It has been reported that the kinetic of BA in the blood depends on the dose, increasing with the amount of BA ingested (Harris et al. 2006). When BA is ingested, the peak of BA in the blood are reached in 20–30 min which is accompanied by paresthesia that is a feeling of tingling on the skin causing by an increases in the sensitivity of the nociceptive responsible for transmitting neuropathic pain and and diminished 1-h after ingestion (Domínguez et al. 2015; Hoffman et al. 2012).

The paresthesia's effects are related to a dose-dependent response that initiates with a dose higher than 10 mg kg^{-1} (Sale et al. 2010). Hence, It is recommended to either divide the dosages with lower doses of 0.8–1.6 g day^{-1} every 3–4 h or using the control release BA capsules for diminishing the appearance of paresthesia (Maughan et al. 2018a; Domínguez et al. 2015).

Collectively, BA supplementation can produce ergogenic benefits when it is supplemented with the dosages from 3.2 to 6.4 g day^{-1} for 4–24 weeks (Perim et al. 2019). Besides, using the maintenance dose of BA supplementation during 6 weeks with 1.2 g day^{-1} followed by 4–6 weeks with a loading protocol of 3.2–6.4 g day^{-1} of can sustain exercise performance (Zandona et al. 2020) through keeping muscle carnosine levels about 30–50% greater than the baseline (Stegen et al. 2014).

3.4 Sodium Bicarbonate Supplementation

Although several studies have reported an ergogenic effect of $NaHCO_3$ supplementation on the training load during resistance exercise session with doses of 300 mg kg^{-1} of $NaHCO_3$ (Carr et al. 2013; Duncan et al. 2014), it has been reported a higher ergogenic effect after 500 mg kg^{-1} compared to 300 mg kg^{-1} of $NaHCO_3$ supplementation during a single high-intensity effort (Doudouros et al. 2006). However, the recommended doses have been proposed on a range of 200–400 mg kg^{-1} of $NaHCO_3$, either in acute or chronic supplementation protocol (Maughan et al. 2018a). These doses have been proposed based on the side effects of ingestion including mainly gastrointestinal discomfort and could be reduced by split lower doses 100 mg kg^{-1} taken over a period of 60–240 min pre-exercise based on individualized experience (Oliveira et al. 2020) or including a portion of food containing carbohydrate and fluid (Naderi et al. 2016) and/or using sodium citrate as an alternative extracellular buffer enhancer with the lower gastrointestinal discomfort characteristic (Maughan et al. 2018a).

3.5 Nitrate Supplementation

Considering that the ergogenic effects of NO_3^- supplementation are positively correlated with the increasing NO levels, it is important to consider this supplement's pharmacokinetics. In this sense, the peak of NO_2^- concentration is obtained within 2–3 h after the supplementation which 2.5 h before exercise has been considered as the optimal timing for this supplement (Domínguez et al. 2017b). Regarding the optimal dosage of NO_3^-, in a study an increase in time to task failure at a severe-intensity exercise after 8.4 mmol (+14%) and 16.4 mmol (+12%), but not after 4.2 mmol of NO_3^- was observed (Wylie et al. 2013). With respect to the comparison between possible differences in acute in comparison with chronic supplementation strategy, in a study was reported that ergogenic effect after a single dose of NO_3^- is maintained during 15 days of chronic supplementation (Vanhatalo et al. 2010) whereas, in another research, it was observed that a single dose of BJ (6 mmol) trend towards increasing cycling economy ($\sim$3%, $p = 0.06$) when the period of supplementation was extended from 7 to 28–30 days ($\sim$3%, $p < 0.05$) (Wylie et al. 2016). In the recent systematic review related to resistance training reported that those studies that used a dose of 6.4 mmol of NO_3^- increased the power output and the maximal number of repetitions with submaximal loads during resistance exercise (San Juan et al. 2019). Although, a single acute NO_3^- supplementation could be insufficient for mediating changes in the content of Ca^{2+} handling or myofibrillar proteins (that need a more extended supplementation protocol), acute NO_3^- supplementation could enhance muscular contraction mediated by cyclic guanosine monophosphate (cGMP) -dependent and independent post-translational modifications of Ca^{2+} handling and myofibrillar proteins (Rodríguez-Fernández et al. 2020). Thus, Resistance trained practitioners can benefit from NO_3^- supplementation with dosages range from 5 to 9 mmol of nitrate for 1–28 days 150 min before every training session (Naderi et al. 2016). In addition, itis necessary to avoid brushing teeth or using mouthwash, chewing gum or ingestion sweets that could contain chlorhexidine, xylitol or other bactericidal substances because the mouth bacteria affectation could affect to the reduction of NO_2^- after NO_3^- intake (Govoni et al. 2008).

4 Filling Gaps

The ergogenic effects of the sport supplements is dependent on the mechanical and metabolic demands of each modality of exercise. Therefore, in the present chapter, it has been reviewed the sport supplements with possible ergogenic aids to increase muscle strength, including the mechanism of action related to increased muscle force production (see Table 1); but it is important to administer these supplements with a correct guideline including the optimal dosage and timing (see Fig. 1).

Table 1 Sports supplements with ergogenic effect on muscle strength

Supplement	Mechanism of action
Caffeine	Enhancement of the mood profile, reduction of RPE, increase recruitment of motor units, improve Na^+–K^+ pump activity and the bioavailability of Ca^{2+} on myoplasm and the number of cross-bridges
Creatine	Increase capacity to resynthesize ATP, aid in shuttling of energy between the mitochondria and cytosol, buffers intracellular H^+, and reduce muscle protein breakdown, alterations in calcium handling potentiates myofibrillar cross-bridge cycling
B-alanine	Increase Ca^{2+} bioavailability on myoplasma and transporting Ca^{2+} at extra muscular space, regulating intramuscular pH
Sodium bicarbonate	Alkalosis on blood that increase the flux of H^+ during activities with high glycolytic demands, regulating intramuscular pH
Nitrate	Favour vasodilation, diminish PCr depletion, increase Ca^{2+} bioavailability in myoplasm and, subsequently the number of actin-myosin cross-bridges

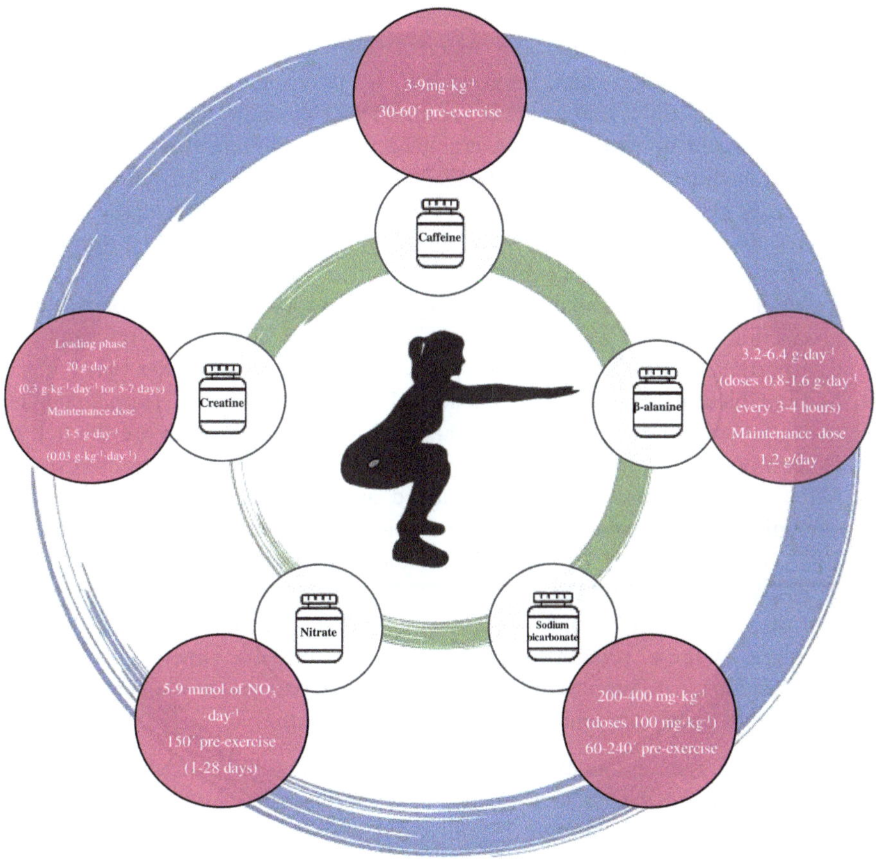

Fig. 1 Optimal dosage and timing of the sport supplement that potentiate muscular strength production

5 Take-Home Messages and Practical Resources

1. Recommendations about the consume of ergogenic sport supplements must be focused mainly on the force production enhancement and, consequently the sport performance.
2. Recommendations about the consume of ergogenic sport supplements for enhancing strength must be focused on the optimization of applied force and, consequently it is not specific of strength sport modalities.
3. The consumption of ergogenic sport supplements cannot be a substitutive of a an unsuitable food selection or an incorrect eating habit.
4. In the selection of ergogenic sport supplements must be considered the risk/benefit relation. A good criteria is attending to the security, efficacy and legalty of the sport supplements.
5. There is a limited number of ergogenic sport supplements that are considered ergogenic for enhancing strength.
6. Caffeine could be an ergogenic sport supplements on strength production mediated by its effect on the mood profile, reduction of RPE, increase recruitment of motor units, improve Na^{+}–K^{+} pump activity and the bioavailability of Ca^{2+} on myoplasm and the number of cross-bridges.
7. Creatine could be considered an ergogenic sport supplement on strength production mediated by an increased capacity to resynthesize ATP, aid in shuttling of energy between the mitochondria and cytosol, buffers intracellular H^{+}, and reduce muscle protein breakdown, alterations in calcium handling potentiates myofibrillar cross-bridge cycling.
8. β-alanine could be considered an ergogenic sport supplement on strength production mediated by an increased Ca^{2+} bioavailability on myoplasma and transporting H^{+} at extra muscular space, regulating intramuscular pH.
9. Sodium bicarbonate could be considered an ergogenic sport supplement on strength production mediated by an alkalosis on blood that increase the flux of H^{+} during activities with high glycolytic demands, regulating intramuscular pH.
10. Nitrate supplements could be considered an ergogenic sport supplement on strength production mediated by vasodilation, diminish PCr depletion, increase Ca^{2+} bioavailability in myoplasm and, subsequently the number of actin-myosin cross-bridges.
11. The timing of sport supplements consume, the appearance of new sports supplements with a scientific backboard as a possible interaction on the co-ingestion of different sport supplements advise to revise the recommendations about sport supplements consume on the future.

It exists different associations, expert meetings and institutions that publish position stand about the efficacy of the sport supplements. Actually, it is distinguished the Australian Institute of Sport (AIS) that renew a position stand every four years approximately, the expert group of the IOC, and the frequent position stands published by the International Society of Sports Nutrition (ISSN).

- Australian Institute of Sport (2019). Sports Supplement Framework 2019 Available from: https://www.sportaus.gov.au/ais/nutrition/supplements
- International Olympic Committee (2018). IOC consensus statement: dietary supplements and the high-performance athlete. Available from: https://journals.humankinetics.com/view/journals/ijsnem/28/2/article-p104.xml
- International Society of Sports Nutrition (2018). ISSN exercise & sports nutrition review update: research & recommendations. Available from: https://jissn.biomedcentral.com/articles/10.1186/s12970-018-0242-y.

References

AIS (2019) Australian Institute of Sport: Sports Supplement Framework 2019. Australian Sports Commission. Available from: https://www.sportaus.gov.au/ais/nutrition/supplements

Alves CRR, Santiago BM, Lima FR, Otaduy MCG, Calich AL, Tritto ACC et al (2013) Creatine supplementation in fibromyalgia: a randomized, double-blind, placebo-controlled trial. Arthritis Care Res 65(9):1449–1459

Baguet A, Reyngoudt H, Pottier A, Everaert I, Callens S, Achten E et al (2009) Carnosine loading and washout in human skeletal muscles. J Appl Physiol 106(3):837–842

Bailey SJ, Fulford J, Vanhatalo A, Winyard PG, Blackwell JR, DiMenna FJ et al (2010) Dietary nitrate supplementation enhances muscle contractile efficiency during knee-extensor exercise in humans. J Appl Physiol 109(1):135–148

Blancquaert L, Everaert I, Missinne M, Baguet A, Stegen S, Volkaert A et al (2017) Effects of histidine and β-alanine supplementation on human muscle carnosine storage. Med Sci Sports Exerc 49(3):602–609

Candow DG, Chilibeck PD (2008) Timing of creatine or protein supplementation and resistance training in the elderly. Appl Physiol Nutr Metab 33(1):184–190

Candow DG, Vogt E, Johannsmeyer S, Forbes SC, Farthing JP (2015) Strategic creatine supplementation and resistance training in healthy older adults. Appl Physiol Nutr Metab 40 (7):689–694

Caraballo I, Domínguez R, Guerra-Hernandez EJ, Sánchez-Oliver AJ (2020) Analysis of sports supplements consumption in young spanish elite dinghy sailors. Nutrients 12(4):E9993

Carr BM, Webster MJ, Boyd JC, Hudson GM, Scheett TP (2013) Sodium bicarbonate supplementation improves hypertrophy-type resistance exercise performance. Eur J Appl Physiol 113(3):743–752

Chilibeck P, Kaviani M, Candow D, Zello GA (2017) Effect of creatine supplementation during resistance training on lean tissue mass and muscular strength in older adults: a meta-analysis. Open Access J Sport Med. 8:213–226

Church DD, Hoffman JR, Varanoske AN, Wang R, Baker KM, La Monica MB et al (2017) Comparison of two β-alanine dosing protocols on muscle carnosine elevations. J Am Coll Nutr 36(8):608–616

Coggan AR, Broadstreet SR, Mikhalkova D, Bole I, Leibowitz JL, Kadkhodayan A et al (2018) Dietary nitrate-induced increases in human muscle power: high versus low responders. Physiol Rep 6(2)

Cooper R, Naclerio F, Allgrove J, Jimenez A (2012) Creatine supplementation with specific view to exercise/sports performance: an update. J Int Soc Sports Nutr 9

De Oliveira LF, Saunders B, Yamaguchi G, Swinton P, Artioli GG (2020) Is individualization of sodium bicarbonate ingestion based on time to peak necessary? Med Sci Sports Exerc 52 (8):1801–1808

Domínguez R, Hernández Lougedo J, Maté-Muñoz JL, Garnacho-Castaño MV (2015) Effects of ß-alanine supplementation on athletic performance. Nutr Hosp 31(1):155–169
Domínguez R, Garnacho-Castaño MV, Cuenca E, García-Fernández P, Muñoz-González A, de Jesús F et al (2017a) Effects of beetroot juice supplementation on a 30-s high-intensity inertial cycle ergometer test. Nutrients 9(12)
Domínguez R, Cuenca E, Maté-Muñoz J, García-Fernández P, Serra-Paya N, Estevan M et al (2017b) Effects of beetroot juice supplementation on cardiorespiratory endurance in athletes A Systematic Review. Nutrients 9(1):43
Domínguez R, Maté-Muñoz J, Serra-Paya N, Garnacho-Castaño M (2018a) Lactate threshold as a measure of aerobic metabolism in resistance exercise. Int J Sports Med 39(03):163–172
Domínguez R, Maté-Muñoz JL, Cuenca E, García-Fernández P, Mata-Ordoñez F, Lozano-Estevan MC et al (2018b) Effects of beetroot juice supplementation on intermittent high-intensity exercise efforts. J Int Soc Sports Nutr 15(2)
Doudouros I, Fatouros I, Gourgoulis V, Jamurtas A, Tsitsios T, Hatzinikolau A et al (2006) Dose-related effects of prolonged $NaHCO_3$ ingestion during high-intensity exercise. Med Sci Sport Exerc 38(10):1746–1753
Duncan MJ, Weldon A, Price MJ (2014) The effect of sodium bicarbonate ingestion on back squat and bench press exercise to failure. J Strength Cond Res 28(5):1358–1366
Dutka TL, Lamb GD (2004) Effect of carnosine on excitation-contraction coupling in mechanically-skinned rat skeletal muscle. J Muscle Res Cell Motil 25(3):203–213
Garthe I, Maughan RJ (2018) Athletes and supplements: prevalence and perspectives. Int J Sport Nutr Exerc Metab 28(2):126–138
Goldstein ER, Ziegenfuss T, Kalman D, Kreider R, Campbell B, Wilborn C et al (2010) International society of sports nutrition position stand: caffeine and performance. J Int Soc Sports Nutr 7
Gonzalez-Badillo JJ, Ribas-Serna J (2019) Fuerza, velocidad y rendimiento físico y deportivo, 1st edn. Librerías deportivas Esteban Sanz S.L, Madrid, p 753
Govoni M, Jansson EÅ, Weitzberg E, Lundberg JO (2008) The increase in plasma nitrite after a dietary nitrate load is markedly attenuated by an antibacterial mouthwash. Nitric Oxide Biol Chem 19(4):333–337
Granier PL, Dubouchaud H, Mercier BM, Mercier JG, Ahmaidi S, Préfaut CG (1996) Effect of $NaHCO_3$ on lactate kinetics in forearm muscles during leg exercise in man. Med Sci Sports Exerc 28(6):692–697
Green AL, Hultman E, Macdonald IA, Sewell DA, Greenhaff PL (1996) Carbohydrate ingestion augments skeletal muscle creatine accumulation during creatine supplementation in humans. Am J Physiol Endocrinol Metab 271(5):34–35
Grgic J, Trexler ET, Lazinica B, Pedisic Z (2018) Effects of caffeine intake on muscle strength and power: a systematic review and meta-analysis. J Int Soc Sports Nutr 15(11)
Grgic J, Mikulic P, Schoenfeld BJ, Bishop DJ, Pedisic Z (2019a) The influence of caffeine supplementation on resistance exercise: a review, vol 49, pp 17–30. Sports Medicine. Springer International Publishing
Grgic J, Sabol F, Venier S, Tallis J, Schoenfeld BJ, Del CJ et al (2019b) Caffeine supplementation for powerlifting competitions: an evidence-based approach. J Hum Kinet 68(1):37–48
Grgic J, Rodriguez RF, Garofolini A, Saunders B, Bishop DJ, Schoenfeld BJ et al (2020) Effects of sodium bicarbonate supplementation on muscular strength and endurance: a systematic review and meta-analysis, vol 50, pp 1361–75. Sports Medicine. Springer
Harris RC, Söderlund K, Hultman E (1992) Elevation of creatine in resting and exercised muscle of normal subjects by creatine supplementation. Clin Sci 83(3):367–374
Harris RC, Tallon MJ, Dunnett M, Boobis L, Coakley J, Kim HJ et al (2006) The absorption of orally supplied β-alanine and its effect on muscle carnosine synthesis in human vastus lateralis. Amino Acids 30(3):279–289
Hernández A, Schiffer TA, Ivarsson N, Cheng AJ, Bruton JD, Lundberg JO et al (2012) Dietary nitrate increases tetanic $[Ca_2+]$i and contractile force in mouse fast-twitch muscle. J Physiol 590(15):3575–3583

Hill CA, Harris RC, Kim HJ, Harris BD, Sale C, Boobis LH et al (2007) Influence of β-alanine supplementation on skeletal muscle carnosine concentrations and high intensity cycling capacity. Amino Acids 32(2):225–233

Hoffman J, Ratamess N, Kang J, Mangine G, Faigenbaum A, Stout J (2006) Effect of creatine and beta-alanine supplementation on performance and endocrine responses in strength/power athletes. Int J Sport Nutr Exerc Metab 16(4):430–446

Hoffman JR, Emerson NS, Stout JR (2012) β-alanine supplementation. Curr Sports Med Rep 11 (4):189–195

Hultman E, Soderlund K, Timmons JA, Cederblad G, Greenhaff PL (1996) Muscle creatine loading in men. J Appl Physiol 81(1)

Jodra P, Lago-Rodríguez A, Sánchez-Oliver AJ, López-Samanes A, Pérez-López A, Veiga-Herreros P et al (2020) Effects of caffeine supplementation on physical performance and mood dimensions in elite and trained-recreational athletes. J Int Soc Sports Nutr 17(1):2

Kaviani M, Shaw K, Chilibeck PD (2020) Benefits of creatine supplementation for vegetarians compared to omnivorous athletes: a systematic review. Int J Environ Res Public Health. MDPI AG 17

Kerksick CM, Wilborn CD, Roberts MD, Smith-Ryan A, Kleiner SM, Jäger R et al (2018) ISSN exercise & sports nutrition review update: research & recommendations. J Int Soc Sports Nutr 15(1):38

Kreider RB, Kalman DS, Antonio J, Ziegenfuss TN, Wildman R, Collins R et al (2017) International society of sports nutrition position stand: safety and efficacy of creatine supplementation in exercise, sport, and medicine. J Int Soc Sports Nutr 14

Lancha Junior AH, de Salles Painelli V, Saunders B, Artioli GG (2015) Nutritional strategies to modulate intracellular and extracellular buffering capacity during high-intensity exercise, vol 45. Sports Medicine

Lanhers C, Pereira B, Naughton G, Trousselard M, Lesage FX, Dutheil F (2015) Creatine supplementation and lower limb strength performance: a systematic review and meta-analyses, vol. 45, pp 1285–94. Sports Medicine. Springer International Publishing

Lanhers C, Pereira B, Naughton G, Trousselard M, Lesage FX, Dutheil F (2017) Creatine supplementation and upper limb strength performance: a systematic review and meta-analysis, Vol. 47, p. 163–73. Sports Medicine. Springer International Publishing

López-González LM, Sánchez-Oliver AJ, Mata F, Jodra P, Antonio J, Domínguez R (2018) Acute caffeine supplementation in combat sports: a systematic review. J Int Soc Sports Nutr

Mata-Ordoñez F, Sánchez-Oliver AJ, Domínguez Herrera R, Villegas García JA (2018) Suplementación en el deporte: directrices desde la responsabilidad profesional. Habilid Mot Rev Ciencias La Act Física y Del Deport 51:4–12

Maté-Muñoz JL, Lougedo JH, Garnacho-Castaño MV, Veiga-Herreros P, Lozano-Estevan MDC, García-Fernández P et al (2018) Effects of β-alanine supplementation during a 5-week strength training program: a randomized, controlled study. J Int Soc Sports Nutr 15:19

Maughan RJ, Burke LM, Dvorak J, Larson-Meyer DE, Peeling P, Phillips SM et al (2018a) IOC consensus statement: dietary supplements and the high-performance athlete, vol 52, pp 439–55. British J Sports Med

Maughan RJ, Shirreffs SM, Vernec A (2018b) Making decisions about supplement use. Int J Sport Nutr Exerc Metab 28(2):212–219

McNaughton LR, Siegler J, Midgley A (2008) Ergogenic effects of sodium bicarbonate, vol 7, pp 230–236. Current Sports Medicine Reports

Medbo JI, Tabata I (1993) Anaerobic energy release in working muscle during 30 s to 3 min of exhausting bicycling. J Appl Physiol 75(4):1654–1660

Mills S, Candow DG, Forbes SC, Neary JP, Ormsbee MJ, Antonio J (2020) Effects of creatine supplementation during resistance training sessions in physically active young adults. Nutrients 12(6):1–11

Mujika I, Padilla S, Pyne D (2002) Swimming performance changes during the final 3 weeks of training leading to the sydney 2000 olympic games. Int J Sports Med 23(8):582–587

Muñoz A, López-Samanes Á, Domínguez R, Moreno-Pérez V, Sánchez-Oliver AJ, Del Coso J (2020) Use of sports supplements in competitive handball players: sex and competitive level differences. Nutrients 12(11):1–14

Naderi A, de Oliveira EP, Ziegenfuss TN, Willems MET (2016) Timing, optimal dose and intake duration of dietary supplements with evidence-based use in sports nutrition. J Exerc Nutr Biochem 20(4):1–12

Outlaw JJ, Smith-Ryan AE, Buckley AL, Urbina SL, Hayward S, Wingfield HL et al (2016) Effects of β-alanine on body composition and performance measures in collegiate women. J Strength Cond Res 30(9):2627–2637

Perim P, Marticorena FM, Ribeiro F, Barreto G, Gobbi N, Kerksick C et al (2019) Can the skeletal muscle carnosine response to beta-alanine supplementation be optimized? vol 6. Frontiers in Nutrition. Frontiers Media S.A.

Persky AM, Brazeau GA (2001) Clinical pharmacology of the dietary supplement creatine monohydrate. Pharmacol Rev 53(2):161–176

Quinn PJ, Boldyrev AA, Formazuyk VE (1992) Carnosine: its properties, functions and potential therapeutic applications. Mol Aspects Med 13:379–444

Rawson ES, Miles MP, Larson-Meyer DE (2018) Dietary supplements for health, adaptation, and recovery in athletes. Int J Sport Nutr Exerc Metab 1–33

Raya-González J, Rendo-Urteaga T, Domínguez R, Castillo D, Rodríguez-Fernández A, Grgic J (2020) Acute effects of caffeine supplementation on movement velocity in resistance exercise: a systematic review and meta-analysis, vol 50, pp 717–29, Sports Medicine. Springer

Rezende NS, Swinton P, de Oliveira LF, da Silva RP, da Eira Silva V, Nemezio K et al (2020) The muscle carnosine response to beta-alanine supplementation: a systematic review with bayesian individual and aggregate data e-max model and meta-analysis. Front Physiol 11

Rodríguez-Fernández A, Castillo D, Raya-González J, Domínguez R, Bailey SJ (2020) Beetroot juice supplementation increases concentric and eccentric muscle power output. Original investigation. J Sci Med Sport

Sahlin K, Harris RC (2011) The creatine kinase reaction: a simple reaction with functional complexity. Amino Acids 40:1363–1367

Sale C, Saunders B, Harris RC (2010) Effect of beta-alanine supplementation on muscle carnosine concentrations and exercise performance. Amino Acids 39(2):321–333

San Juan AF, López-Samanes Á, Jodra P, Valenzuela PL, Rueda J, Veiga-Herreros P et al (2019) Caffeine supplementation improves anaerobic performance and neuromuscular effciency and fatigue in Olympic-level boxers. Nutrients 11(9)

San Juan AF, Dominguez R, Lago-Rodríguez Á, Montoya JJ, Tan R, Bailey SJ (2020) Effects of dietary nitrate supplementation on weightlifting exercise performance in healthy adults: a systematic review. Nutrients 12:1–16

Steenge GR, Simpson EJ, Greenhaff PL (2000) Protein- and carbohydrate-induced augmentation of whole body creatine retention in humans. J Appl Physiol 89(3):1165–1171

Stegen S, Bex T, Vervaet C, Vanhee L, Achten E, Derave W (2014) β-Alanine dose for maintaining moderately elevated muscle carnosine levels. Med Sci Sports Exerc 46(7):1426–1432

Stellingwerff T, Anwander H, Egger A, Buehler T, Kreis R, Decombaz J et al (2012) Effect of two β-alanine dosing protocols on muscle carnosine synthesis and washout. Amino Acids 42 (6):2461–2472

Swietach P, Youm JB, Saegusa N (2013) Coupled Ca2þ/Hþ transport by cytoplasmic buffers regulates local Ca2þ and Hþ ion signaling. Proc Natl Acad Sci USA 110

Syrotuik DG, Bell GJ (2004) Acute creatine monohydrate supplementation: a descriptive physiological profile of responders vs nonresponders. J Strength Cond Res 18(3):610

Thomas DT, Erdman KA, Burke LM (2016) American College of Sports Medicine Joint Position Statement. Nutrition and Athletic Performance. Med Sci Sport Exerc 48(3):543–568

Vanhatalo A, Bailey SJ, Blackwell JR, DiMenna FJ, Pavey TG, Wilkerson DP et al (2010) Acute and chronic effects of dietary nitrate supplementation on blood pressure and the physiological

responses to moderate-intensity and incremental exercise. Am J Physiol Regul Integr Comp Physiol 299(4)
Volek JS, Kraemer WJ (1996) Creatine supplementation: its effect on human muscular performance and body composition. J Strength Cond Res 10(3):200–210
Wallimann T, Schlösser T, Eppenberger HM (1984) Function of M-line-bound creatine kinase as intramyofibrillar ATP regenerator at the receiving end of the phosphorylcreatine shuttle in muscle. J Biol Chem 259(8):5238–5246
Weber A, Herz R (1968) The relationship between caffeine contracture of intact muscle and the effect of caffeine on reticulum. J Gen Physiol 52(5):750–759
Wylie LJ, Kelly J, Bailey SJ, Blackwell JR, Skiba PF, Winyard PG et al (2013) Beetroot juice and exercise: pharmacodynamic and dose-response relationships. J Appl Physiol 115(3):325–336
Wylie LJ, Ortiz De Zevallos J, Isidore T, Nyman L, Vanhatalo A, Bailey SJ et al (2016) Dose-dependent effects of dietary nitrate on the oxygen cost of moderate-intensity exercise: acute vs. chronic supplementation. Nitric Oxide Biol Chem 57:30–39
Yamaguchi GC, Nemezio K, Schulz ML, Natali J, Cesar JE, Riani LA et al (2020) Kinetics of muscle carnosine decay after β-alanine supplementation: a 16-week washout study. Med Sci Sports Exerc
Zandona BA, Ramos RA, de Oliveira CDS, McAnulty SR, Ferreira LHB, Smolarek AC et al (2020) Reduced dose of beta-alanine is sufficient to maintain performance in repeated sprints. J Strength Cond Res

Raúl Domínguez is an extraordinary doctoral thesis award on Sport Sciences and is an pProfessor at the University of Seville (Spain). Raúl Domínguez leads a research line on sport nutrition with a scientific production that exceed 65 articles published in journal with impact factor JCR/SJR.

Alireza Naderi is a master graduated in exercise physiology and a researcher in the field of sport nutrition with 13 published scientific papers. As a performance nutritionist, he works with elite athletes.

Antonio Jesús Sánchez-Oliver has a Ph.D. on Sport Nutrition Sciences and is a full professor in the Department of Motricidad Humana y Rendimiento Deportivo at the University of Seville (Spain). Antonio J. works in a research line on sport supplement use with a scientific production that exceed 63 articles published in journal with impact factor JCR/SJR.

www.ingramcontent.com/pod-product-compliance
Ingram Content Group UK Ltd.
Pitfield, Milton Keynes, MK11 3LW, UK
UKHW021315130926
15127UKWH00002B/9
9783030819880